T0138462

A Nice Derangement of Epistemes

A Nice Derangement of Epistemes

Post-positivism in the Study of Science from Quine to Latour

John H. Zammito

THE UNIVERSITY OF CHICAGO PRESS
Chicago and London

JOHN H. ZAMMITO is the John Antony Weir Professor of History at Rice University. He is the author of *The Genesis of Kant's Critique of Judgment* and *Kant, Herder, and the Birth of Anthropology,* both published by the University of Chicago Press.

The University of Chicago Press, Chicago 60637
The University of Chicago Press, Ltd., London
© 2004 by The University of Chicago
All rights reserved. Published 2004
Printed in the United States of America

13 12 11 10 09 08   2 3 4 5

ISBN: 0-226-97861-3 (cloth)
ISBN: 0-226-97862-1 (paper)

Library of Congress Cataloging-in-Publication Data

Zammito, John H., 1948–
    A nice derangement of epistemes : post-positivism in the study of science from Quine to Latour / John H. Zammito.
    p. cm.
Includes bibliographical references and index.
ISBN 0-226-97861-3 (alk. paper) — ISBN 0-226-97862-1 (pbk. : alk. paper)
    1. Science—Philosophy.   2. Science—History.   3. Progress.   I. Title.

Q175 .Z25 2004
501—dc21

2003011970

⊗The paper used in this publication meets the minimum requirements of the American National Standard for Information Sciences—Permanence of Paper for Printed Library Materials, ANSI Z39.48-1992.

To Albert Van Helden

Contents

Acknowledgments

This was no easy book to compose. Without the encouragement of Albert Van Helden, I doubt I would have undertaken it. He convinced me that my historical way of coming to terms with theoretical issues made them more comprehensible for him and might do so for others as well. I dared to make the effort, and I hope the end product still fulfills the terms of Al's original commission. Neither he nor any of the others I acknowledge below bear responsibility for errors or excesses to which I have succumbed, though, to be sure, they have rescued me from many.

I began my study of philosophy with Michael Polanyi, and I want to take this opportunity to express my reverence for this early teacher, whose influence must somehow have worked its way across the decades to this project. Larry Laudan and Thomas Haskell read and provided me with detailed commentary on an early version of the manuscript. Larry argued that I was being too kind to Kuhn; Tom argued I was being too critical. I have tried in the ultimate version to find my way between. I thank each of them for many helpful suggestions. Larry's influence on this whole work is self-evident. Even where I part company from him, I can think of few indeed whose overall approach I find so thoroughly compatible with my own. Tom shares with me the project, to which I now turn my full attention, of working out what a moderate historicism should mean for empirical practice. Our collaboration is one of the linchpins of my intellectual orientation. Richard Grandy read a draft of the manuscript and responded with remarkable generosity to one who so fundamentally disputed philosophical and historical matters with which he was intimately associated. I am grateful for his friendship and for his rigor. Andy Pickering read an early draft of the chapter on social constructionism in science studies, and I was pleased that he found my grasp of the matter civil and . . . well, constructive. Our points of convergence as well as our points of contestation mark his work out for me as indispensable. Without constant dialogue with Steven Crowell, this book would never have come to completion. Steve challenged every step of my argument, and the upshot has been to force me to clarify my philosophical claims substantially, even if inevitably they have not achieved ideal clarity, to say nothing of conclusiveness! There is nothing more salutary for a novice naturalist than to have to answer to a fiercely transcendental inquisitor. I look forward to more rounds and better arguments.

I began the discussions for this book with Susan Abrams, whose achievement in building a list in science studies is incomparable. I thank her for her interest in my project and hope for her all the best. David Brent, my editor for two earlier books with the University of Chicago Press, stepped in to bring the project to completion. It was a thorough pleasure to work with him on this new endeavor, as it had been with projects in the past. I thank him and all the others at the press for their smooth and professional management of this, my third book with them. In particular I thank the anonymous readers for the press: first, for grasping with such discernment what I was attempting; second, for finding it worthwhile; and, finally, for offering me such constructive suggestions for improvement. I have not been able to incorporate every suggestion, but the book is immeasurably stronger for these interventions. Thanks, too, to the copyeditors—Catherine Howard here at Rice and Clair James for the press—for their meticulous efforts.

And thanks, last but not least, to Katie.

The contemporary tendency to regard interpretation as something second class reflects, I think, not a craving for objectivity but a craving for absolutes—a craving for absolutes and a tendency which is inseparable from that craving, the tendency to think that if the absolute is unobtainable, then "anything goes." But *"enough is enough, enough isn't everything."*

— HILARY PUTNAM, "THE CRAVING FOR OBJECTIVITY"

Introduction

This study is intended as an introductory consideration of post-positivist philosophy and theory of science since 1950 for scholars in the empirical human sciences, especially those in history and science-studies. It explores and questions new "dogmas of (anti)empiricism" at the epistemological core of contemporary "theory."[1] It is not a neutral book, though it will strain toward objectivity.[2] What I wish to suggest is that some "theorists" have drawn upon post-positivism to initiate an attack upon the *practice of empirical inquiry* itself.[3] I consider and contest certain extravagant gestures in philosophy of language which, when taken seriously, threaten to undermine indispensable canons of empirical inquiry. Probably the most extravagant theses in philosophy of language have been the "theoretical" uptakes of Thomas Kuhn's notion of "incommensurability," of Willard van Orman Quine's dogma of "indeterminacy of translation," of Donald Davidson's modulation of that into "indeterminacy of interpretation," and—in a different tradition—of Jacques Derrida's disquisitions on *différance*.[4] The quantity of exegesis that has been devoted to explicating these notions is only matched by its inconclusiveness. It is hardly in the compass of my endeavor to resolve even one of these conundrums, but there may be an advantage in stringing them together: the parallelism in the debates they occasion and the aporia into which they conduct commentators may itself evoke a rather Gordian response to their convolutions. At any rate, there is no prospect of defining and defending empirical inquiry theoretically without confronting these hydras.

Recent theorists have typically asserted their claims rather dramatically—one might say hyperbolically—in order to galvanize attention. The ensuing debate has typically reduced the original claims to far more modest terms, though rarely has the result been a return to the *status quo ante*. In short, such "theory" needs to be deflated. I propose both to document this deflation and to take seriously whatever residual of warranted implication remains. Thus my work will be narrative in structure, telling the story of the proposal of claims, their reception, and the resulting stabilization of understanding (if, indeed, such stabilization has been achieved). I seek to provide, in a narrative frame, an account of how these ideas were developed and what remains of them in the wake of the debates which have raged over them. In setting the narrative in just this form, I offer two critical vantages: an immanent confrontation with the claims themselves, and a mediated appraisal based on the critical reception which has greeted them. On occasion, I will gesture to the wider cultural and political context of the original formulation or of the reception history, but I do not pretend to offer more than a "history of ideas." My purpose is unequivocally presentist; I am interested exclusively in what these ideas mean for those who practice any of the empirical human sciences today.

I will operate with the presumption that "theory" in current parlance embraces three distinct but not always distinguished domains: methodology, epistemology, and rhetorical reflexivity. By methodology I mean what relates to the concrete production of accounts—how and what one can write. By epistemology I mean what relates to the *validity* of the claims of an account. The epistemological borders on one side on the ontological and on the other on the methodological. The rhetorically reflexive, finally, has to do with the inextricably rhetorical and linguistic form of all argument; it is a form of hypertrophy of epistemology in a shifted key or a "jumped frame." To "jump frames" is to take as one's *topic* one's target's *resource*.[5] Malcolm Ashmore identifies this as the strategic meaning of *deconstruction:* "What is meant by deconstruction? Essentially it means the subversion of a Participant level self-understanding by a superior Analyst level meta-understanding."[6] The disputed question is the scope and authority of "theory," in particular in this third, hyperbolic form, but this cannot be clarified unless the other two domains are explored. To what extent does "theory" trade on (inverted) foundationalist concerns with "first philosophy" which exceed the capacities and fall outside the concerns of empirical inquiry? What is the consequence for *empirical knowledge,* both of nature and of man? My claim is that we need to know what "incommensurability," "indeterminacy," "underdetermination," "theory-ladenness of data," and other such daunting negativities betoken for the pursuit of empirical inquiry. I propose that once we dispel the extravagant, what remains in most of these recent "theoretical" claims is fully assimilable into—*not* preemptive of—empirical inquiry. This is, I recognize, hardly a dramatic result, but I do believe that a little sobriety is salutary after all the sound and fury. Sometimes there are differences that make a difference; sometimes claims are false. This study cautions unwary practitioners of empirical inquiry against submission to extravagant and irresponsible "readings" of philosophy by self-proclaimed "theorists." Philosophers—the real ones—have increasingly taken a deflationary view of their authority over the empirical disciplines.

Post-positivism has been a preponderantly analytic—hence Anglo-American—enterprise, though its origin must be traced to the Vienna Circle. It emerged in philosophy of science and it escalated into philosophy of language. The story begins with the crisis of logical positivism/empiricism in the 1950s sparked by Quine's rebellion. The decisive turning point was the publication of Kuhn's *Structure of Scientific Revolutions* in 1962.[7] Problems Quine and Kuhn discerned in theorizing empirical science seemed to require "semantic ascent" into philosophy of language for resolution. Here the "causal theory of reference" of Hilary Putnam and others assumed salience. But reflections upon the general "linguistic turn" led some prophets (e.g., Richard Rorty) to proclaim the "end of epistemology." The linguistic turn prompted two further impulses: the "historicization of reason" and the "social construction of knowledge." That is, to grasp science it was felt necessary, first, to situate it in historical process, and second, to situate it in social context. The first impulse led to what came to be called the "marriage" of the history of science with the philosophy of science, and the second to what came to be called the "sociology of scientific knowledge" (SSK). Imre Lakatos, Larry Laudan, and "naturalized epistemology" pursued the first agenda. The "Strong Program" of David Bloor and Barry Barnes instantiated the second. Neither of these developments has had unequivocally happy outcomes. The "marriage" of the history and the philosophy of science has been rocky, and some would say it has ended in divorce.[8] The story of the sociology of scientific knowledge is even more lamentable, for in the all-out version of radical social constructivism, it has occasioned a literal *reductio ad absurdum*. That is the overview of my argument. What follows are the details of the specific chapters.

Chapter 1 considers the term *positivism* and appraises both its historical burden and the question of what post-positivism has repudiated of this legacy. Two objectives of this chapter are, first, to dispute the postmodern tendency to lump anything associated with empirical inquiry with a global and pejorative sense of the term *positivism,* and, second, to recognize the recent historical findings that logical positivism had already begun the philosophical deflation of its own claims for which post-positivism is generally credited. I make the case that there is still quite a substantial set of problems that can legitimately be conceived under the rubric post-positivism.

Chapter 2 considers Quine's three "indeterminacies"—of translation, of reference (also called inscrutability and sometimes ontological relativity), and of theory (also called underdetermination). Just fixing his claims will prove difficult, for, clear as he believed himself to be, the reception suggests that he was all too cryptic, if not simply inconsistent. In any event, no single analytic philosopher has had more impact on the current parameters of epistemology.[9] The most extravagant claims are bandied about today under the ostensible warrant of Quine's arguments.[10] To grasp what Quine said, I will routinely discriminate several versions, on a spectrum from impossibly strong to innocuously weak. Both immanently and with recourse to the reception, I will deflate these to what can be warranted in Quine's positions. The result, in my view, discredits a substantial number of the invocations of Quine by later "theorists."

Chapter 3 takes up Kuhn's *Structure of Scientific Revolutions* and his efforts thereafter, in response to the torrent of confutations with which he was greeted, to explain what he really meant. I will subject two of Kuhn's key concepts—paradigm and incommensurability—to the same immanent and reception-historical examination as Quine's indeterminacies. Significantly, Kuhn invoked Quine to explain himself. For a time he assimilated his views to Quine's, but ultimately he distinguished his position through contrast with Quine's. I will pursue the sharp "semantic ascent" which resulted from the philosophical reception of Kuhn's work (often under the shadow of Quine), in terms of the debate about meaning variance and conceptual change and the so-called causal theory of reference. The displacement of philosophy of science into philosophy of language (in which Quine and Kuhn both represent major stages) proves, upon a consideration of this narrative, to have been less than illuminating for science, suggesting that the linguistic turn here has quickly run into a dead end.

Chapter 4 examines a second and ostensibly more positive reception of Kuhn's "revolution"—the effort to bring philosophy of science more critically into interaction with the history of science. This effort dominated the philosophy of science in the 1970s, involving such figures as Stephen Toulmin, Imre Lakatos, Ernan McMullin, Larry Laudan, Richard Burian, and others. But the enterprise ultimately reached an impasse over the reconciliation of naturalist and normative notions of epistemology as applied to science. Nevertheless, the impulses toward a naturalized or evolutionary epistemology hold out prospects for a coherent and viable approach to the methodology of empirical inquiry which may well rescue us from current theoretical impasses.

Chapter 5 takes up the emergence of the so-called Strong Program in the sociology of scientific knowledge (SSK) as the disciplinary conscription of Kuhn into sociology, a conscription he both incited and resented. SSK explicitly adopted Kuhn at his philosophically most extravagant, and radicalized that stance still further. The Strong Program was an aggressive effort to displace philosophy of science with sociology of science. Ironically SSK represented yet another philosophy of science, and not a very good one.[11] But it linked up fruitfully with a methodological revolution in the social sciences, launched by Peter Winch under the auspices of the later Wittgenstein and taken up in the so-called rationality debate in anthropology. This debate was complemented by the rebellion within sociology itself not merely against the sociology of science of Robert Merton but against the overarching function- alism of Talcott Parsons. This wider rebellion organized itself under the rubrics of symbolic interactionism and, most importantly, ethnomethodol- ogy. In short, with the rise of SSK the debate in philosophy of natural science fused with a debate in philosophy of social science.[12]

Chapter 6 considers the successors of SSK in science studies. The Strong Program was subjected to a withering barrage of external criticism in the 1980s, but that did not prevent it from proliferating a rapid succession of progeny. Laboratory studies and controversy studies, the turn to "microso- ciology" and to practice characterized this shift, but so did an even more drastic commitment to constructivism "all the way down." Practitioners of science studies swiftly became concerned about the cogency of their own procedures: while they embraced "constructivism," they worried about how they could trust the "social." All they seemed to have left was "discourse."

Chapter 7 pursues the proliferation of science studies in three veins: the development of the so-called actor network theory (ANT), the intervention of feminism in the interpretation of science, and the conception of a "man- gle" of practice developed by Andrew Pickering. In each elaboration, the ef- fort to face the discursive impasse that undercut a conventionally "social" constructivism led to the invocation of various "hybrid" forms. Not only did these approaches weave into post-positivism a considerable amount of French poststructuralism, they each essayed to become "post-human(ist)."

Chapter 8 takes up the last and most ironic twist in the narrative. The harvest of the misappropriation of philosophical ideas, together with the confusion of ideas from other discourses (preponderantly poststructuralist), has resulted in a series of absurd postures, culminating in radical "reflexiv- ity." The Sokal affair made it clear just how clueless some of the fashionable new constructivists had become.[13]

Hyperbolic postmodern "theory," whether in the vein of philosophy of language or in the vein of constructivism "all the way down," requires defla- tion. This does not mean that we have not learned something important from each of these "theoretical" extravagances. But it does mean that it is time to take up a more moderate historicism.

"Positivism" has today become more of a term of abuse than a technical term in philosophy. The indiscriminate way in which the term has been used in a variety of different polemical interchanges in the past few years, however, makes all the more urgent a study of the influence of positivistic philosophies in the social sciences.

— ANTHONY GIDDENS, "POSITIVISM AND ITS CRITICS"

From Positivism to Post-positivism

To elucidate post-positivism one must both specify what *positivism* was—in both its general historical and its logical forms—and then spell out clearly what displaced it. Positivism is a long tradition of thought about science in the West, emanating out of the nineteenth century.[1] Its effect has been to induce *scientism*, the idealization and privileging of natural-scientific method, conceived monolithically, for accessing truth and reality.[2] As Hilary Putnam has put it, "part of the problem with present day philosophy is a scientism inherited from the nineteenth century—a problem that affects more than one intellectual field."[3]

Positivism can be construed so broadly that it loses purchase as a conceptual discrimination. This is all the more true since it has come to be used pejoratively to signify whatever is distasteful about an opponent's position. Thus, all too frequently in postmodern discourse, positivism is used to dismiss any concerns with warrant or evidence, any recourse to empirical inquiry or to the rational adjudication of disputed questions.[4] It is essential that we resist the tendency to identify empirical inquiry generally with positivism. Neither the pursuit of empirical knowledge nor the reflection on its procedures (methodology) or standards (epistemology) can simply be lumped with positivism. Things are vastly more complex. There is no intellectual justification for the rhetorical misuse of *positivism*.

The two key moments explicitly identified with positivism in recent Western intellectual history were the philosophy of history developed by Auguste Comte in the early nineteenth century and the philosophy of science developed by the Vienna Circle in the early twentieth century. It was Comte who coined the term *positivism* and gave it its two central theoretical dimensions—as an epistemology and as a theory of progress. The theory of progress involved both a theory of the overall advancement of the human race as well as a theory of the specific development of particular sciences. According to Comte, human thought originally deployed theological structures to order the world of experience: everything from primitive animism to sophisticated monotheism, in the measure that it postulated a transcendent agency governing the behavior of the human environment, counted on this scheme as theological. Gradually, and differentially by content, human thought progressed from theological to metaphysical conceptions of the world. Immanent but essential properties *in* the world came to explain the nature and function *of* the world. Yet these speculative entities ultimately proved inaccessible to critical appraisal. Finally, and again sequentially by field, human thought advanced to the ultimate stage of positive knowledge. This form was achieved first in mathematics and mathematical physical science in the seventeenth century, followed by chemistry in the late eighteenth and early nineteenth century. It would be achieved, Comte was confident, in biology, and ultimately in the study of human conduct, which he dubbed "sociology." This sequential achievement of the positive stage by the hierarchy of sciences propelled the progress of the human race. To Comte the pragmatic warrant for science was its incontestable and cumulative contribution to human flourishing. Positive science offered a more certain predictive grasp of reality and concomitantly such knowledge was utility in a Baconian sense: prediction was power. The conquest of ever new domains, ultimately the sphere of human conduct itself, by the methodology of positivism was both inevitable and welcome. For Comte "the coming into being of sociology is supposed to mark the final triumph of positivism."[5]

Thus, the theory of history served as a powerful pragmatic reinforcement, in Comte's scheme, for the epistemological premises of positivism—relentless phenomenalism, antimetaphysicalism, and unitarian hierarchy of knowledge. Comte's epistemology built upon the critical thrust of Enlightenment thought, perhaps most centrally David Hume's combination of empiricism and skepticism. Thus, Comte maintained radically the traditional empiricist stance that there could be nothing in the mind that was not first in the senses, and he drew the epistemological consequence that there could be no evidential warrant, no claim to validity, which did not trace the claim back to sensory data. Consequently, nothing was more odious to Comte than recourse to transcendent or metaphysical categories, in other words, to anything which postulated the reality of what could not be confirmed by sensory observation. That was, from Comte's historical vantage, an unacceptably prescientific and obsolete approach to reality. Finally, Comte subscribed utterly to the view that all science fell under the authority of a unitary model and method, first established in the mathematical-physical sciences, and destined to triumph in all the other sciences in turn. The two elements of this commitment were the demarcation of science from all nonscience (which came nigh being labeled nonsense) and the effort to achieve a principled "reduction" or unification of all particular sciences into a singular theory grounded in physics.

Comte's ideas were taken up in the balance of the nineteenth century by a group of thinkers in both the natural and the social sciences. They attenuated Comte's schematic philosophy of history, but retained the association of science with progress, and they elaborated the epistemological dimension. Key figures here were Emile Durkheim in the social sciences and Jules Henri Poincaré and Pierre Duhem in the natural sciences. The implicit positivism of these thinkers involved their affirming a singular method for attaining valid knowledge, namely, the method of natural science, and a reductive ontology of science, namely, unity grounded in monistic materialism as articulated in physical science. This compound of epistemic and ontological positivism placed the social sciences in a position of inescapable inferiority vis-à-vis the natural sciences, and it relegated the humanities to a phantom realm of subjectivity.[6] It was against this consequence of positivism that Wilhelm Dilthey and the German neo-Kantians struggled to develop an alternative conception of the human sciences grounded in hermeneutic interpretation, only to run afoul of the epistemological scruples of the preponderantly positivist culture.[7] Logical positivism in the Vienna Circle was one of the direct responses to this quandary.[8] It is aptly termed neo-positivism, for it endeavored to restate, with the full force of dramatically enhanced symbolic logic and semantic theory, all the core epistemological tenets of the positivist tradition.

One of the essential points about this reassertion of positivism is its equivocal attitude regarding the place of philosophy itself vis-à-vis science. On the one hand, logical positivism professed to see philosophy reduced "to expressing the emergent synthesis of scientific knowledge," a role somewhat akin to that which Locke claimed for himself in relation to the Newtonian synthesis—the "underlaborer."[9] But at the same time, logical positivism held out for a strong sense of philosophy's sovereignty in the *justification* or legitimation of scientific claims to knowledge. Thus, against the well-publicized phenomenal empiricism of their approach, we must stress the primacy of their logicism. The Vienna Circle emerged precisely "to develop a view of science which would recognize the vital significance of logic and mathematics in scientific thought as systems of symbolic representations."[10] By contrast, post-positivism can be construed as the belated fulfillment of logical positivism's professed subordination of philosophy to science, the "naturalization" of epistemology.

The story goes—and my account would appear to conform to it—that the post-positivist theory of science was the result of the challenge posed primarily by Quine and Kuhn to the dogmas of logical positivism articulated by the Vienna Circle early in the century and still dominant in the philosophy of science after World War II. Recently, revisionists have demonstrated compellingly that by the 1930s logical positivism had articulated internally many crucial insights which post-positivism has been credited with achieving *against it* only decades later.[11] For example, a series of penetrating essays has demonstrated in Kuhn's work pronounced affinities to his ostensible *bête noire,* Rudolf Carnap.[12] Indeed, Kuhn's work appeared under the auspices of the Vienna Circle series *International Encyclopedia of Unified Science* and was also received by Carnap with cordially expressed approval, not out of politesse but from substantial philosophical concurrence. While Kuhn failed to grasp the invitation to further accommodation until long after Carnap's death, he did come increasingly to find points of tangency with the theories of one of Carnap's most important associates, Carl Hempel.[13] Much of the work of Kuhn *after The Structure of Scientific Revolutions* appears to have been an effort to achieve rapprochement with the (positivist) philosophy of science he seemed to have repudiated *in* it.[14]

Is, then, the whole idea of post-positivism misguided?[15] I think not. But we must be much more discriminating in formulating it. That logical positivism anticipated key tenets of post-positivism does not really undercut Kuhn's claim that in 1960 interpreters still labored under a misleading image of science, nor does it erase remaining important philosophical differences that Quine, Kuhn, and post-positivism generally *did* introduce in philosophy of science. Quine, for example, did not seek to *historicize* the a priori so much as to eliminate it. That went well beyond what Carnap and company envisioned; they clung to the analytic-synthetic distinction well into the 1950s. Kuhn's dissolution of the distinction of the context of justification from the context of discovery was similarly more drastic than anything Hans Reichenbach and the logical empiricists were prepared to countenance.[16] And Kuhn's introduction of a more radical historicism, though it had some affinities to what Karl Popper advocated in rivalry to logical positivism, went beyond anything Popper could countenance. As Evandro Agazzi has noted, the "Popperian philosophy of science" was "very far from an historicist way of thinking."[17]

The mention of Popper raises perhaps the most important point. By 1960 it was not logical positivism/empiricism alone which prevailed in the philosophy of science, for it faced the sharply critical rival view of Popper, whose *Logic of Scientific Discovery* was reissued in English with considerable éclat in 1959.[18] Kuhn's *Structure* might better be conceived historically in juxtaposition to Popper, not Carnap.[19] At least it was the Popperians who represented the most energetic opposition to Kuhn over the next decade.[20] Thus, the discernment of continuities between Carnap and Kuhn, important as this has been, does not confute the revolutionary impact of Kuhn's text; one must only reconsider its immediate context. The key idea was the historical contingency of scientific growth and change, something neither Carnap nor Popper would consider. Moreover, there were decisive features which Popperian philosophy of science shared with the logical positivists and empiricists: "Popper shared with the logical empiricists two essential tenets. The first is the conception of theories as deductive systems . . . ; the second was that comparison between theories, and thus the problem of justifying and interpreting theory change, was to be approached according to a deductive model focused on the relation of logical deducibility between the axioms of the theories and some single sentences belonging to them."[21] Post-positivism set about overthrowing just this commitment to the logical entailment or "sentential view of theories." The notion of semantic holism which undergirded post-positivism denied that the meaning of terms used in two different theories (languages) remained identical, and with this meaning variance all strict logical deducibility was undermined.

A central tenet of logical positivism/empiricism was the theory/observation distinction. It was only because observations were independent of theories that they could serve as evidential warrants to appraise the adequacy of theories, to ground theory comparison. "The observational vocabulary provided an objective ('theory independent') basis for interpreting and judging single theories."[22] But post-positivism would demonstrate definitively that this essential distinction could not be upheld. The most common formulation of this post-positivist principle is the "theory-ladenness of observation." The point is simply that "'observation terms' . . . are *not* . . . completely free from 'theory.'" There is *no* "neutral observational vocabulary." Instead, "*what counts* as an observation, and the *interpretation* or *meaning* of observation terms is at least partly [theory-] dependent."[23] In this sense, post-positivism picks up a line of criticism launched by an early antipositivist, Friedrich Nietzsche, who wrote famously: "facts are precisely what there is not, only interpretations."[24] The suggestion that observations always entail a theoretical frame for their discernment played a central role in Pierre Duhem's argument that "crucial experiments" could never definitively refute theories: adjustments could always be made regarding the background assumptions to excuse the failure of an experiment. This "Duhem thesis" is the centerpiece upon which post-positivism in science rests, as we will have occasion to explore in detail.

What post-positivism did was to generalize Duhem's principle into philosophy of language as a claim for "semantic holism." In other words, words only mean in sentences; sentences only mean in languages; and therefore one can grasp a specific semantic content only from the vantage of the entire language. As a corollary, a change in any element will ripple across the entire web of belief. These are decisive interventions associated with Quine's philosophy of language, and they are constitutive for post-positivist philosophy of science.

Quine wreaked further havoc with two other demarcation projects dear to logical positivism/empiricism: the analytic/synthetic distinction and the science/nonscience demarcation. From the vantage of Quine's analysis, neither could stand up, and Carnap's carefully crafted categorization of "internal" versus "external" questions fell into acute quandary. Kuhn and Feyerabend would contest one further demarcation, that between facts and values, by arguing for the theory-dependence of standards. The result was to make the very idea of a theory thoroughly problematic. For the logical positivists/empiricists it had signified primarily an interpreted axiomatic system. But Kuhn's historicization drastically extended the idea of theory to encompass background knowledge and to allow internal mutation within theories, rendering them thus difficult to recast in terms of formal logic. *Theory* became a term without a concept. "There is today," Shapere wrote in 1984, "no completely—one is almost tempted to say remotely—satisfactory analysis of the notion of a scientific theory."[25]

Two further consequences of the dissolution of the theory/observation distinction need to be noted. First, the question arises whether the logical positivist/empiricist model of logical entailment characteristic of formal-mathematical systems combined with protocol sentences for evidential warrant could serve as an adequate analogue for theories of causal interaction of real-world scientific process. That adequation has been fundamentally repudiated by post-positivists, and we now know that some of the logical positivists/empiricists themselves came to discern the inadequacy of this agenda. Second, the theory/observation distinction played a central role in the antimetaphysical disposition of logical positivism/empiricism, in other words, its decided preference for antirealism or instrumentalism. The key issue, here, is the ontological status of theoretical entities. For logical positivism, "it was all too easy to view the distinction between observational and theoretical as paralleling a distinction between existent and non-existent entities."[26] The more insistent their phenomenalist epistemology, the more glaring became the incongruity between the philosophy of logical positivism and the practice of empirical science. As Grover Maxwell summed up the state of affairs around the beginning of the 1960s, "the fact remains that the referents of most (not all) of the statements of the linguistic framework used in everyday life and in science are not sense contents but, rather, physical objects and other publicly observable entities."[27] That opens up a central dimension of philosophical contestation for post-positivism, the relation of ontology to epistemology. Much of the confusion and the distortion of arguments over the last half century has arisen out of the obscurity of this relation. As Robert Nola maintains, "ontological/semantic considerations of reference, extension and sense should be kept distinct from epistemological considerations."[28] Hilary Putnam makes the essential point: "a statement can be (metaphysically) necessary and epistemologically contingent."[29] For Nola, this was key to the proper assessment of Kuhn's claims: "Kuhn avoids considerations of ontology and semantics and resorts to considerations of epistemology."[30] These considerations bring us to a final, crucial issue between positivism and post-positivism, the distinction between the context of discovery and the context of justification.

The contestation of the distinction between the context of discovery and the context of justification will prove crucial for my interpretation of post-positivism in science studies. Upholding that distinction was central to logical positivism/empiricism's claims for the autonomy and indeed the a priori authority of philosophy vis-à-vis the empirical sciences. The sciences, or empirical inquiry more generally, set about discovering by hook or by crook how the world might be, but whatever results emerged had to be submitted to the sovereign judgment of philosophy to be warranted as valid or justified knowledge. Philosophy presumably possessed unimpeachable standards for such appraisal, typically the a priori categories of logic, whether timeless in the Kantian sense or relativized, in the sense of Carnap and others.[31] Shattering that posture, rejecting the possibility of a "first philosophy," is essential to post-positivism. That the logical positivists/empiricists contributed to the undermining of the posture which warranted their own enterprise is an irony that does not lessen the significance or finality of the rupture.

What conceptual work does the discrimination of the context of discovery from the context of justification do? The term *context* itself occasions some bafflement because it is remarkably vague for so crucial a demarcation. Does it discriminate temporal regions, such that the context of discovery is temporally distinct from and prior to that of justification? Or does it discriminate activities, such that certain actions constitute the one context and certain other actions the other? Or is it a question even more specifically of the governing motives or purposes of agents embroiled in the respective contexts? *Discovery* has also come under scrutiny, and the idea that science has more "invention" than "discovery" about it will be central to much postpositivist theorizing. But the most controverted notion is "justification." How is it to be understood? Is it even possible?

The debate over the distinction between the context of discovery and the context of justification compounded many controversies, all of them central to the post-positivist rupture. First, the whole question of temporal sequence—did discovery precede justification?—raised important issues about the actuality of scientific practice, which could hardly be so neatly divided. There were even more questions about the implicit division of labor that seemed to animate the distinction, namely, that scientists would be involved in the first task while philosophers would take over in the second. This was particularly vivid in the logical positivist/empiricist penchant to envision philosophy of science as the temporally subsequent but timelessly authoritative appraisal of "finished" theoretical systems of statements. Again, it raised the question of the role and right of the scientists themselves in the appraisal of their theories. Not only was the discrimination in this sense descriptively false, it also assigned authority over the appraisal of scientific achievements in a way that flouted the competence and the actual decisiveness of scientific practitioners. But only in this way, it appeared, could philosophy carve out a legitimate role for itself in relation to science.

The distinction had further senses. One of the most important was the implicit and sometimes quite blatant denial of rationality—of logic—to discovery. It was taken by the logical positivists/empiricists as inherently random, serendipitous, mysterious. Genius, luck, trial and error—anything but a logic of induction—seemed proper to its actual process.[32] Hence philosophers of science gladly relinquished that context to historians, sociologists, psychologists, or any other inquirers who felt prepared to work in the chaotic realm of happenstance.[33] The scientists who achieved determinate theories with truth-claims were themselves merely human, achieving their breakthroughs with luck and chance at play, with the vagaries of social and historical circumstance always pitted against the imaginative power of their genius and the rigor of their theoretical and experimental methods. But the theories they produced could legitimately be construed as truth-claims that could be appraised with the thorough lucidity and definitiveness of formal logic.

Analysis of the context of discovery, like that context itself, had to be irreducibly empirical. But the analysis of the context of justification, like that context, privileged formal logic. "[T]he point of Reichenbach's distinction is that information relevant to the *generation* of a scientific idea is irrelevant to the *evaluation* of that idea."[34] This was yet another form of the distinction of the "internal" from the "external," in which the internal was identified with the logical, and the external with the empirical or even the irrational. The thrust of post-positivism would be to argue both for the pursuit of a logic in discovery and for the importance of generative considerations in justification. In doing so, the contention would itself be empirical: that logic (induction or abduction, for example) did function in discovery and that social and psychological elements and, above all, culturally constructed values played a role in justification.

With the introduction of the concept of "values," the ultimate issue in the discrimination of the two contexts emerges: the distinction between the factual and the normative. The use of the distinction between discovery and justification in this instance underscored the essential autonomy of values for appraisal—normative standards—from empirical facts or theories. Norms could not be derived from empirical results; the abyss was utter. The fact/value discrimination, indeed, is the greatest of all the dogmas of positivism/empiricism. One of the most important elements in post-positivism is denying this abyssal divide. That is what "naturalized epistemology"—perhaps the most fruitful harvest of post-positivism—is all about.

Post-positivism in science studies deflates the pretense of philosophy to stand above and to dictate to the empirical sciences. But in doing so it thrusts not only philosophy but empirical inquiry more generally into profound perplexities about standards of appraisal. The challenge that post-positivism must face is how to reestablish rationality immanently out of human learning, discovery, and invention. That is the thrust of what has come to be called "evolutionary epistemology" or "naturalized epistemology."

As a preliminary formulation of what post-positivism challenged in the positivist legacy, the list composed by Clifford Hooker can serve admirably.[35] He noted ten challenges:

1. Theories cannot be reduced to observations;
2. Scientific method is not merely logical entailment;
3. Observation is not theory-neutral;
4. Theories do not cumulate historically;
5. Facts are theory-laden;
6. Science is not isolated from human individuals;
7. Science is not isolated from society;
8. Method is not timelessly universal;
9. Logic should not be privileged;
10. There is no gulf between fact and value.

CHAPTER 2

> We do not deny the possibility that the world is such that equally viable,
> incompatible theories of it are possible. We do not deny the possibility of
> the world's being unamenable to epistemic investigation and adjudica-
> tion, beyond a certain level. But whether or not the world is like that is
> itself an empirical question open to investigation. The answer cannot be
> preordained by a transcendent, epistemic skepticism.
> — LARRY LAUDAN AND JARRETT LEPLIN, "EMPIRICAL EQUIVALENCE
> AND UNDERDETERMINATION"

The Perils of Semantic Ascent: Quine and Post-positivism in the Philosophy of Science

Empirical inquiry in the human sciences appears to be beset by philosophical indeterminacies.[1] Their foremost exponent has been Willard van Orman Quine. Dealing with Quine, then, becomes incumbent upon those who wish to establish a space for empirical inquiry in the new philosophical world of post-positivism.[2] In a study devoted to the philosophical parameters within which historical practice as empirical inquiry must proceed, Murray Murphey dwelled long on the philosophy of Quine.[3] The issues in which Murphey embroiled himself appeared, for a very knowledgeable reviewer, to be remote from the concerns of disciplinary history.[4] Tacitly, what this reviewer assumed was a distinction between methodology and metaphysics. What Murphey realized was that this line could only be drawn by reckoning what lay beyond it. It is one of Hegel's most subtle and supple insights that the idea of limit or boundary requires a projection beyond it for its own constitution. To dispute the hegemonic claims of "first" philosophy upon empirical inquiry, one must first take up those claims.

A sophisticated and directly relevant consideration of the implications of Quine's philosophy for historical practice is Gary Hardcastle's essay "Presentism and the Indeterminacy of Translation." It brings Quine's conceptions of indeterminacy directly to bear upon the methodological disputes between presentist and reconstructionist approaches to history of thought and science.[5] Contending that Quine's concept of "translation" bears tellingly upon historical reconstruction, Hardcastle raises questions about access and warrant in historical interpretation, targeting in particular Quentin Skinner's methodological precepts.[6] Thus, the obligation of historians to answer just the sorts of questions Hardcastle raises impels them to a careful study of Quine and post-positivist philosophy in general.

Raymond Weiss more than twenty years ago discerned the potential in Quine's philosophy for a postmodern reading, which Weiss termed, somewhat confusingly, "historicism."[7] He wrote, "[T]he seeds of historicism are present in his theory of knowledge—in spite of the positivist vestiges still there. He is, as it were, in a half-way house between positivism and historicism, thus exemplifying an ever growing tendency in philosophy to move toward historicism. This movement in America brings it within hailing distance of continental philosophy."[8] Taking up Quine's philosophy of language, Weiss maintained that "language cannot reach the thing in itself," hence "empirical evidence cannot provide a secure starting point" and "in the final analysis we know only what we have constructed, the objects we ourselves have posited."[9] Because Quine's epistemology is "infested with historicism," it "fosters cultural relativism." Indeed, Weiss proclaimed, "Quine's theory of knowledge . . . heralds the demise of reason."[10] Since "Quine contends that a choice can be made among different theories [only] on pragmatic grounds," and since a "pragmatic standard . . . can justify the most fantastic conceptual schemes," Weiss maintained, "reason is now impotent to pass judgment upon the end," and "the once hallowed distinction between science and myth is no longer tenable."[11] Thus, "homeric gods have the same cognitive status as physical objects."[12] Of course, this is a caricature of Quine, but the point of caricature, after all, is that it catches *some* semblance even in its distortion. A lot of theory taken today as established (what "everyone knows or ought to know," in the idiom of all too many) is the result of inept reception of philosophical arguments.[13] Weiss illustrates nicely this sort of misappropriation based on partial comprehension.

What is beyond dispute is that Quine's ideas have exerted an enormous influence upon the philosophy of the post-positivist period. He is responsible for introducing what he recently termed "three indeterminacies": the underdetermination of theory by evidence, the inscrutability of reference, and the indeterminacy of translation.[14] For Quine, positivism in its final form of logical empiricism came to a dead end when it recognized "the hopelessness of grounding natural science upon immediate experience in a firmly logical way."[15] In his epoch-making essay "Two Dogmas of Empiricism" (1950), Quine targeted two prime considerations which had governed logical positivist thinking, the analytic-synthetic distinction, on the one hand, and reductionism, the notion that "each sentence has its own separate empirical content," on the other.[16] If, after Gottlob Frege, philosophy had taken the sentence as the proper unit of meaning and reference, logical empiricism had to recognize that even "a statement about the world does not always or usually have a separable fund of empirical consequences that it can call its own."[17] That is, "the empiricist is conceding that the empirical meanings of typical statements about the external world are inaccessible and ineffable."[18] Thus, finally, "the Cartesian quest for certainty," the quest for an absolute foundation, "was seen as a lost cause."[19] There can be no first philosophy, no foundationalism. Quine concludes that we are left only with linguistic or semantic "holism" and "ontological relativity."[20]

The strategy of this chapter is to reconstruct each element of Quine's ensemble of indeterminacies and to assess their independent warrant, their interrelation, and their mutual support. The place to begin is with the "Two Dogmas" essay and the so-called Duhem-Quine thesis.

The Duhem-Quine Thesis

The debate surrounding the Duhem-Quine thesis has been intense and central to the philosophy of science in the post-positivist era.[21] Allegedly articulated by the French philosopher of science Pierre Duhem in the first decade of this century, the thesis only attained prominence when it was taken up and extended by Quine in the 1950s as part of his decisive attack on the "dogmas of empiricism."[22] Recent historical work has established that Otto Neurath played an important intermediary role between Duhem and Quine by elaborating Duhem's observations on physics into a general linguistic holism.[23] This is a significant piece of the larger archaeology exposing logical positivism's own internal disassembly.[24] But the rise of *post*-positivism set in only with Quine's evocation of Duhem in his 1950 essay.

Ironically, the Duhem-Quine thesis has become a *new* dogma of empiricism, or a *dogma* of the new empiricism.[25] The challenge is to clarify what exactly the thesis signifies. Moreover, how does it relate to two other core elements of the new empiricism, Quine's famous doctrines of the underdetermination of theory by data and the inscrutability of reference, especially in connection with what has been called the theory-ladenness of facts, the *second* dogma of the new empiricism?[26] Do they indeed nest together into a network of mutually implicating ideas, or must they be disentangled? Are their merits disguised or highlighted by their elision into one another? How and why has the unified thesis come to be interpreted—indeed, extended further—by those who have taken up the issue in the wake of Quine's essay? What is the *current status* of the thesis? These must be our concerns.

Quine first enunciated what came to be known as the Duhem-Quine thesis in his "Two Dogmas" essay, using his own terms and ascribing the idea to Duhem in a footnote reference. Quine wrote, "our statements about the external world face the tribunal of sense experience not individually but only as a corporate body."[27] His footnote says merely: "This doctrine was well argued by Duhem."[28] In "Two Dogmas in Retrospect" (1991) Quine observed, as a "matter of curiosity," that he knew little of Duhem when he composed the 1950 essay and added the footnote only in response to comments from Carl Hempel and Philip Frank.[29] Adolph Grünbaum then initiated the controversy by challenging what he called The Duhemian Argument but was in fact Quine's new reading of it.[30] In his second essay on the question, Grünbaum coined the phrase *Duhem-Quine thesis* itself.[31] Others—notably Larry Laudan and Imre Lakatos—took up the issue, in the latter instance drawing the Duhem-Quine thesis into the debate about the other great fountainhead of post-positivism in the philosophy of science, Thomas Kuhn's doctrine of incommensurability.[32]

With the Duhem-Quine thesis now a matter of contemporary philosophical debate, philosophers began to ask whether Duhem actually said what the Duhem-Quine thesis imputed to him. In other words, is there *one* Duhem-Quine thesis or are there, rather, two theses, that of Duhem and that of Quine, with distinctly different purport?[33] To try to access what a philosopher actually argued is, of course, one of the more disputed practices of historical reconstruction.[34] Nonetheless, historians of philosophy persist in doing this, and it has not proven fruitless in disciplinary clarifications, as is witnessed by this instance. Laudan initiated the historical question by claiming that the thesis which Grünbaum disputed, whatever its merits, was *not* the thesis Duhem had advocated: "he has misconstrued Duhem's views on falsifiability. . . . [T]he logical blunder which he discussed should not be ascribed to Duhem, but rather to those who have made Duhem's conventionalism into the doctrine which Grünbaum attacks."[35] Laudan distinguishes a strong from a weak form of the Duhem-Quine thesis, claiming that only the weaker form was in fact advocated by the historical Duhem. As Laudan understands the strong form, it goes as follows: "Any theory can be reconciled with any recalcitrant evidence by making suitable adjustments in our other assumptions about nature."[36] Grünbaum's attempt at refutation bore on only the strong form, and thus Laudan claimed that the "authentic" version of Duhem's argument was "untouched by Grünbaum's critique."[37] But what exactly was this authentic version?

It is worthwhile to juxtapose the respective theories in their simplest terms. Duhem wrote: "an experiment in physics can never condemn an isolated hypothesis."[38] The central argument is this: no theoretical idea can be tested without the inclusion of a set of auxiliary assumptions necessary to specify its observational consequences for testing. Therefore, the outcome of the experiment bears not upon the simple theoretical claim but upon the ensemble of that claim plus all the auxiliary notions necessary to undertake the test. Perhaps the clearest articulation Duhem offered of what he was claiming is the following passage: "The only thing the experiment teaches us is that among the propositions used to predict the phenomenon and to establish whether it would be produced, there is at least one error; but where this error lies is just what it does not tell us."[39] Quine writes: "our statements about the external world face the tribunal of sense experience not individually but only as a corporate body." Duhem is decidedly more concrete in the domain under consideration, namely physics, and also quite determinate in his claim that isolated hypotheses cannot be refuted by experiment. Quine talks about statements as "a corporate body" (linguistic "holism") and also invokes "the tribunal of sense experience" (which raises the question of Quine's peculiar strain of behaviorist empiricism). This is not to say that such a juxtaposition settles anything. It simply equips us to turn to the exploration of the difference between the historical Duhem thesis and the "Quinery" (in the endearing phrase of Ian Hacking) that has surrounded it.[40]

Philip Quinn has tried to establish that difference in an essay entitled "What Duhem Really Meant."[41] He claims that Duhem presumed that the "separability" thesis—the idea that "no single or individual theoretical hypothesis *by itself* has *any* observational consequences"—implied the "falsifiability" thesis—namely that such an individual theoretical hypothesis could not be falsified by any observations.[42] Quinn doubted this implication, but his ensuing exchange with Nancy Tuana forced him to admit that the inference *was* entailed. As Quinn notes, "Duhem never doubted that experiments can refute whole theoretical systems; he was only concerned to argue that no single theoretical hypothesis can be falsified by any observations."[43]

For our purposes, the essential issue here is what exactly distinguishes such an "individual theoretical hypothesis" from a theory as a whole. It is important to note that one of the decisive provocations inciting Duhem's argument was the doctrine of the crucial experiment. It dominated late nineteenth-century philosophy of science as the model of rationality in theory choice. A crucial experiment juxtaposed two rival theories in terms of their mutually exclusive predictions of an experimental outcome. Thus, the outcome adjudicated unequivocally between them. One could logically conceive this as a disjunctive syllogism—if theory A is false, then theory B must be true. Through such crucial experiments, a theory could be refuted or at least shown decisively inferior to its surviving rival. But Duhem emphatically denied that there could be such crucial experiments.

If Duhem held both that crucial experiments cannot occur and that (whole) theories can be refuted, the issue arises: *how* are such theories refuted? And what makes a theory *whole;* in other words, what are the limits of a theory? As Quinn notes, "if the only theoretical system large enough to be falsifiable as a whole is the totality of our knowledge and beliefs," the consequences for falsification are drastic.[44] Quinn's discussion of crucial experiments seems to suggest that Duhem was already operating with a form of the underdetermination thesis, since Duhem's argument against the disjunctive syllogism was that "the class of all possible alternatives to a given hypothesis is, at least potentially, infinite" (hence there is always the possibility of theory C or D, and the disjunct is never complete).[45] That seems very close to the claim that, for any given base of empirical data, logically distinct alternative theories can always be developed, which epitomizes Quine's notion of the underdetermination thesis. The mutual implication of the Duhem-Quine thesis and the underdetermination thesis accordingly becomes a crux of current theoretical dispute.

Nancy Tuana, in her brief but powerful emendation to Quinn, adds further important perplexities by raising the question of "the truth value of observational statements" for Duhem.[46] For Tuana, Duhem clearly asserted that all such "data" must entail and consequently be compromised by, theoretical presuppositions. As Duhem himself put it, "without theory it is impossible to regulate a single instrument or to interpret a single reading."[47] All experimental results are always *interpreted* in terms of the theories of the apparatus employed in the experiment; these constitute a massive component of the auxiliary assumptions necessary to test the hypothesis under consideration. Consequently, experimental data or "descriptions of observational results involve inductively based propositions . . . [and] can never [provide], as a matter of principle, conclusive evidence."[48] They are clearly "amongst the challengeable presuppositions of the experiment" for Duhem.[49] Tuana has made the notion of the theory-ladenness of data central to Duhem's thesis. If in Duhem's original argument there were anticipations of the underdetermination of theory by data or the theory-ladenness of facts, then Quine's appropriation of Duhem for his own purposes and the emergence of the full-blown Duhem-Quine thesis becomes even more difficult to assess.

Gary Wedeking concludes that Quine's notorious line, "any statement can be held true come what may, if we make drastic enough adjustments elsewhere in the system," simply extrapolates vastly beyond Duhem's position.[50] So does Carlo Giannoni: "Essentially Quine is proposing an extension of Duhem's thesis of conventionalism in physics to include the truths of logic as well as *all* of the laws of science."[51] Like Quinn, Giannoni emphasizes the holism of Quine's reformulation: "'The unit of empirical significance is the whole of science.' Only science as a whole can be verified or falsified."[52] That is, "only science as a whole, including the laws of logic, is empirically testable."[53] This is a drastic claim, and, Giannoni notes reasonably, "We anticipate from Quine at this point a defense as strong and as hard hitting as his critique of the analytic-synthetic distinction. But our expectations are not fulfilled."[54] That issue will concern us anon. The important point is that the undertaking described is a vast generalization beyond Duhem's original claim.

Jules Vuillemin argues that one should segregate Duhem's thesis from Quine's on four counts.[55] First, in terms of their respective methodologies of approach, Duhem was distinctively historical in his construction of argument, while Quine is logical and formal. Second, there is the difference in scope; Vuillemin insists that Duhem restricted his claims to physical science, explicitly excluding physiology from his argument, whereas Quine's argument encompasses everything: "mathematics and logic on one hand, linguistics on the other, are here on a par with . . . physical theory."[56] That entails a third difference: in the "articulation [of] the body of scientific laws."[57] Where Duhem took seriously differences among sciences, and between science and other domains of knowledge or experience, Quine appears to annul any such categorical divides. For Vuillemin, this universalism brings the indeterminacy of translation of language (formal or natural) into intimacy with the "empirical underdeterminacy" of physical theory. Quine's holism entails that "mathematics, language, and physics belong to the same fabric of human knowledge."[58] Thus, "a sentence of physics has the same underdetermined relation to experiments as a sentence in natural language has to the things and events which it is about," as Vuillemin understands Quine.[59] Moreover, "the apparatus of quantification and of identity is only parochially determined," that is, ontology is always relative to theory, or reference is always inscrutable ("somewhat arbitrary and subjective," in Vuillemin's formulation).[60] Finally, Vuillemin articulates the different metaphysical/ontological commitments that these views entail for Duhem and Quine, respectively. Duhem's "instrumentalism" was tied, historically, to his religious and metaphysical commitments; Quine's "pragmatism" entails ontological commitments (to natural science) with epistemological consequences: what has come to be called "naturalized epistemology."

Henry Krips goes so far as to insist that "Quine and Duhem radically disagree . . . [and hence] there is no univocal doctrine which can be identified as *the* Quine-Duhem epistemology."[61] Krips raises questions about the wholeness of theory and about the quality of the observational evidence which challenges it, like Quinn and Tuana. He also explores the important question of the strategy of theory revision: in other words, what rules or guidelines motivate the selection of particular revisions of the theory in light of the negative evidence. Krips stresses an important criterial word—*rational*—in the views of both Quine and Duhem. The strategy of adjustment turns on *rational* response to negative results. Quine, according to Krips, thinks it "*not* rational to continue . . . ignoring recalcitrant experiences indefinitely," despite his "holism."[62] Yet Quine could clearly be suspected of holding the rejected view, based on the key passage from his "Two Dogmas" essay: "Any statement can be held true *come what may*, if we make drastic enough adjustments elsewhere in the system" (my emphasis).[63] Which posture is the true Quine? Krips claims, fairly, that Quine has been on both sides of the issue.

Krips finds Duhem, on the other hand, quite concrete in confining the problem within the experimental exploration of a physical hypothesis. In that particular setting, according to Krips, Duhem emphasized that "the observation statements describing the experiment and its result presuppose some theory—in particular, an instrument theory to interpret meter readings in terms of the values taken by quantities which are referred to in the observation statements. Also, for any such experimental situation there is an error theory—which estimates by how much the meter readings may *diverge* from the values of the quantities they are supposed to measure."[64] The materiality of Duhem's way of articulating his thesis was not inconsequential for its scope and persuasiveness. Indeed, Duhem's concreteness ties in profoundly with recent efforts to get at science as "practice."[65] On questions of strategy in the face of adverse results, Krips finds that Duhem wrote only of "good sense," and he elaborates, "What Duhem seems to be saying here is that the reasons of scientists are inscrutable: their 'knowledge' of how to pick hypotheses is (in more modern terms) 'tacit.'"[66]

Like Krips, Roger Ariew insists that Duhem would not subscribe to the Duhem-Quine thesis. Indeed, "Duhem himself would not have been able to recognize what is attributed to him."[67] Moreover, for Ariew, "Duhem's actual thesis is much superior to the thesis normally attributed to him."[68] Employing Quinn's categories, Ariew argues: "Duhem is claiming that the falsifiability thesis is a result of the separability thesis, and that the separability thesis is an empirical thesis depending upon factors that do not govern all sciences."[69] Ariew suggests that Duhem explicitly restricted it to physical science. That means that "Duhem does not claim that no statement of science *qua* science ever confronts observation directly."[70] He even admitted cases in physiology where some *did*. This sort of restriction of the claim *never* appears in the contemporary applications of the Duhem-Quine thesis. Along the same lines, it is physical science, not any and all science, that Duhem conceived as a whole. He wrote: "Physical science is a system that must be taken as a whole; it is an organism in which one part cannot be made to function without the more remote parts coming into play, some more than others, but all to some degree."[71] For Ariew, this holism is unwarrantedly generalized and transposed into the "underdetermination thesis," which Duhem never held or conceived of holding.

Two technical discussions of the Duhem-Quine thesis explore more concretely the question of strategies of revision. Ulrich Gähde and Wolfgang Stegmüller contend that Quine's approach would seek to preserve both "(1) those statements, which describe experience directly [what Quine calls "observation sentences"], and (2) those which, intuitively speaking, are furthest away from these 'observational statements,' i.e., the most fundamental and general laws."[72] That is, adjustments will tend to be made in the mediations between those general laws and the concrete observational data. The authors also ascribe to Quine a commitment, in the strategy of revision, to a "correction which 'disturbs the total system as little as possible,'" a rule which has been labeled by other commentators as "conservatism."[73] Quine himself accepts this "hierarchy of options of revision, ranged according to disruptiveness" as an "objective" formulation of his view.[74] In later essays, he dubs it the "maxim of minimal mutilation."[75]

In another essay Yuri Balashov draws upon Popper and Lakatos to try to specify this strategy of revision more tightly and in that measure heighten the falsification potential of experiment. His thrust is to emphasize the possibility of "independent support" for elements in the "inferential string" of assumptions linking the hypothesis under investigation to its observational test result. "What matters is the degree of independent corroboration enjoyed by the corresponding elements of a theory."[76] The more such support, the less likely these elements should bear the burden for the failure. In addition, Balashov invokes Lakatos's notion of progress and degeneration in research programs to suggest that multiple testing and modification can identify proper targets for revision in terms of their fruitfulness for future predictions that can be tested.[77] Quine's own later writings show strong resonances of just such Popperianism.[78] Balashov, however, concludes that these principles provide substantial escape from the holistic indeterminacy that seems to be the thrust of the Duhem-Quine thesis.

Finally, John Greenwood, in an excellent critical appraisal of the "dogmas of neo-empiricism," including the Duhem-Quine thesis, makes the case that a rationally warranted version must be construed quite narrowly. He argues that while a given observation might be dismissed, a series of such anomalous observations is not so easily evaded. Similarly, auxiliary hypotheses are not always so easily tampered with. As he puts it, "there is usually a set of exploratory theories about interventive techniques and instruments, which cannot be seriously questioned without undermining the prior support for the explanatory theory [the one under scrutiny], based upon precisely these techniques and instruments."[79] Thus, Greenwood urges, "the 'web of belief' tends to reduce rather than increase the room for intellectual maneuver when faced with recalcitrant observations."[80] Greenwood is clear that modification of auxiliary hypotheses to salvage a theory *has* occurred in the history of experimental science, but he denies "the common assumption that exploratory theories can always be modified in the face of anomalies to preserve the evidential equivalence of an explanatory theory with respect to its rivals."[81] Instead he urges a more dynamic model of the dialectical interplay of exploratory and explanatory theories. New exploratory theories make possible the testing of new explanatory hypotheses and these, in turn, stimulate the elaboration of new exploratory theories to test them.

What Greenwood's argument portends is a narrowing of focus, once again, onto the concrete circumstance of scientific practice in experimentation and the dialectic of theory and experiment, as against the linguistic-logical universalism of Quine. What it also suggests is that there is resistance intrinsic in both observational data and the accumulated basis of auxiliary hypotheses (exploratory theory) which constrains the experimenter such that "it simply cannot be presumed that there will always be some logically possible formal accommodation of an anomaly which preserves evidential equivalence between competing explanatory theories."[82]

This tallies with Ian Hacking's insistence that mature laboratory sciences "lead to an extraordinary amount of rather permanent knowledge." Hacking elaborates that while it is "utterly contingent" how these sciences got this way—in other words, the *path* is ineradicable from the *result* (or, history matters)—it is very hard to change them. "[A]s laboratory science matures, it develops a body of types of theory and types of apparatus and types of analysis that are mutually adjusted to each other. . . . They are self-vindicating in the sense that any test of theory is against apparatus that has evolved in conjunction with it—and in conjunction with modes of data analysis."[83] Duhem's original thesis seems to fit more closely such empirical assessments of scientific practice in terms of experimental behavior than would Quine's semantic globalization. In turn, this makes the repeated invocations of the Duhem thesis that have appeared in recent years seem all the more plausible. Nonetheless, in embracing Duhem they also often unwittingly embrace Quine's amplification.

What we may gather from this consideration of the Duhem-Quine thesis is that Quine has extended the original proposition of Duhem in two crucial manners. First, he has introduced the radical phrase "come what may" into the considerations, so that he can be taken to have endorsed "drastic" rearrangement of the web of our beliefs not merely as possible but as plausible. Second, he has practiced a "semantic ascent" upon Duhem's empiricism, transposing an issue of theory and experiment into an issue of language and, indeed, situating the issue at the most *holistic* level within this scheme of semantics.[84] Thus, Quine offers phrases like: "the unit of empirical significance is the whole of science."[85] At that level of abstraction, there could be no real distinction of the whole of science from the whole of language. This twofold extrapolation of Duhem carries us into utterly new terrain in which it is not clear that Quine's holism leaves any room for his empiricism. That becomes all the more problematic when we see how Quine and his inter-preters take the Duhem-Quine thesis to be synonymous with Quine's under-determination thesis.

Conversely, we have seen that Quine's critics have tried to circumscribe, to contain the penchant toward hyperbole in Quine's extrapolations, to bring the discussion back to the terrestrial haunts of scientific practice of the original Duhem thesis. This contest of constraint against hyperbole will be-come quite familiar as we proceed. It is, in this light, highly noteworthy that in his essay "Two Dogmas in Retrospect," Quine writes: "Looking back on it, one thing I regret is my needlessly strong statement of holism. 'The unit of empirical significance is the whole of science. . . . Any statement can be held true come what may.' This is true enough in a legalistic sort of way, but it di-verts attention from what is more to the point. . . . In later writings I have in-voked not the whole of science but chunks of it, clusters of sentences just inclusive enough to have critical semantic mass."[86] That is, Quine himself repudiates extreme holism. On the question of the adjustment of the theory cluster to recalcitrant results, Quine recognizes a "maxim of minimal muti-lation" that particularly exempts mathematical principles, since adjustment there "would reverberate excessively through the rest of science."[87] In gen-eral, Quine acknowledges that adjustments in the theoretical cluster aim ra-tionally "to maximize future success in prediction."[88] On the question of how holistic Quine's moderate holism in this revised form still remains—in other words, how "global" the theory or language must be that is testable—Quine writes: "a testable set or conjunction of sentences has to be pretty big, and such is the burden of holism."[89] Even in this revised form, it should be noted, Quine appears to view holism not only as the upshot of Duhem's claims but also as the essential point of his own notion of the underdetermi-nation of theories by empirical data. We must turn to the relation of the Duhem-Quine thesis to the underdetermination thesis.

The Underdetermination of Theory Thesis

Once again, it will prove crucial to ascertain as crisply as possible exactly what the underdetermination doctrine maintains. Again, ambiguities abound, both from Quine's own pen and from those trying to parse his meaning. The upshot of a review both of Quine's texts and those of his commentators is that there are *at least* two versions of the underdetermination thesis, a mod-est and a drastic one, a weak and a strong one, with radically different im-plications.[90]

The simplest and most explicitly articulated position is what Laudan has called the "nonuniqueness" of theories.[91] Paul Roth articulates this succinctly: "it is possible to formulate empirically equivalent but logically incompatible scientific theories."[92] That is, Quine maintains that one can always conceive at least one other theory which is compatible with the same body of evidence but is logically distinct from any given theory. Quine initially took this to be relatively unproblematic.[93] In trying to defend the even more controversial thesis of "indeterminacy of translation," Quine confidently invoked his notion of underdetermination of theory: "Let us put translation aside for a while and think about physical theory. Naturally it is underdetermined by past evidence; a future observation can conflict with it. Naturally it is underdetermined by past and future evidence combined, since some observable event that conflicts with it can happen to go unobserved."[94] To this point, underdetermination would only seem to betoken the fallibilism of empirical research, that is, of *local* scientific theories undergoing elaboration in real time. Thus, the reason that Quine expects "wide agreement" is that it is intuitively clear that the theories we actually employ are drawn on limited data and can always be disconfirmed. But Quine in fact has a slightly different notion, namely, that there are always elements in a theory that are at once necessary as integument but not directly linked to any observation. One may formulate that in terms of a distinction between observation and theoretical sentences.[95] The underdetermination thesis thus maintains that a change in some theoretical sentence would leave every observation sentence unaffected.

Initially Quine appeals to matter-of-fact, actual scientific practice as we grasp it in real time. But then he insists that philosophy is a question of *principle*.[96] And so he radicalizes the argument:

> Moreover many people will agree, far beyond all this, that physical theory is underdetermined even by all *possible* observations. Not to make a mystery of this mode of possibility, what I mean is the following. Consider all the observation sentences of the language: all the occasion sentences that are suited for use in reporting observable events in the external world. Apply dates and positions to them in all combinations, without regard to whether observers were at the place and time. Some of these place-timed sentences will be true and the others false, by virtue simply of the observable though unobserved past and future events in the world. Now my point about physical theory is that physical theory is underdetermined even by all these truths. Theory can still vary though all possible observations be fixed. Physical theories can be at odds with each other and yet compatible with all possible data even in the broadest sense. In a word, they can be logically incompatible and empirically equivalent.[97]

Quine's passage now asserts a *global* thesis of underdetermination. It is global in two crucial senses that need to be attended. First, it is global in conceiving "physical theories" as total "systems of the world," in the language of Quine's 1975 essay.[98] Second, it is global in that the evidential base that is postulated is *all possible observations*. That is, the claim of the strong underdetermination thesis is that, even if all conceivable data were available, it would still not be possible to establish a uniquely correct theory of the physical world.

Does *any* form of the Duhem-Quine thesis warrant global underdetermination?[99] It is certainly not obvious that it does. Paul Roth places a great deal of weight upon discriminating between the Duhem-Quine thesis and the underdetermination thesis.[100] Yet Quine is not very cooperative in keeping these distinct. Roth recognizes: "since Quine does not clearly distinguish in his early writings between (DT) [the Duhem thesis] and (UT) [the underdetermination thesis], he assumes that whatever (DT) implies, (UT) does also."[101] This is particularly the case in "Two Dogmas of Empiricism," where the holism of the former thesis is articulated in terms of its underdetermined implications.[102] But Roth insists that Quine does undertake to distinguish between the two theses in later writings, and he highlights as decisive the revisionist essay "On Empirically Equivalent Systems of the World" (1975).

Clearly, Quine engaged in considerable "backtracking" (to take up Laudan's uncompromising terminology) in that 1975 essay.[103] There Quine did distinguish between the two theses, terming the Duhem thesis "holism" and elaborating a new formulation of his underdetermination thesis.[104] But he did not wish at all to sever the logical connection between the two. As he put it, "the holism thesis lends credence to the underdetermination thesis."[105] Laudan refuses to accommodate Quine's backtracking in the 1975 essay, "because Quine's holism is often (and rightly) seen as belonging to the family of underdetermination arguments, and because it has become customary to use the term underdetermination to refer to Quine's holist position."[106] Unfortunately, Laudan takes holism to signify the *strongest* formulation of Quine's underdetermination thesis, which he labels "egalitarianism" of theories, and which he distinguishes from the weaker form of "nonuniqueness" of theories.[107] In glossing Quine's "holism" this way, however, I believe Laudan makes an interpretive error. What Quine appears to have meant by "holism" in 1975 is, rather, what Roth claims, namely, an argument about the limitations entailed in the verification of a particular hypothesis within a theory.[108]

Roth discerns that by holism Quine was distinguishing the Duhem thesis, about which he remained confident, from the underdetermination thesis, especially in its *global and in principle form*, about which he had developed substantial doubt. Roth takes Quine's revised position as follows: "As Quine now sees the matter, underdetermination does *not* concern the intra-theoretic dependence of terms and sentences, as does (DT), but emphasizes the methodological/logical problems of the possible compatibility of a set of observations with theories which are, in turn, not compatible with one another. . . . (DT) is concerned with the internal relation of theoretical terms and sentences; (UT) is concerned with the relation of distinct theories compatible with the same body of evidence."[109] Thus, Roth claims that "the Duhem thesis is consistent with either the truth or the falsity of (UT)," and that "at best, Quine believes, (DT) makes plausible the claim that (UT) is true."[110] Like Roth, Carl Hoefer and Arthur Rosenberg identify Quine's "holism" of 1975 with the Duhem thesis in order to distinguish it *from* and thereby clarify different forms *of* the underdetermination thesis.[111] Yet Quine himself, in his reply to Roth's paper, asserts that the relation between the Duhem-Quine thesis and the underdetermination thesis is stronger than Roth wishes to suggest, that indeed, the former *does* imply the latter.[112]

Certainly, Quine made gestures repudiating a global understanding of holism. Thus, he wrote: "It is an uninteresting legalism . . . to think of our scientific system of the world as involved *en bloc* in every prediction" (as if this "legalism" were someone else's work!).[113] In the pivotal essay of 1975, Quine wrote: "We can . . . appreciate how unrealistic it would be to extend a Duhemian holism to the whole of science, taking all science as the unit that is responsible to observation. Science is neither discontinuous nor monolithic. It is variously jointed, and loose in the joints to varying degrees. . . . Little is gained by saying that the unit is in principle the whole of science, however defensible this claim may be in a legalistic way."[114] But it is not clear that Quine salvages the coherence of his overall position or even grasps the full implications of his repudiation of this legalism as impractical while still insisting it is defensible. His holism and his verificationism are in acute tension, if not outright contradiction.[115]

Roth notwithstanding, it is not clear that Quine ever extricates himself from the early confusion between the weak and the strong forms of the Duhem thesis and their embroilment in the various underdetermination theses. Even in the 1975 essay, as Laudan correctly charges, "despite his concession [in his 1962 letter to Grünbaum] that the D-Q [Duhem-Quine] thesis is untenable in its nontrivial version, Quine was still defending his holistic account of theory testing."[116] That is, Quine promoted a holistic version of the Duhem thesis in which the only unit that could be tested was *global science*, a total theory.[117] Such an extreme holism, we have seen, is by Quine's own admission an egregious basis for evaluation of theory in real science.[118] Trimming holism down to the original terms of the Duhem thesis should bar Quine's recourse to the extravagant global version of his underdetermination thesis in any but the most vacuous sense of logical possibility.

It is important to pause a moment and reflect on the *sort of claim* involved in global underdetermination and to ask what this has to do with science as it is actually practiced. Murray Murphey has made a very important point in this connection:

> [I]n setting up the strong form of the thesis, Quine converts the indefinitely extended process of scientific inquiry into a finished state in which all possible observations have been made. . . . Quine's "limiting case" of scientific knowledge is not really a case of scientific knowledge at all, but a case of postscientific knowledge in which all inquiry has ceased. . . .
>
> But ruling out further inquiry is illegitimate: it supposes the completion of an indefinitely extended process of scientific inquiry in a finite time . . . and it supposes that there is a determinate set of all possible observation sentences that can be expressed in our language despite the fact that the language itself is constantly changing as a result of scientific inquiry.[119]

Along Murphey's lines, two objections to global underdetermination immediately present themselves.

First, there is no compelling reason to accept Quine's argument for the inevitability of a global underdetermination. As A. C. Genova puts it: "Everyone would presumably agree that theory is underdetermined by available evidence, but why should we be obliged to believe that it is underdetermined by all *possible* evidence?"[120] Quine himself seems to have grown more doubtful in the years after writing the cited passage. In Murphey's view, Quine's 1975 essay offers only an "exceedingly qualified" assertion; in other words, it is "an unproven hypothesis he believes to be true, although he admits it is unprovable."[121] Quine says the same thing, in so many words, "This, for me, is an open question."[122] Under the pressure brought to bear upon his position, his latest word on these matters is revealing: "The fantasy of irresoluble rival systems of the world is a thought experiment out beyond where linguistic usage has been crystallized by use."[123] An unkind gloss would be: it is arrant speculation. Laudan and Jarrett Leplin make the essential case in the epigraph cited at the head of this chapter. There is no point to the global assertion; it is an a priori claim without the necessary transcendental warrant. Empirically, only the contingency and fallibility of inquiry in real time have bearing. Science has long since acknowledged this, and it takes such local underdetermination robustly in stride.

Second, it is not clear what implications for actual science—or more broadly, for empirical epistemology—follow even if we postulate Quine's thesis. For the global thesis can have two readings. First, it can be the reflection that it is logically possible that we may never get a definitive, total theory of the world, simply because our empirical science is too weak an instrument. That appears to be the upshot of Quine's 1975 view.[124] Roger Gibson notes that this "amounts to reducing the doctrine of under-determination from the status of a *theoretical* claim . . . to that of a *practical* claim about what is humanly possible."[125] One can surely countenance such a sobering *conjecture* without either having to assume it *must* necessarily turn out this way or despairing of the construction of partial theories in the meanwhile. Such, I maintain, is the preponderant attitude among practicing researchers.

On the other hand, one *could*, with Cartesian abandon, mount this into a reinvocation of "evil demon" skepticism and argue that it renders all theories suspect as the night proverbially renders all cows dark. But is there any merit in that position? Laudan evidently does not think so. With Leplin he urges that the argument is "out of proportion to the conceptual credentials of the basic idea of empirical equivalence."[126] And in his own voice he is far harsher, terming radical inferences from underdetermination "hollow, anti-intellectual sloganeering."[127] Clearly, something inflammatory is involved here. As Laudan presents matters, the underdetermination thesis "is presupposed by many of the fashionable epistemologies of science of the last quarter century" as the "central weapon" in undermining *any* form of empirical epistemology.[128] "Sloppy formulations of the thesis of underdetermination have encouraged authors to use it—sometimes inadvertently, sometimes willfully—to support whatever relativist conclusions they fancy."[129] The stakes, it would appear, are enormous.

As Laudan understands the position, the radical extrapolation from the strong underdetermination thesis offers the following line of contention. If one grants that there is underdetermination of theory by evidence *in principle*, discrimination among theories becomes *in principle* arbitrary, hence all theory is equal. Or, in his terminology, there is "cognitive egalitarianism"; in other words, *"every theory is as well supported by the evidence as any of its rivals."*[130] As it stands, this has the form of an entirely a priori argument. Laudan ascribes this position to Quine, though he acknowledges that Quine neither articulated it nor would welcome its ascription to him.[131] As Laudan sees it, it follows nonetheless from the postures Quine has (at least at times) assumed.[132]

There are, then, two issues here. First, is this a fair ascription to Quine? One could well hedge here, but Quine has written some extraordinarily hyperbolic lines. In my view, Laudan's intuition about the ultimate *significance* of Quine's deployment of holism in 1975 turns out to be far more plausible than his specific rendering of Quine's *term*. Second, whether or not it is fair, has Quine's position *been taken to entail literally these consequences* not merely adversarially by Laudan, but applaudingly in other quarters? The answer to this second question is uneqivocally yes.[133] Quine would abjure any responsibility for the drastic appropriation of his ideas. He wraps himself in the mantle of his naturalism.

It is left to others to mend the damage. Laudan launches two arguments against the strong thesis of underdetermination. First, he challenges the notion that the logical-semantic notion of entailment of evidence necessarily exhausts the rational grounds for discriminating among empirically equivalent theories.[134] Laudan argues that there are "ampliative" forms of inference which are central to current scientific practice yet they do not receive adequate recognition from Quine. Quine has been peculiarly equivocal on this question. In *The Web of Belief,* Quine and J. S. Ullian made full acknowledgment of the canonical rational criteria for theory choice, even suggesting that these might well suffice to discriminate among rival theories.[135] But elsewhere Quine contended that, nevertheless, underdetermination was possible, even taking these supervening criteria into consideration.[136] Quine's supporters have striven to uphold this stronger form of the thesis and with it the global version of underdetermination. What they have resorted to, in defending this view, are elaborate logical permutations in formal languages. Therefore, in his second argument, Laudan challenges the notion that merely logical possibilities may count as rivals to actual scientific theories.

Together with Leplin, Laudan offers three distinct arguments that the underdetermination of theory by evidence does not entail the equivalence of all rival theories. There is, as Quine himself would affirm, the set of what we might call coherence constraints, for example, simplicity or conservatism, which would promote one theory over another, even given the same evidence base.[137] Second, there is the Bayesian probability model for which, given differing prior probabilities, the same evidence base will result in different posterior theoretical probability.[138] Finally, other evidence, not logically entailed by the thesis, can nevertheless bear upon its validity by confirming or disconfirming elements in the background of the theory or confirming or disconfirming a consequence that *is* entailed tangentially by the theory. One critic of their view maintains that this third form of argument was refuted in advance by Carl Hempel.[139] It is nevertheless true that their general observation holds: "Much sophisticated reasoning in the natural sciences would be vitiated by restricting evidence relevant for assessing a theory to the entailments (via auxiliaries) of the theory."[140] Once we take seriously the full scope of actual scientific practice, there is little reason to maintain *in principle* that the relative superiority of one theory cannot be established. Practice, not principle, governs here, as it should in all naturalized epistemology. And scientific practice strongly suggests that superiority actually *can* be established. Here is one of the domains where philosophy of science is called upon to assert its independence from philosophy of language in order to grasp science in real time.

Hoefer and Rosenberg concede that Laudan and Leplin are correct to argue that there are no serious issues for scientific practice entailed by the underdetermination of any local theories.[141] But these are the only sorts of theories that Laudan and Leplin think worth defending. Laudan insists that "theory choice is generally a matter of comparative choice among extant alternatives."[142] He elaborates: "[T]o say that a theory can be rationally retained is to say that reasons can be given for holding that theory, or the system of which it is a part, as true (or empirically adequate) that are (preferably stronger than but) at least as strong as the reasons that can be given for holding as true (or empirically adequate) any of its *known* rivals."[143] Laudan invokes *rationality*, not merely logical possibility, in strategies of revision in the Duhem quandary. Similarly, he and Leplin insist upon the rationality, not merely the logical possibility, of rival theories in assessing the underdetermination thesis. Too much of what has been offered in support of alternative theories, Laudan and Leplin complain, boils down to little more than "logico-semantic trickery."[144] In his comment on the Laudan and Leplin essay, André Kukla observes: "the whole philosophical dispute between [Quineans] and Laudan and Leplin comes down to the issue of distinguishing genuine theoretical competitors from logico-semantic tricks."[145] In their reply, Laudan and Leplin affirm this stance: "What we require for a case of competition is that there be alternative—and incompatible—theoretical accounts of a common body of phenomena or independently established facts, which there is antecedent reason to regard as subject to unified explanation. It is not trivial that such an alternative will always be forthcoming. Indeed, in contemporary theoretical physics it is often difficult to come up with *one* theory meeting well-established desiderata, let alone two."[146] As they insist (see the epigraph to this chapter), it is an empirical matter whether underdetermination of theories occurs.[147] It cannot be established a priori. (Nor can it be refuted, but the burden of proof surely lies with those who wish to make the contention that any given theory can always be confronted by a rival equally adequate to the evidence.)

Above all, Laudan and Leplin fault Quine for generally "conflating epistemic and semantic relations."[148] For them, Quine has unwarrantedly shifted the entire epistemic issue concerning warranted belief in theories (evidence) into a semantic issue of their logical entailments (truth). This has an air of paradox to it, and that is enhanced when they cite Quine's words from the conclusion of "Epistemology Naturalized": "epistemology now becomes semantics. For epistemology remains centered as always on evidence; and meaning remains centered as always on verification; and evidence is verification."[149] Laudan and Leplin take this to mean that Quine is practicing a "semantic ascent" here. But in fact, this is a passage in which Quine is invoking his most unregenerate *naturalistic* or "empiricist" commitments.[150]

Still, Laudan and Leplin have a sounder point at the more general level, for indeed, "semantic ascent" *is* a principle of Quine's philosophy, and it defines the "linguistic turn" in philosophy.[151] Certainly one can see Quine's re-fashioning of Duhem's thesis in precisely this light. Laudan and Leplin have a case that a linguistic notion of "truth" has gained ascendancy over an epis-temic notion of "knowledge"—in other words, that "syntax, logical struc-ture, and semantics" have displaced "epistemic and pragmatic dimensions of confirmation and explanation" or "judging the reasonableness of belief." One can sympathize with their blunt objection: "We dispute the ability of se-mantic considerations to resolve epistemic issues."[152] The richest deploy-ment of this argument is Ian Hacking's *Representing and Intervening,* espe-cially the chapters on "Internal Realism" and "Observation."[153]

What I wish to maintain—in line with Laudan and Leplin, Hoefer and Rosenberg, and Hacking—is that one can affirm the narrow Duhem thesis and the sort of underdetermination it implies without at all being compelled to embrace the wider underdetermination thesis. Decoupling the two argu-ments is essential, because the wider arguments for underdetermination have become the epistemological gateway to hyperbolic skepticism in the philosophy of science. Yet when these hyperbolic postures get queried, their strategy has been to invoke the narrow Duhem thesis as their warrant.[154] That must be blocked. In the words of Hoefer and Rosenberg, "Holism is a license neither for scientific willfulness nor for relativism."[155] Moreover, if radical skeptics can appeal only to local, fallibilist versions of theory rivalry, then Laudan is right to argue that "the nonuniqueness thesis will not sustain the critiques of methodology that have been mounted in the name of under-determination."[156] As Laudan and others have noted, empirical fallibility is universally recognized, but it has none of the spectacular skeptical conse-quences that are usually associated with Quine's thesis.[157]

The Inscrutability of Reference

What Quine presents in his major works is a theory of linguistic learning, of increasing sophistication in semantics, which in fact dissolves our natural, instinctual identification of bodies with what is "real" about the world.[158] Quine has persuaded Ian Hacking that the phrase "Nature can be carved at the joints" is infamous.[159] What, then, of our propensity for "natural kinds"?[160] Quine explains this in terms of his theories of language acquisi-tion and cultural progress. In both instances the *naturalistic* basis of his ar-gument is decisive. There is a biological predisposition, probably tied to evolutionary survival, to discern in the environment "bodies" of fixed and solid contour: "We and our fellow mammals have a robust sense of the real-ity of gross bodies around us. It hinges on salient, integrated patches and sharp edges in the visual field, reinforced by correlated tactual and olfactory stimulations, and subject mostly to only gradual distortion over time."[161] Quine relativizes our "innate flair" for categorizing by natural kinds, since this is a capacity that we share with animals. "Interestingly enough, it is characteristically animal in its lack of intellectual status."[162] That is, "natural kind" categorization is irretrievably *crude*.

It interests Quine that this natural disposition to form categories of natural kinds—in other words, to develop expectations of the recurrence of similar patterns—*works*. "[W]hy does our innate subjective spacing of qualities accord so well with the functionally relevant groupings in nature as to make our inductions tend to come out right? Why should our subjective spacing of qualities have a special purchase on nature and a lien on the future?"[163] The only tentative answer Quine sanctions is evolutionary advantage. "Without some such prior spacing of qualities, we could never acquire a habit; all stimuli would be equally alike and equally different."[164] Learning by induction is a "natural rationality" (to borrow a phrase from some crucial later extrapolators of Quine).[165]

But he marvels still more that we *outgrow* natural kinds. What Quine celebrates in human cognitive development is that we have gone beyond such natural patterns of order to more abstract and powerful ones. Language acquisition overlays this natural disposition to categorize by an artificial system of words. "[L]earning to use a word depends on a double resemblance: first, a resemblance between the present circumstances and past circumstances in which the word was used, and second, a phonetic resemblance between the present utterance of the word and past utterances of it."[166] This capacity for abstraction evolves further. "We revise our standards of similarity or of natural kinds on the strength, as Goodman remarks, of second-order inductions. New groupings, hypothetically adopted at the suggestion of a growing theory, prove favorable to inductions and so become 'entrenched.'"[167] Thus, "it is a mark of maturity of a branch of science that the notion of similarity or kind finally dissolves, so far as it is relevant to that branch of science."[168] He fastens on our natural predilection for color as an instance; color turns out to be a category that has no deep interpretive force in natural science. As science matures, we relegate color to insignificance.

Thus, "our ontological preconceptions have a less tenacious grip on the deliberate refinements of sophisticated science."[169] "Natural kind" terms mutate into structures defined by science and logic. "Bodies are our paradigmatic objects, but analogy proceeds apace; nor does it stop with substances."[170] Sophistication leads from body to abstract "physical object" as a duration in space/time. That, in turn, gets transfigured by the introduction of numbers, sets, classes, and ultimately *proxy functions*. "[W]e can drop the space-time regions in favor of the corresponding classes of quadruples of numbers according to an arbitrarily adopted system of coordinates. We are left with just the ontology of pure set theory."[171] While each of these is a growth in sophistication, *any* is an arbitrary notation. Crisply formulated, "the lesson of proxy functions is that all the technical service rendered by our ontology could be rendered equally by any alternative ontology."[172] Demonstrating how our ordinary language can be reformulated in terms of set theory and proxy functions, Quine observes, "made nonsense of reference." And he insists: "reference *is* nonsense except relative to a coordinate system."[173] "What particular objects there may be is indifferent to the truth of observation sentences, indifferent to the support they lend to the theoretical sentences, indifferent to the success of the theory in its predictions."[174] Thus, "what one takes there to be are what one admits as values of one's bound variables," and "reference and ontology recede thus to the status of mere auxiliaries."[175] That is, "what is empirically significant in an ontology is just its contribution of neutral nodes to the structure of the theory."[176] That is ontological relativity.

The primordial form of all Quine's indeterminacies is this inscrutability of reference, the claim that language "fits loosely" on the world, and hence the manner in which any given term or statement links up with the world is indeterminate. That is what must be gleaned from Quine's notorious barbarism, "gavagai."[177] Gavagai is both an *observation sentence*—"(Lo, a) rabbit!"—and a *term*. Observation sentences may be as veridical as language gets, but their *terms* remain ultimately *inscrutable*. Quine presumes observers will assent to "Gavagai." as an observation sentence in the presence of the same stimulation (which is all the evidential support Quine thinks one can get). But as a term, *gavagai* can be translated as rabbit, or, incorrigibly, as "undetached rabbit part" or "temporal slice of rabbithood," and so on. That is because language is an ultimately arbitrary (loose, slack) mapping of the world. "Confusion of sign and object is original sin, coeval with the word."[178] The problem of "gavagai" is not just a problem of translation from an alien language; it is endemic to language as such. It "cuts across extension and intension," reference and meaning. "The inscrutability of reference runs deep, and it persists in a subtle form even if we accept identity and the rest of the apparatus of individuation as fixed and settled; even, indeed, if we forsake radical translation and think only of English."[179] As Quine puts it, "radical translation begins at home."[180] Quine's illustration via "gavagai" could just as well have been one from the home language. Substitute, for example, the English word *rabbit* for *gavagai*. Assume two speakers of English are trying to understand one another. One English manual might substitute "undetached rabbit part" as a translation of *rabbit* and satisfy all the observational evidence. Only, for each English speaker, *within* either manual, the distinction between *rabbit* and "undetached rabbit part" would be intelligibly sufficient to deny their synonymy, so that the two manuals would be incompatible.

The looseness of fit of language to world, the reduction of "natural kinds" to a vanishing minimum, and the denial that there are any "joints" at which nature can be "cut," leaves us with the inescapable condition that language participates in the construction of objects: in other words, there is no theory-neutral language. All observations are theory-laden, to cite the most common articulation of this post-positivist insight. At the basis of everything else in Quine is his realization of the inscrutability of reference even *within* a given language: "any characterization of the stimulus meaning occurs *within* the theory (language) we have learned to articulate, and there is no path back to the sensory starting points."[181] "It has been objected that what there is a question of fact and not of language. True enough. Saying or implying what there is, however, is a matter of language; and this is the place of the bound variables."[182] Ontology is always *relative to a translation manual,* in other words, dependent upon a given language scheme.[183] "To say what objects someone is talking about is to say no more than how we propose to translate his terms into ours; we are free to vary the decision with a proxy function."[184] As Quine puts it, we can fix "the objects of the described theory only relative to those of the home theory; and these can, at will, be questioned in turn."[185] This suggests a dangerous regress, but it does get halted: "in practice we end the regress of background languages, in discussions of reference, by acquiescing in our mother tongue and taking its words at face value."[186] While we are free to switch standpoints, we have to stand *somewhere.* "[I]t is a confusion to suppose that we can stand aloof and recognize all the alternative ontologies as true in their several ways. . . . It is a confusion of truth with evidential support. Truth is immanent, and there is no higher."[187]

But what about "evidential support"? In what sense is *it* not immanent as well? The question is whether it is possible to be a linguistic holist of Quine's variety and still conceive empirical inquiry in any sense that saves the very idea of evidence and hence of epistemology. Quine's notion of ontological relativity puts drastic pressure on the naturalized epistemology he professes, for the inscrutability of reference which it entails makes it difficult to discern in what sense one might verify a claim. As Paul Roth puts it, "Quine's Duhemian outlook effectively eliminates the view that we are able to unambiguously identify the data independent of our scientific outlook."[188] In other terms: "holism precludes the possibility of looking past the veil of theory." Despite himself, perhaps, "Quine has shown that, once one forsakes the old empiricism, there is no basis for assuming semantic determinacy within an austerely empirical theory of knowledge."[189] Indeterminacy of *language* simply obviates evidence. We have come to the most problematic of all Quine's indeterminacies, the indeterminacy of translation.

The Indeterminacy of Translation

What exactly does Quine mean by indeterminacy of translation and what warrant does he offer for his view? At the risk of sounding cavalier, one might well say there is very little in philosophy that is more indeterminate than Quine's indeterminacy thesis itself. One of his most sympathetic interpreters, Dagfin Föllesdal, has observed: "In spite of all that has been written . . . there is considerable disagreement not only as to whether the thesis is true, but also as to what the thesis is."[190] As Miriam Solomon puts it, "We need to know what Quine means when he claims that methods of translation are not, and cannot be made, objective: we need to know what is meant by 'objective' in such claims."[191] But she notes immediately that "no consensus has been reached on the nature of Quine's point of view."[192] Moreover, Solomon states flatly what others amount to saying: "the arguments for indeterminacy are different—even contrary—on different occasions."[193] Quine himself equivocated. He ventured a number of claims, and he was forced to reconsider a number. The challenge is to grasp what he was after and what he has left.

Quine's position is keyed to *translation,* in other words, to the problem of rendering sentences from one language into another. His claim is that there is no determinate or unique procedure to achieve such rendering, so that *many* plausible translations may be offered for any given sentence in the object language which will *conflict* in the metalanguage. Anyone who has dabbled at all in translations will hardly find it surprising that translations should show a measure of uncertainty.[194] Still, some translations appear superior to others. We make judgments and we offer reasons. But actual translation problems do not at all capture Quine's intent. "Quine's discussion, rather than arguing that there are no rational criteria for translation, seems to presuppose it."[195] Richard Rorty and Miriam Solomon have tried to garner from Quine's writings all that he had to say about the problems of actual translation and choice among translation manuals. It proves a disappointingly thin gleaning.[196] Though Quine gestures to such problems, he does not really take them seriously. His "principle of charity" is a trivial element in negotiating the indeterminacy thesis, in striking contrast with Davidson's.[197] Quine does not care about success in translation; he cares about what trying to translate *reveals about language.*

Quine argues that the question of language needs to be conceived in its most extreme circumstance, namely, the acquisition of linguistic competence without any mediating resources. This is the situation of the "radical translator" that Quine invents in one of the most influential thought experiments in modern philosophy. He argues that a field translator confronted with a jungle language for which there is no bilingual speaker to aid in translation is left only with the device of testing for dispositions to assent to, or dissent from, sentences presumed to respond directly to the situation of observation. He argues that this is similarly the situation of the child in first-language learning. Working from such patterns of disposition to assent or dissent to observation sentences, the field linguist develops analytical hypotheses about the elements and structure of the observation sentences. She then tests these by uttering other sentences not directly keyed to observation but employing the elements and structures derived from observation sentences and seeking assent or dissent from the natives to these sentences. Quine imagines that this would be an onerous iterative procedure; more, he argues that the desired outcome of fluent discourse in the native language could be achieved by myriad paths of iterative analytical hypotheses, none of which could claim for itself any specific warrant. In that such paths of analytical hypothesizing would result in translation manuals in some home language, the notations would be entirely arbitrary. Inevitably these manuals would generate mutually inconsistent sentences in the home language. Nevertheless, any one manual's analytical hypotheses would achieve all the congruence available with the evidence from the native language, namely, the correlation of observation sentences, holophrastically considered, between the native and the field linguist. Each such manual, *ex hypothesi*, would be "adequate" in the measure that it enabled fluent dialogue. But none could be "true," because there would be literally nothing apart from such global success to warrant the entire structure of the translation manual or analytical hypotheses in use. Hence, indeterminacy of translation.

Translation is a matter not of linguistics but of *ontology* for Quine. He wants to demolish the idea of meanings as mental entities. Furthermore, he wants to demolish the ontology of mental states altogether. Miriam Solomon gets the context exactly right: "Quine treats the question of indeterminacy and the question of getting rid of a mentalistic semantics interchangeably."[198] In the tradition of logical positivism, Quine takes language and theory as synonymous terms, and he is interested in treating language as theory, as a semantics of truth.[199] "The whole point of indeterminacy," writes Föllesdal, "is just that inseparability" of "meaning and information, language and theory."[200] Hence, Joseph Levine has argued that Quine's indeterminacy thesis is ultimately a "thesis concerning the ontological status of belief."[201] In other words, "the whole point of the indeterminacy thesis is to demonstrate the ontologically soft nature of intentional attributions."[202]

In response to Jaako Hintikka, Quine indicated the *motivation* for the indeterminacy thesis was "to undermine Frege's notion of proposition or

Gedanke."[203] That is, the driving force in Quine's philosophical project was the campaign against the "'idea' idea," the "myth of the museum," or mentalist semantics.[204] What was the nature of this opposition? One can imagine it as an epistemological scruple: how could one know for sure that such mentalistic forms existed? But Quine seems far less tentative than this. For him, it is an ontological posit: such things just don't exist. But how does Quine justify such a claim? How, to reinvoke epistemology, can he know his ontological premise is warranted? We must distinguish an epistemological from an ontological sense of Quine's claims for indeterminacy of translation. The former has to do with the possible evidence we might have to prefer one translation manual to another. That is properly a matter, Michael Friedman and others have suggested, of the underdetermination of theory by data with reference to empirical linguistic theory. Quine does not care about empirical linguistics, in the final analysis. Rather, Quine is operating from a prior conviction that linguistic theories—theories of meaning and reference—have a special ontological defect relative to other theories. While, as Friedman notes, "it is hard to find any passage in Quine's writings which is clearly an argument for the ontological version of the indeterminacy thesis," the core of Quine's robust sense of naturalism lay in his confidence that he could discriminate a "fact of the matter" for physical theory over against its utter absence for semantics.[205] Roger Gibson earlier earned Quine's gratitude for clarifying, in a way Quine thought ultimately successful, what warranted Quine's view that physical theory diverged from semantic theory precisely because the former had and the latter lacked the touchstone of a "fact of the matter."[206] Can such a distinction be upheld?

One striking feature of Quine's discussion of indeterminacy of translation is that for a considerable time, starting with *Word and Object,* he acted as though he did not need to offer a proof of the thesis, as though it were already established.[207] Indeed, there are strong grounds for the supposition that he believed that the necessary argument which established indeterminacy of translation had been made, without any reference to the notion, in "Two Dogmas of Empiricism." That is, he presumed indeterminacy of translation followed from the collapse of the analytic/synthetic distinction insofar as that demolished synonymy and with it any grounds for mentalist semantics. But in the face of repeated challenge over the last thirty years, Quine has had to mount several efforts to support his thesis.

Indeterminacy of translation is not, in Quine's view, to be reduced either to underdetermination of theory by data or to inscrutability of reference. It has a force over and above either, though Quine has on occasion maintained that he could argue to indeterminacy of translation from "above," via the underdetermination thesis, or from "below," via inscrutability of reference.[208] What *is* indeterminacy of translation, then? And what are its independent arguments, as well as those ostensibly from "above" and from "below"? How does it relate to these other theses of Quine? Indeterminacy of translation is more than inscrutability of reference in that it deals with sentences, not terms.[209] Indeterminacy of translation is also more than underdetermination of theory by data, in Quine's view. In his harsh "Reply to Chomsky," Quine wrote: "indeterminacy of translation is not just inherited as a special case of the underdetermination of our theory of nature. It is parallel but additional."[210] Quine elaborated in the key essay "On the Reasons for Indeterminacy of Translation" (1970):

> As always in radical translation, the starting point is the equating of observation sentences of the two languages by an inductive equating of stimulus meanings. In order afterward to construe the foreigner's theoretical sentences we have to project analytical hypotheses, whose ultimate justification is substantially just that the implied observation sentences match up. But now the same old empirical slack, the old indeterminacy between physical theories, recurs in second intension. . . .
>
> The indeterminacy of translation is not just an instance of the empirically underdetermined character of physics. The point is not just that linguistics, being a part of behavioral science and hence ultimately of physics, shares the empirically underdetermined character of physics. On the contrary, the indeterminacy of translation is additional.[211]

How is this supplementary nature of indeterminacy of translation to be understood? And is Quine entitled to the claim that one can nonetheless infer the indeterminacy of translation from the underdetermination of theory? As Michael Dummett has cautioned, "one should not too lightly pass from the underdetermination theory to the indeterminacy of interpretation."[212] While it makes perfect sense to argue that theories of linguistics, as an empirical science, are underdetermined, it remains to be proven that they suffer a more serious flaw. As Michael Friedman has contended, "Quine has not provided us with a reason for thinking that linguistic theory is different from any other higher-level theory."[213] It is here that some powerful challenges to Quine's position have been registered by Noam Chomsky, Rorty, Friedman, and Louise Antony.

Interpreters have over the course of thirty years attempted to come up with arguments for the indeterminacy of translation. Quine has himself been more forthcoming. But the result has been remarkably desultory. Early on, Dagfin Föllesdal conceived Quine to have two arguments for indeterminacy of translation.[214] The first sought to achieve a proof of indeterminacy from underdetermination, his so-called argument from above. The second argument Föllesdal discerned was grounded in the ontology of language, and this he took to be Quine's most important argument. Quine himself claimed this was "the real ground of the doctrine, . . . different and deeper" than inscrutability of reference, though linked to it by the so-called argument from below.[215] Solomon, some twenty years later, also identified two forms of argument to the indeterminacy thesis.[216] The first reflected upon the problems in actual translation practice to demonstrate substantial underdetermination. This, she noted, could not really distinguish, as Quine demanded, the indeterminacy of translation from underdetermination of other empirical theories. She identified the second, essential argument, accordingly, with high-level reflections on theory of language—in effect, Quine's a priori views on the ontology of language.

As it turns out, the argument from above does not work. As Quine elaborated it in "Epistemology Naturalized," it has two components, the Duhem thesis (holism) and the Peircean idea of verificationism. "If we recognize with Peirce that the meaning of a sentence turns purely on what would count as evidence for its truth, and if we recognize with Duhem that theoretical sentences have their evidence not as single sentences but only as larger blocks of theory, then the indeterminacy of translation of theoretical sentences is the natural conclusion. And most sentences, apart from observation sentences, are theoretical. This conclusion, conversely, once it is embraced, seals the fate of any general notion of propositional meaning."[217] As Dagfin Föllesdal argued early on, the verificationism was the weak link in this argument. The problem, as we have seen, is that inscrutability of reference wreaks havoc on the compatibility of verificationism and holism, especially in terms of all-out linguistic holism.[218]

Neither the argument from above nor the argument from below yields indeterminacy of translation. The *motive force* for Quine is an ontological conviction against mental states. The question we are left with, as Miriam Solomon has urged, is whether there is any *argument* uniquely upholding the indeterminacy of translation. If not, we are left with Quine's a priori intuitions about the ontology of language. Many interpreters have urged that we must understand Quine in terms of fundamental background commitments of his "naturalized epistemology." Three key notions have been introduced along those lines: *physicalism, empiricism,* and *behaviorism.* Quine has unequivocally endorsed each of these in various contexts. In what measure do his endorsements bear upon, and do his arguments justify—uniquely or conjointly—his ontological premise about mentalist semantics and accordingly about indeterminacy of translation? We must consider each of these three concepts, both for what each professes and for the warrant Quine offers for endorsing it. We must see, yet again, how they interrelate and whether they are mutually reinforcing. And, finally, we must bring back the whole consideration to the epistemological and the ontological forms of Quine's claim for indeterminacy of translation, and we must see whether and how they feature in the arguments he has, over the course of his career, offered for that thesis.

As with matters hitherto, it will often be his commentators rather than Quine himself who flesh out the nature of his arguments and the point of his thesis.

By postulating the ontology of current natural science, Quine believes he can anchor epistemology as much as it can be anchored, since there is no first philosophy, no foundationalism. There is a "fact of the matter" about the world for Quine, because he adopts as his ontology the views of contemporary natural science. Physicalism postulates that all there is, ultimately, is the world of physics, the world physics as a research science has unearthed, or the world according to the latest theory of physics. But is such physicalism a warranted adoption? Is it more than a stipulation, an "article of philosophical faith"?[219] As Quine idiosyncratically holds it, physicalism entails further a commitment to a (strong or weak) reductionism of all other scientific inquiries into physics, and an epistemological concomitant, namely, that only in the measure that evidence can be produced in terms of the latest standing theory of physics is there warrant for any claim in second-order science.[220] Chemistry and biology maintain natural scientific status for Quine by virtue of their ostensible reduction to physics. Psychology is not yet in a position to do so, though a research program in causal neurophysiology is at work on that project. The social sciences and the humanities are in no position even to conceive such a possible (strong or weak) reduction, in Quine's view, and therefore they do not have full scientific status.[221]

Linguistics, the discipline at issue, unquestionably falls among the latter. Worse still, in Quine's view, the very facts it seeks to theorize have no physical concomitants whatever. There are *no* facts there. "The point is not that we cannot be sure whether the analytical hypothesis is right, but that there is not even . . . an objective matter to be right or wrong about."[222] As Antony observes, Quine "is arguing that linguistics is indeterminate—nonobjective—because it is underdetermined *with respect to physics*," and since "reality is circumscribed by physics, that the only real facts are facts describable in the language of physics."[223] Michael Friedman sees it the same way: for Quine, linguistic theories "unlike other higher-level theories . . . fail to be (strongly or weakly) reducible to physics," that is, "translation is not determined by the set of truths of physics."[224] But as Antony protests, this restriction of facts is an a priori stipulation. "A truly naturalized epistemology must, it would seem, take it as an open and *empirical* question what constitutes an objective domain."[225] Hence, "Quine's 'in principle' case against a theory of language—the set of arguments for the indeterminacy of translation—is by no means empirical . . . the indeterminacy thesis is in serious conflict with the notion of a *naturalized epistemology*."[226] This is the ultimate judgment, as well, of Christopher Hookway.[227]

If *fact* means what Quine maintains under his physicalist commitment, then his claim that there is no "fact of the matter" in linguistics follows trivially. But physicalism is by no means required by naturalized epistemology. Physicalism may well be a useful methodological commitment in some natural sciences. It is not clear that it is equally constructive for—and more important, it is not clear that it is methodologically incumbent upon—the human sciences.[228] Part of the naturalism that ostensibly motivates Quine's epistemology is the criterial value of fruitfulness for future research of any standing theory, as we have noted in our discussion of the underdetermination thesis. Quine is, as Friedman has suggested, making a bet on the future of the science of psychology (more specifically, psycho-linguistics). But such a bet cannot *in principle* carry the force of prescription which Quine wishes to impose. Moreover, "the central issue . . . whether there is a fact of the matter about translation is an empirical one—it is not something that can be settled by philosophical argument."[229]

Bluntly stated, naturalized epistemology does not warrant the specific form of Quine's physicalism. Indeed, it militates against the sort of a priori ontological posits Quine undertakes. The essential issue becomes the choice between the moderate form of the underdetermination thesis, the idea of fallibility, and the unrestricted form of the indeterminacy of translation thesis, which militates for an all-out semantic holism. One way leads to the naturalized epistemology of empirical inquiry; the other, to the fancies of an uncontrollable skepticism. But Quine did not foresee such a choice. Why not? What motivated Quine's commitment? We must look beyond Quine's physicalism to his empiricism. Quine is an empiricist in the technical epistemological sense, namely, that there is nothing that can serve as evidence but what appears in sense. On this score, Quine remains an unregenerate positivist. Thus, Quine insists upon his "sensory receptors" and his "neural endings" in establishing the intractable empiricism of his epistemology:

> Two cardinal tenets of empiricism remained unassailable [even after Hume's and subsequent critiques], and so remain to this day. One is that whatever evidence there *is* for science is sensory evidence. The other . . . is that all inculcation of meaning of words must rest ultimately on sensory evidence. . . .
>
> The stimulation of his sensory receptors is all the evidence anybody has to go on, ultimately, in arriving at his picture of the world."[230]

All human experience can draw upon is sensory input, and hence all evidence must be assigned ultimately to stimulus. The linchpin of Quine's "epistemology naturalized" is the conviction that "it is a finding of natural science, however fallible, that our information about the world comes only through impacts on our sensory receptors."[231]

Donald Davidson believes that Quine must abandon what is left of his empiricism because his concept of evidence is incoherent. "Quine does not, I think, ever directly answer the question in what the evidence consists on which our theory of the world depends."[232] According to Davidson, "what is needed is a description of *how* stimulations determine the meaning—the content—of observation sentences."[233] But "nothing, it seems, is properly called the evidence."[234] Quine's residual empiricism is culpable for this skeptical impasse. According to Davidson, skepticism is parasitical upon this "general idea that empirical knowledge requires an epistemological step between the world as we conceive it and our conception of it."[235] That is, "to base meaning on evidence necessarily leads to the difficulties of . . . truth relativized to individuals, and skepticism."[236] That is, one lands back in the Cartesian box. The problem, as Davidson sees it, is that Quine wants to interpose a theoretical language at the very point where he needs direct contact with observation to have any epistemological, any evidential purchase. Language, the very language of observation sentences, takes the world prima facie in terms of the objects of that language, the everyday objects of ordinary natural language. Davidson calls this the "distal" sense of observation sentences. The thrust of Davidson's claim is that a "distal" theory of observation sentences should displace Quine's "proximal" theory of stimulus and sensation in order to keep the relation between language and world immediately available for interpretation. That alone, in Davidson's view, escapes the threat of skepticism.

Quine recognizes the danger; he strongly resists being lumped with the radical appropriation of linguistic holism for incommensurability or all-out relativism, as he understands the position of Norwood Russell Hanson, Thomas Kuhn, and Paul Feyerabend. It appears that Quine's last stronghold is his theory of observation sentences. The recent essay "In Praise of Observation Sentences" represents one of Quine's most explicit attempts to distance himself from Hanson, Feyerabend, and Kuhn. He holds that they "overreacted" against traditional epistemology with their formulation of the "theory-laden" nature of all observation.[237] Quine insists that "we must recognize degrees of observationality."[238] Observation sentences are "Janus-faced"—they face "outward to the corroborating witnesses and inward to the speaker." They must be seen as working simultaneously "holophrastically" and "piecemeal." Quine wishes to hold that the holophrastic usage of the observation sentence is the vehicle *both* for language learning and for evidential support, but he recognizes that the outward, the intersubjective use of observation sentences, is piecemeal: "Piecemeal is how the sentence relates to scientific theory, where its words recur in new combinations and contexts."[239] Quine wants to believe that the ostensions of the observation sentences, as holophrastic, will serve adequately "as experimental checkpoints for theories about the world."[240] That is, he believes that the terms within the given sentences do not entail any ineradicable indeterminacies for the sentence taken as a whole. "Seen holophrastically, as conditioned to stimulatory situations, the sentence is theory-free; seen analytically, word by word, it is theory-laden."[241]

Quine believes he can anchor his epistemological position because of the actuality of language acquisition. Given that language is acquired, and that acquisition has an undeniable element of ostension via "observation sentences," Quine can assert with confidence, "The observation sentence, situated at the sensory periphery of the body scientific, is the minimal verifiable aggregate; it has an empirical content all its own and wears it on its sleeve."[242] But here Quine's naturalism seems to beg the question of his holism, for this last assertion seems fairly explicitly to contradict the Duhem thesis. To rescue himself, Quine asserts that the Duhem thesis only holds for "theoretical sentences."[243] By contrast, "observational sentences" are determinate, not only for theory but for translation. But it is questionable whether he can hold that line. Davidson queries, what guards against one observer gesturing with the observation sentence "Gavagai" to what another perceives as a warthog? There is no security that the terms of one individual's ostension might not conflict with those of another, and thus collapse epistemic support toward solipsism.[244] All-out semantic holism *obliterates* the determinacy of reference.[245] With Davidson's revisions of Quine's philosophy of language, reference passes from inscrutability to obscurity.[246] At best, "the existence of an empirical (sensory) correlate—the thing in itself—becomes a *formal* (necessary) condition for there being a language."[247]

All along, Quine had in mind that the "intended notion of matter of fact . . . be taken *naturalistically within our scientific theory of the world.*"[248] Still, "facts of the matter" are *immanent* or "internal to our theory of nature."[249] It is impossible to take a stance outside a theory (language), and hence *any* statement about evidence is *within* a theory. As Gibson puts it, "there are no unique evidential relations to be found between sensory evidence and the theories it supports."[250] There is no theory-neutral observational language, to use the terminology of Mary Hesse.[251] That suggests that there can never be access to these facts apart from the theory/language in which they are formulated. It becomes impossible for Quine to establish a "distinction between the conditions that justify belief in a theory and the conditions that warrant the attribution of truth to a theory."[252] Genova concludes: "So just as meaning is relativized to a translation manual, truth is relativized to a background theory."[253] Quine's struggle to keep his notions of epistemology and ontology distinct fails: "what he construes as a categorical distinction between an epistemological as opposed to an ontological context, is really only a generic distinction, a distinction of degree within his naturalized epistemology."[254] The problem is that empirical content is "given" always and only *within* some language. Quine is clear that "truth is an immanent notion."[255] But it appears that his linguistic holism makes *evidence* equally immanent. Epistemology thereby appears to lose its purchase.

To be sure, Quine *wants* to be an empiricist. To be sure, physicalism is a perfectly plausible *pragmatic* recourse. But, as Quine once put it, we are "making a philosophical point."[256] Or rather, three decisive extrapolators of Quine have made this point: Mary Hesse, Donald Davidson, and Richard Rorty. Critics are pressing to dissolve Quine's last distinction between physics and language, which would swallow everything up in indeterminacy of translation, in semantic holism. Gibson now argues that Quine's maneuvers have robbed him of the distinction he once possessed: "Quine is left without a way to differentiate under-determination and indeterminacy."[257] That is, "if we accept the indeterminacy thesis, we cannot make objective sense of the Underdetermination thesis. A proponent of indeterminacy can at most give objective sense to the possibility of two theories which exhibit *syntactic* differences."[258] As Gibson puts it, the long course of the debate about underdetermination "over the last twenty-five years has landed [Quine] in a contradiction."[259]

One of the most difficult things to grasp in Quine's philosophy is the relation between ontology and epistemology, between truth and evidence. What makes disentangling these notions particularly difficult is that Quine assimilates theory to language, science to semantics.[260] For Quine, theory is just a distillation of ordinary language, and *epistemologically* there is no separating their warrants. "The evidence relation and the semantical relation of observation to theory are coextensive."[261] Scientific theory is just a specialized language, that is, the language of a specialized community: "What counts as observation sentences for a community of specialists would not always so count for a larger community," but *mutatis mutandis* the same principles apply.[262] In Gibson's terms: "from the point of view of *epistemology*, underdetermination of physical theory and indeterminacy of translation *are* on a par: Just as alternative *ontologies* can be erected on the same observational basis, so alternative *translations* of a native expression can be erected on the same observational basis. All are equally warranted by the evidence."[263] While underdetermination of theory is *about* ontology—"it is the thesis that different systems of objects (or systems of sentences about objects) may link past and present sensory stimulations to future ones"—it remains a thesis *in* epistemology—"it is a statement about *evidence* for theory, not about *truth* of theory."[264]

Following his path of semantic ascent, Quine's emphasis on indeterminacy leads to the paradoxical conclusion, fully exploited by Richard Rorty, that there simply is no place left for epistemology, indeed for philosophy in the old sense.[265] Rorty notes the asymmetry which results from Quine's empiricist theory of language: "Since linguistic behavior is a datum for psychology, economics, sociology, intellectual history, and the like, presumably in all these fields 'indeterminacy of translation' applies. Since we can trade off theories of what people believe and desire against theories of what their utterances mean, the special indeterminacy Quine attributes to translation will infect all fields in which humans *qua* believers and desirers are studied. This means that in the whole field of *Geisteswissenschaften* 'there is no fact of the matter' in the way there are facts in the *Naturwissenschaften*. I find it hard to imagine Quine welcoming this result—but I cannot see how he might avoid it."[266] But in fact Quine is quite happy to make this discrimination. "In softer sciences, from psychology and economics through sociology to history (I use 'science' broadly), checkpoints are sparser and sparser, to the point where their absence becomes rather the rule than the exception. Having reasonable grounds is one thing, and implying an observation categorical is another. Observation categoricals are implicit still in the predicting of archaeological finds and the deciphering of inscriptions, but the glories of history would be lost if we stopped and stayed at the checkpoints."[267] Quine clearly entertains the old positivist hierarchy of the sciences. It is Rorty who will wash away all discrimination between the *Geisteswissenschaften* and the *Naturwissenschaften,* though in a manner diametrically opposed to Quine—by dissolving the claim of the natural sciences to any "fact of the matter" of their own. Everything dissolves into indeterminacy.

"The Indeterminacy thesis advises us of the existence of too many (incompatible) translations of alien theoretical sentences into our own, consistent with all the evidence."[268] The only way out seems to "require us to look at practicing scientists from some point of view which is *outside* their practices, and to see the theories they work up, and work with, as (ultimately) 'make-believe,' or 'a put-up job.'"[269] That is precisely what the so-called sociology of scientific knowledge has done, but to establish this connection we will have to consider the other great fountainhead of post-positivism, Thomas Kuhn's *The Structure of Scientific Revolutions.* Before we make that fateful turn, it behooves us to ask whether there is not a *philosophical* recourse to rescue naturalized epistemology from Quine's contradictions.

Rescuing Naturalized Epistemology from Its Founder

Quine is clear that the theoretical account for the assimilation of sensory excitation into linguistic utterance is not yet available. He believes that it is nonetheless possible to imagine how in principle it must work, namely, as a causal structure. In any event, there is an epistemological surrogate that obviates for the present the need to resort to such a theory, though should it become available it would obviously have pride of place. The surrogate is *disposition* toward observation sentences. This much psychology Quine appears ready to accept.[270] Disposition, however, is a theoretical concept. At its simplest, Quine conceives it as consistent assent or dissent to a sentence in the context of stimulus situations as identical as possible. Initially Quine was concerned with the dispositions of a single individual, but increasingly in his thought what has mattered are the dispositions of the sum of individuals constituting a "language community." Yet he recognizes that assent and dissent and their observable indicators are not accessible except via a theory. Thus the verification of even observation sentences depends upon what Quine calls "analytical hypotheses." This creates at least one portal for skepticism. Still, Quine wishes to adhere to the evidential warrant that can be supplied by dispositions to assent or dissent to observation sentences. In it consists, in his view, our ultimate access to physics, to "facts of the matter," to "what there is." Observation sentences taken holophrastically are the only checkpoints we can access. Quine summarizes this: "Grant that a knowledge of the appropriate stimulatory conditions of a sentence does not settle how to construe the sentence in terms of existence of objects. Still, it does tend to settle what is to count as empirical evidence for or against the truth of that sentence."[271] Theory has access to facts in such sentences, and that is all there is to say about evidence. But just for this reason, "the thesis of indeterminacy of translation follows fairly directly from Quine's empiricist view of evidence," in the words of Föllesdal, for there are clearly myriad sentences in any language which do not fit tightly to such "facts."[272]

In such cases, all there is to observe in language is the behavior of speakers. Quine's empiricism in epistemology carries him over to *behaviorism* in linguistics. Not only is Quine conceiving epistemology as *empiricist* ("sensory experience"), he is also intent on its *social* character ("learned language"). "Each of us learns his language by observing other people's verbal behavior and having his own faltering verbal behavior observed and reinforced or corrected by others."[273] Language is socially inculcated and controlled; the inculcation and control turn strictly on the keying of sentences to shared stimulation. Internal factors may vary *ad libitum* without prejudice to communication as long as the keying of language to external stimuli is undisturbed."[274] For Quine, this is relentlessly *behavioristic:* "We have been beaten into an outward conformity to an outward standard; and thus it is that when I correlate your sentences with mine by the simple rule of phonetic correspondence, I find that the public circumstances of your affirmations and denials agree pretty well with those of my own."[275] Language is a *social* form: "Language is where intersubjectivity sets in. Communication is well named."[276] "I hold . . . that the behaviorist approach is mandatory. In psychology one may or may not be a behaviorist, but in linguistics one has no choice. . . . Our mental life between checkpoints is indifferent to our rating

as a master of the language."[277] It is commonplace in the secondary literature to insist that Quine not be lumped with the now quite discredited views
of behaviorism as a general theory of psychology or human action. There are
grounds, I would counter, to question such an exemption for Quine. In any
event, he clearly professes to discriminate between behaviorism as a theory
of psychology and behaviorism as a theory of linguistics, and he demands to
be judged on the merits of his case for the latter, which he thinks is insuperable. This claim is at the core of indeterminacy of translation. If it is upheld,
much that Quine claims cannot be withstood. If it falls, his whole position
becomes susceptible to considerable revision. Quine's strongest defenders
believe they can rescue indeterminacy of translation from obscurity or from
fallacy on that basis. In my view, his case fails just there. We have reached the
crux.

I question whether Quine's thought experiment of radical translation
could really even get off the ground. Quine himself marvels that such "mea-
gre input" should generate such "torrential output." It is not clear that
Quine's field linguist could, in fact, construct a translation manual in the situation in which Quine has placed her. His gesture to the actual practices of
field linguists such as Kenneth Pike begs the question, for there are substantial grounds, some of which were invoked by John Searle, to see the procedure that Pike and other practical field linguists employ as vastly more
complex than Quine has postulated.[278] And this is just the point I wish to
stress: the analytical hypotheses which a field linguist brings to bear upon an
alien language, in just the measure that she postulates that what is involved
is a *language, must* go beyond the "austere" empiricism (and a fortiori, behaviorism) Quine postulates. Only because the field linguist brings to bear a
whole tacit or active theory of language as a structure—with phonemics,
syntax, and semantics, with a congeries of lexical units, and with certain
overarching communicative goals—is it possible to venture even the crudest
approximation of an analytical hypothesis. Without the whole of the home
language—a *natural* language, it should be stressed—the field linguist would
be helpless in the endeavor.

What, one can imagine a Quinean immediately protesting, about the child
learning its original language? "What, indeed?" is the appropriate response.
It is Quine who has postulated that the child's learning of a first language
must follow the analogy of the field linguist. It is for Quine to *prove*, not
simply postulate that.[279] And he cannot.[280] The entire weight of empirical
evidence, Quine notwithstanding, bars that path.[281] Quine suggests that
first-language learning begins with ostensions. That is plausible. What is not
plausible is his view of how the rest of language can be acquired. In *From
Stimulus to Science,* Quine ventures some ideas about the sequence of analytic hypotheses through which a natural language might have been phyloge-

netically and ontogenetically acquired, offering both lucid characterizations of the structures required and candid admissions that he has no idea how they might have been put in place either for the species or for the child learning its first language. But this is a rather formidable question to beg for a naturalized epistemologist. I agree with Nobuharu Tanji: "How can we learn a language bit by bit? . . . [I]t is very natural to invoke observation sentences. . . . But even if we could fully understand observation sentences . . . the question would remain untouched: How can we learn *the other parts* of language bit by bit?"[282] It will not do for Quine to disclaim responsibility for this empirical question. His ontological dogmatism already embroils him in claims about it.

Translation is the central concept in Quine's theory of language. Since he is primarily concerned with the question of the appraisal of claims in language, this concentration on translation suits his epistemological purposes quite well. Yet it downplays the crucial matter of first-language learning. The point is that we are never in the position of a field linguist with reference to our first language. The metaphor of translation distorts primary-language learning. If the way we acquire our first natural language and become empowered in this extraordinary capacity gets flattened out in Quine's pervasive invocation of translation, it seems also that his doctrine of the indeterminacy of translation is robbed of most of its sinister implications once we reassert the complexity and yet pervasive success of primary-language learning. Instead of prescribing to, we must *learn from,* children's primary-language learning. Instead of prescribing to, we must learn about, natural language in its difference from formal logic. There is still too much "first philosophy" in Quine. We must rescue naturalized epistemology from its own founder.

The child, originally, then the field linguist, and ultimately all philosophers of language in their train must all draw upon a far more complex order of natural language if they are ever to attain the requisite fluency to be accepted in a speech community. It is highly likely that this can only be accounted for on psycholinguistic premises fundamentally at variance with Quine's. *Theorizing* the facts of *that* matter means conceiving the situation of linguistic behavior in terms of a far more complex theory than simply one of observable dispositions. Cognitive science is an empirical science working to unearth the mechanisms through which natural language constitutes itself. That account has had to recognize the indispensability of mental states, of beliefs, if it is ever to become adequate to the problem. All the more clearly, in the discourses of culture, we are compelled to seek a more robust naturalism that does not succumb to reductionist physicalist or behaviorist presuppositions. We must, in the terms of Davidson, advance from translation to interpretation, and what that requires is the overthrow of Quine's empiricist/behaviorist strictures on evidence. We must work, considering each language learner, with the triangulation of beliefs, meanings, and truth and, among speakers, with the triangulation of two speakers and a situation whose contours appear relatively stable.[283] If philosophy of language were to suspend its a priori posturings and hearken, as naturalized epistemology mandates, to the best hypotheses of current empirical science—here, the cognitive sciences—its strictures on interpretation in cultural and historical studies might well lose some of their imperious "radicalism" and thus make peace with contingent, fallible, but progressive empirical inquiry.

The irony and the tragedy is that, in spite of official honours and genuine attempts at reconciliation by both Kuhn and others, he himself was never truly at home in any of these disciplines [i.e., philosophy, history, or sociology], nor in their intersection. The majority of historians and philosophers of science never permitted Kuhn to feel genuinely comfortable in their professional associations. The sociologists tried, but Kuhn himself was not comfortable in their company. He died professionally homeless.

— RONALD GIERE, "KUHN'S LEGACY FOR NORTH AMERICAN PHILOSOPHY OF SCIENCE"

It might be said that *Structure* has a philosopher's sense of sociology, a historian's sense of philosophy, and a sociologist's sense of history.
— STEVE FULLER, *THOMAS KUHN: A PHILOSOPHICAL HISTORY FOR OUR TIMES*

Living in Different Worlds? Kuhn's Misadventures with Incommensurability

The Structure of Scientific Revolutions began with the assertion: "History, if viewed as a repository for more than anecdote or chronology, could produce a decisive transformation in the image of science by which we are now possessed."[1] That is, Thomas Kuhn charged we were somehow misguided in our sense of science, and he proposed to correct that image through the invocation of history. This opening sentence offered a clarity of intent the whole work failed to sustain. But there remains even in this first sentence one very large ambiguity: who was this "we" that Kuhn invoked? Who was Kuhn's intended audience?

As it turns out, the *actual* audience for the book has been monumentally larger than he could have imagined; no other work in the history or philosophy of science has been so widely read, discussed, or appropriated by the entirety of the literate public. The extraordinary reception of *Structure* beyond the sphere of philosophy of science and natural science itself can most readily be construed against the backdrop of an ambient idolatry of science. The social sciences, for example, were so caught up in this mystique of "scientism" that *Structure* sent them scurrying to establish possession of disciplinary paradigms attesting to their scientific maturity.[2] Yet simultaneously it unleashed, both in them and in the more general literate public, a festering resentment toward natural science's cultural privilege.[3]

But the question still remains: for whom was Kuhn writing? Was it for scientists, to correct a faulty self-perception? Was it for historians of science, to correct a faulty methodology in the representation of actual scientific achievements?[4] Or was it for philosophers of science, to correct their ostensibly monolithic notion of scientific method? After forty years, we can judge that Kuhn primarily sought to reach the third of these groups, and hence that *The Structure of Scientific Revolutions* was intended as a contribution to the *philosophy of science*. Sergio Sismondo observed correctly: "[W]e should remember that Kuhn's discipline when he wrote *Structure* was history of science. . . . [I]t was his dissatisfaction with philosophy of science in the light of historical work that prompted his change of fields from physics to history."[5] I would carry Sismondo's point further, namely, to the claim that his dissatisfaction carried Kuhn from history of science into *philosophy* of science. In 1968, delivering the Isenberg Lecture at Michigan State University, Kuhn avowed: "I stand before you as a practicing historian of science. . . . I am a member of the American Historical, not the American Philosophical, Association." But his address was entitled "The Relations between the History and the Philosophy of Science," and he was clear that his goal was to impact the second via the first.[6] In the last extended interview in which he discussed his career, Thomas Kuhn observed: "And I thought of *Structure*, when I got to it finally, as being a book for philosophers."[7] More generally, Kuhn professed that his "objectives in [his approach to science studies], throughout, were to make philosophy out of it."[8] Later in the interview he commented: "I like doing history. . . . The philosophy has always been more important."[9] Especially in his later life, this philosophical pursuit was central to his identity: "I've gone on now in the last ten, fifteen years, really trying now to develop this philosophical position."[10]

Yet by 1980, it is fair to say, Kuhn had become a marginal figure for the philosophy of science.[11] Despite his assiduous efforts to win acceptance by the philosophical community, this marginality persisted to his death.[12] Ronald Giere's obituary comment cited as our epigraph illuminates this sad marginality that proved to be Kuhn's fate.[13] Though it proved disheartening for Kuhn, it should not be surprising that philosophers of science represent the audience most antagonistic to his message.[14] Kuhn's suggestion that they were "possessed" by an "image" would obviously provoke some psychological resistance. But this resistance was not due simply to disciplinary pride. Kuhn proposed such profound revision that his own terminology applies: it represented a "paradigm shift" for the philosophy of science.[15] As Giere has observed, it was the difficult achievement of Kuhn's *Structure* to "shame" philosophers of science into "dealing with *real* science."[16] As Kuhn put it in

1968, "the philosophers' reconstruction is generally unrecognizable as science to either historians of science or to scientists themselves."[17] What was the "image" of science that had kept philosophers from it? Perhaps we might term what Kuhn wished to dispel "scientism" rather than "positivism." Yet it remains historically important to retain the latter term alongside, and as an explanation for, the former. Even the most sophisticated philosophy of science propagated scientism. This conventional wisdom, or "Received View," found expression, above all, in *reductivist* agendas for science within philosophy and in *presentist* scenarios of progress in standard science textbooks.[18] Perhaps the key feature was "logicism," which Stephen Toulmin phrases thus: "an unquestioned presupposition for philosophers of science that the *intellectual content* of any truly scientific theory formed a timeless 'propositional system.'"[19]

The judgment of philosophers of science proved hostile to Kuhn from the outset, and it scarcely relented. The two major contexts for the interrogation of Kuhn's ideas were the proceedings of the 1965 International Colloquium in the Philosophy of Science held in London, July 11–17, 1965, and the symposium on the structure of scientific theories held in Urbana, Illinois, March 26–29, 1969. The first resulted, eventually, in the publication of *Criticism and the Growth of Knowledge,* and the second, in the publication of *The Structure of Scientific Theories.*[20] Even before these major conferences, the hostile reception of Kuhn by philosophers was clear from the initial reviews of *Structure.* From the outset, Dudley Shapere and Israel Scheffler condemned Kuhn either as hopelessly self-contradictory or as recklessly irrationalist, idealist, or relativist.[21] They lumped him with Norwood Russell Hanson and Paul Feyerabend and denounced all of them for attempting to relativize claims of scientific validity drastically to particular cultures or communities.[22] Kuhn seemed to hold that no rational mode of adjudication between rival theories could be achieved.[23] Thus, in the first and very influential review of Kuhn's work among philosophers, Shapere drew all these ideas together: "If one holds, without careful qualification, that the world is seen and interpreted 'through' a paradigm, or that theories are 'incommensurable,' or that there is 'meaning variance' between theories, or that all statements of fact are 'theory-laden,' then one may be led all too readily into relativism with regard to the development of science."[24] Shapere insisted that philosophy of science need by no means be driven to this extremity by the historical record of science. Instead, the fault lay with Kuhn, who was philosophically inept, finding a problem of "meaning" in what should only be a matter of "application."[25]

The Popperian reception of Kuhn's work documented in *Criticism and the Growth of Knowledge* showed a similarly harsh repudiation. Karl Popper himself, with Alan Musgrave, made it clear he found little to his liking in

Kuhn's approach.[26] Popper had already developed his response to Kuhn in an important 1965 essay, "The Myth of the Framework," which he revised in 1972.[27] There, Popper challenged the plausibility of any drastic relativist implications from Kuhn's work. At the 1965 conference, Popper concentrated on his distaste for Kuhn's account of "normal science." Imre Lakatos set himself the task of reconstructing Popperian theory to offer a better version of whatever insight Kuhn might have been gesturing toward.[28]

And philosophers *remained* unprepared to take Kuhn seriously. Kuhn's efforts to clarify his position over the next decade were met with continued sharp resistance.[29] The publication of *The Essential Tension* (1977) did nothing to allay the philosophical disapproval.[30] And the publication of *Black-Body Theory* (1978) was received with little enthusiasm, because Kuhn failed to use it as an exemplar of his theoretical stance.[31] He spent the balance of his career endeavoring to persuade philosophers that the outlandish things he was taken to have said he never really meant. Ironically, the result was rarely to his liking, and for two reasons. Either he failed to persuade, and was taken (positively or negatively) to endorse "radical" views notwithstanding his disavowals, or he succeeded, only to be considered either renegade or trivial in his later, more reasonable postures.[32]

Paradigm, Normal Science, and Revolution: Kuhn's Theory of Scientific Development

Kuhn's *Structure* essentially offers a theory of scientific development.[33] *Theory*, as distinct from history: history is, as Kuhn puts it at the outset, a "repository," not a conceptualization.[34] Kuhn utilizes incidents from the history of science for his philosophical purposes, but these actual occurrences are evoked not for their own sake or even to constate a narrative. Rather, they exemplify "structures," that is, patterns with theoretical significance. Kuhn is after "what *all* scientific revolutions are about" (6; my emphasis).[35] Particular sciences achieve "maturity" by constituting themselves around a distinctive achievement and exploiting its example to undertake an ongoing, focused practice ("paradigm" and "normal science," respectively). But such mature science routinely generates anomalies (failures of articulation or prediction) and episodically suffers discontinuous change ("scientific revolutions" involving "incommensurability"). A first reading would lead one to think that Kuhn is concerned strictly with massive ruptures in the course of science—in other words, famous and transfigurative moments in the history of the modern physical sciences on the order of the Copernican revolution.[36] As one reads more closely, and certainly if one attends to all that Kuhn wrote in explication of *Structure*, *any* change which revises a paradigm is termed revolutionary.[37] Not surprisingly, critics have retorted that *revolution* seems too grand a term for such recurrent ruptures and that the question of discontinuity might need to be nuanced.[38]

In the preface to *Structure*, Kuhn offers the reader his first characterization of paradigms: "universally recognized scientific achievements that for a time provide model problems and solutions to a community of practitioners" (viii). Notoriously, Kuhn's term, *paradigm*, was overdetermined.[39] That it had some twenty-two different senses in *Structure* signifies not only the philosophical slack in Kuhn's notion, but also its cultural potency.[40] A paradigm is a *specific achievement* of insight into natural process. Second, it is *recognized* (soon or late, but only upon such recognition can it *be* a paradigm). Finally, it is recognized precisely as *exemplary*, or capable of emulation. Its status as paradigmatic only follows from its inauguration of a subsequent, ongoing, and fruitful practice; hence, it is a *historical* concept, accessible only retrospectively.

"Normal science" was one of the most difficult notions to appraise in *Structure*. Many readers took Kuhn to be disparaging toward it, representing it as mere "hackwork."[41] Accordingly, despite their differences, both Popper and Feyerabend felt that normal science should not be taken as science at all.[42] Normal science is normal in two senses: the conventional sense that this is how science usually takes place and the connotational sense that it is governed by norms, that is, that there is a determinacy to the practice, a disciplinarity. Kuhn writes: " 'normal science' means research firmly based upon one or more past scientific achievements, achievements that some particular scientific community acknowledges for a time as supplying the foundation for its further practice" (10). The emergence of normal science is both constraining and enabling, and it is enabling *through* constraint.[43] Specialization, while it segments intellectual life, allows for a deepening of inquiry which is not otherwise imaginable and which is precisely what has given science its historically distinctive record of efficacy. "When the individual scientist can take a paradigm for granted, he need no longer, in his major works, attempt to build his field anew, starting from first principles and justifying the use of each concept introduced. That can be left to the writer of textbooks. Given a textbook, however, the creative scientist can begin his research where it leaves off and thus concentrate exclusively upon the subtlest and most esoteric aspects of the natural phenomena that concern his group" (19–20). *Why* is disciplinarity, conceived in Kuhnian terms as paradigm-governed normal science, enabling? To work under a paradigm is to think analogically. A paradigm is a specific achievement which "is rarely an object for [mechanical] replication. Instead, like an accepted judicial decision in the common law, it is an object for further articulation and specification under new or more stringent conditions" (23).[44]

Repeatedly Kuhn points out that normal science does not seek "major substantive novelties" but rather seeks those problems for which, in accordance with the paradigm, there *ought to be* a solution. Kuhn likes to think of such work as "puzzle solving."[45] The challenge is, then, "solving a puzzle that no one before has solved or solved so well" (38). This entails not just theorizing but experimentation: "problems of paradigm articulation are simultaneously theoretical and experimental" (33). What Kuhn finds essential is that "paradigms guide research by direct modeling as well as through abstracted rules" (47). Scientists, Kuhn asserts, "can . . . agree in their *identification* of a paradigm without agreeing on, or even attempting to produce, a full *interpretation* or *rationalization* of it. . . . Indeed, the existence of a paradigm need not even imply that any full set of rules exists" (44). The theory that articulates the substantive and methodological implications of the paradigm "need not, and in fact never does, explain all the facts with which it can be confronted" (18). Kuhn writes of discovering historically "what isolable elements, explicit or implicit, the members of that community may have *abstracted* from their more global paradigms and deployed as rules in their research" (43). He insists on the essential role of the *implicit* or, in the terminology of Michael Polanyi, the "tacit."[46] It is here, perhaps more than anywhere else, that Kuhn's work has effected a long-term shift of inquiry in the philosophy of science: from logic to practice.[47] "Scientists, it should already be clear, never learn concepts, laws, and theories in the abstract and by themselves. Instead, these intellectual tools are from the start encountered in a historically and pedagogically prior unit that displays them with and through their applications" (46). Thus, Kuhn places a great emphasis on the process of enculturation or socialization into the scientific community and the character of the *learning process* whereby one becomes a scientist. But he also insists that this "tacit dimension" is an *ongoing and essential* component of the disciplinary community, securing its efficacy.

Though normal science has no ambitions to be revolutionary and its goal is simply "the steady extension of the scope and precision of scientific knowledge" (52), the actuality is that "new and unsuspected phenomena are . . . repeatedly uncovered by scientific research." Consequently, "research under a paradigm must be a particularly effective way of inducing paradigm change" (52). The very rigor of paradigm-directed inquiry is a dynamo of change. Kuhn offers a precise theory of innovations in science: "Produced inadvertently by a game played under one set of rules, their assimilation requires the elaboration of another set" (52). Novelty is always inadvertent because it violates the paradigm out of which it emerges. But following paradigms inevitably leads to such novelties. "Novelty ordinarily emerges only for the man who, knowing *with precision* what he should expect, is able to recognize that something has gone wrong" (65). Hence, "In science . . . novelty emerges only with difficulty, manifested by resistance, against a background provided by expectation" (64).

First, of course, every effort will be made to resolve the anomaly, to restore the coherence of the paradigm-governed practice. Initially the presumption must always be an inept execution, not a fallacious premise. Normal science always sets out with the presumption that *there ought to be a solution* to any puzzle its game postulates. And usually the anomaly can be reassimilated. But not always. One of the features of real-time science, science as practice, is its robust tolerance of margins of error. Such tolerance, Kuhn observes, explains the persistence in a theory even when anomalies have not all been resolved. There are always things that organizing inquiry in any given way leaves out. Such costs can generally be borne. A measure of cognitive dissonance is tolerable in exchange for the ongoing fruitfulness of a paradigm-governed practice.

However, some anomalies are more insufferable than others. Perhaps this is because they violate a very pervasive rule of the given game, or because they cluster in their resistance to resolution. Kuhn avoids reducing it to a formula: "We . . . have to ask what it is that makes an anomaly seem worth concerted scrutiny, and to that question there is probably no fully general answer" (82). In any event, after the initial self-criticism of the scientist, persistent anomalies solicit not only obsessive attention as "*the* subject matter of their discipline" but also an increasing bending of the rules of the game to make the recalcitrant element fit (83). Scientists will "devise numerous articulations and *ad hoc* modifications of their theory in order to eliminate any apparent conflict" (78). The scene is set for revolution: "with continuing resistance, more and more of the attacks upon it will have involved some minor or not so minor articulation of the paradigm," and consequently, "the rules of normal science become increasingly blurred" (83). "Crisis loosens the rules of normal puzzle-solving in ways that ultimately permit a new paradigm to emerge" (80). That is, the crisis itself has a structure. As the contours of the existing paradigm weaken and as a given problem originating from the anomaly becomes the obsessive key to disciplinary self-conception, the possibility for the genesis of an alternative paradigm presents itself. To reinvoke Kuhn's phrase, "produced inadvertently by a game played under one set of rules, their assimilation requires the elaboration of another set" (52).

In the postscript of 1969 Kuhn sought to specify more precisely both the empirical characteristics of his scientific community and the sense of paradigm that he had imputed to it in *Structure*. The second of these endeavors preponderated. Kuhn discriminated two senses of paradigm (from the twenty-two variants Margaret Masterman had distinguished) as decisive.

The sense he took to be "the most novel and least understood aspect" in his thinking was that of paradigm as *exemplar,* a concrete achievement of practice to be emulated (hence my exposition above). But he acknowledged that he also used *paradigm* to evoke something more global, the "constellation of group commitments" of his scientific community; for this sense of paradigm he offered the alternative conceptualization, "disciplinary matrix."[48]

He articulated four components of this global sense, one of which was the idea of the exemplar behind his original use of *paradigm.* The other three components—symbolic generalizations, metaphysics, and values—seemed to impel Kuhn to a "sociological" elaboration of his theory. By symbolic generalizations Kuhn meant those formulations that "function in part as laws but also in part as definitions" (183), such as Newton's $f = ma$, through which paradigms as exemplars occasion *theory* in the determinate sense. Kuhn stressed that there was a tenuous balance between the legislative and the definitional component in these symbolic generalizations, a balance that could shift as the theory came under duress (183). Moreover, he retained his strong emphasis on the tacit dimension connected with direct modeling of the exemplars, so that a symbolic generalization had always to be conceived far more as a "law-sketch" (189) than as a law, in other words, a *schema* whose application to any given case had always to be worked out. The idea of metaphysics in Kuhn's elaboration remained the most hodgepodge, for it included not only what we might term background beliefs but also investigative hunches, "the spectrum from heuristic to ontological models" out of which the community not only established its "roster of unsolved puzzles" but what would be "accepted as an explanation and a puzzle-solution" (184). Indeed, this metaphysical element "suppl[ied] the group with preferred or permissible analogies and metaphors" (184).

In short, Kuhn used his first two categories of the disciplinary matrix to articulate what traditionally would have been called *theory* and *methodology* but also to blur them with *metaphysics*—anathema to the "demarcation"-minded Received View. His third new elaboration was the notion of values, which extended beyond any narrow scientific research specialty to "natural scientists as a whole" (184). It was the dimension of values that allowed scientists both to make choices within a given theory and to compare rival theories. Kuhn disputed the allegation that he had eliminated any prospect of rationality in theory choice here, by emphasizing the role of values. What he was trying to assert, he clarified, was that "Debates over theory-choice cannot be cast in a form that fully resembles logical or mathematical proof" for which "premises and rules of inference are stipulated from the start" (199). Neither in the pursuit of a given paradigm nor in the evaluation of competing paradigms could one apply any fixed, rule-driven algorithm.[49] Yet, he went on, it was misguided to presume that this was the only sense in which rationality figured in science. Kuhn insisted that he fully upheld the conventional value criteria for theory choice: "accuracy, simplicity, fruitfulness, and the like" (199). But these values, while shared by individuals in a scientific community, were construed differently in isolation and ranked differently ensemble by each member. Not only was this inevitable, it was beneficial, serving "functions essential to science" by "distributing risk" and protecting "the long-term success of its enterprise" (186).[50]

Incommensurability: Kuhn's Most Provocative Concept

What made Kuhn's theory of scientific development appear so radical in 1962? What conventional wisdom did it so outrageously flout? There are two overarching points. First, Kuhn suggested by the discontinuity he asserted between episodes of normal science, and by his invocation of the notion of incommensurability to describe their mutual relation, that there could be no simple idea of "progress" as the continuous accumulation of scientific knowledge. More specifically, there was no warrant to be found in the history of scientific change for any realist metaphysic; science was not getting closer to what was "really there."[51] Second, by stressing incommensurability and discontinuity, Kuhn seemed to be suggesting that there was nothing like rational choice between theories. In explaining how one paradigm displaced another, Kuhn invoked terms such as "Gestalt switch" and "conversion experience." In explaining the discontinuity, he frequently wrote of scientists of different paradigms working or "seeing" in "different worlds." The upshot seemed to be a challenge to the very idea of the rationality of scientific progress or, at a minimum, of theory choice. Thus, Imre Lakatos grasped matters as follows: "For Popper scientific change is rational or at least rationally reconstructible and falls in the realm of the *logic of discovery*. For Kuhn scientific change—from one 'paradigm' to another—is a mystical conversion which is not and cannot be governed by rules of reason and which falls totally within the realm of the (*social*) *psychology of discovery*. Scientific change is a kind of religious change."[52] For Kuhn almost all of this represented severe misunderstandings of his claims, though he did take responsibility for some rhetorical carelessness in exposition in the original version of *Structure*.[53] Kuhn was trying to grasp the historical circumstance of scientists with different theoretical orientations failing to communicate, "talking past one another." Perhaps no one seriously doubts that this happens; certainly no one who has ever studied intellectual exchanges *should* doubt it. What created consternation in the minds of philosophers presented with this rather obvious historical point was the implication that scientists would therefore not be able rationally to resolve their differences. The issue of the rationality of theory choice was at the core of the philosophical reception of incommensurability.

Repeatedly in *Structure*, Kuhn wrote of scientists operating under different paradigms as working (or seeing) in "different worlds" (110, 117, 119, 134, 149). No other rhetorical phrase has occasioned stronger reaction: charges of "relativism" or "idealism" largely revolve around these passages. Kuhn was cautious always to qualify the phrase, to admit that he meant this only in some sense he could not fully specify. Nonetheless, the phrasing was so frequent and woven so thoroughly into the exposition of his very ideas of paradigm, scientific revolution, and incommensurability that critics were not altogether unjustified in taking the phrase as more than a dispensable metaphor. Indeed, it was. One of the most meticulous expositors of Kuhn's theory, Paul Hoyningen-Huehne, has tried extensively to discriminate two senses of "world" in Kuhn along Kantian lines to get at this idealist element.[54] It has been claimed that Kuhn was never interested in contesting the *rationality* of science but rather full-blown scientific *realism*.[55] What makes this a bit less perspicuous than it might otherwise appear is that Kuhn professed (in my view with no little irony) to be as much a realist as Richard Boyd.[56] What he was implying, I believe, is that what science takes to be real *with reference to a theory* is as real as science ever need (can) be: that the metaphysical issues between scientific realism and varieties of instrumentalism have no epistemological consequence.[57]

In my view, Kuhn is more important for his theory of scientific development than for his theory of incommensurability, for his philosophy of science than for his philosophy of language, even if he longed to be a philosopher of language. I would not go so far as Jack Meiland to claim that "clearly the idea of 'anomaly' is the main idea" of *Structure*, but I do agree that it is central to "explaining the occurrence and nature of scientific revolutions" and that "the very existence of anomalies shows that, according to Kuhn, nature provides a 'check' for each paradigm."[58] That is, Kuhn developed a model of scientific change which had a claim to be considered rational and objective—even potentially "realist" in view of the idea of natural *constraint*. Nevertheless, the Kuhn of incommensurability has preoccupied philosophical consideration.

The thesis of incommensurability was enunciated simultaneously by Thomas Kuhn and Paul Feyerabend, colleagues at the time at Berkeley, in 1962.[59] It arose out of conversations between them, but it does not follow that it ever had synonymous meaning even for the two of them. Indeed, there are grounds for the suspicion that they *never* shared the same idea of incommensurability. Moreover, it can be conjectured that the initial idea of incommensurability was undertheorized in both thinkers' formulations in 1962, only to suffer increasing obfuscation in their efforts to address their critics' queries and confutations. Subsequently, figuring out what incommensurability could sensibly mean has become a massive and less than fulfilling enterprise. It has generated an enormous literature but remarkably little consensus even as to what is at stake in the concept. There were serious ambiguities about incommensurability from the outset. Did incommensurable mean incomparable? How could it, and still allow for rivalry? But if it did not, what *did* it mean?[60] Second, was the claim total or only partial? Did every term change when a theory changed, or only some? Was this a claim about language or about science? Were the spheres invoked *theories* or *languages*, and did that matter? What occasioned incommensurability? Was it the terms a particular theory used or the problems it pursued that represented the driving consideration?

One thing is clear: the context that elicited the concept. Both Feyerabend and Kuhn were striving to formulate a critique of the logical positivist distinction between theory and observation. Both of them were committed to the idea that theory inevitably participated in the constitution of data, and therefore the very idea of a theory-neutral observation language was misguided. All observations were theory-laden. The way to bring that home, each of them realized, was to evoke the situation occasioned by theory change. Whereas the logical positivist view was that the observation language would remain unchanged and therefore supply the basis for theory comparison, Kuhn and Feyerabend argued that theory change would occasion change also in the observations, resulting in a loss of direct contact between the two theories—incommensurability.

Kuhn's idea of incommensurability was formulated somewhat drastically in the original form of *Structure*. In invoking the idea that under a new or different paradigm scientists "worked in different worlds" or belonged to separate linguistic communities, Kuhn assimilated the idea of meaning variance as it had already been articulated by Norwood Hanson and Paul Feyerabend.[61] He used it to highlight the idea that there could be no theory-neutral language through which to compare alternative theories. The "theory-ladenness of data" (Hanson) entailed "meaning variance" (Feyerabend) and accordingly "incommensurability" (Feyerabend), which rendered theory comparison problematic. But perhaps the most important philosophical source for Kuhn was Willard Quine.

The connection between Kuhn and Quine proved particularly important for the philosophical reception of Kuhn's thought in the 1970s, especially in the context of Kuhn's own explicit association of incommensurability with Quine's indeterminacy of translation. That is, Kuhn was thoroughly complicit in the assimilation of his ideas in philosophy of science into the domain of philosophy of language. In the preface to *The Essential Tension* (1977) Kuhn wrote: "I am now persuaded, largely by the work of Quine, that the problems of incommensurability and partial communication should be treated in another way. Proponents of different theories (or different paradigms, in the broader sense of the term) speak different languages—languages expressing different cognitive commitments, suitable for different worlds. Their abilities to grasp each other's viewpoints are therefore inevitably limited by the imperfections of the processes of translation and of reference determination."[62]

In an interview on November 26, 1979, Kuhn told Daniel Cedarbaum that his "entire philosophical perspective had a distinctly 'Quinean' cast."[63] Cedarbaum notes that Kuhn and Quine spent the academic year 1958–59 at the Center for Advanced Study in the Behavioral Sciences at Stanford. While Kuhn was working on *Structure,* Quine composed his seminal second chapter of *Word and Object* that dealt with radical translation, and he explicitly acknowledged Kuhn's comments on the draft in the preface to the finished work. Kuhn derived still more from the interaction. Cedarbaum concludes that "Quine's work on both the underdetermination of theories and the indeterminacy of translation . . . had a profound impact on Kuhn."[64] Unfortunately, the strong forms of Quine's theses seem to have been the ones that Kuhn assimilated and, as Keekok Lee has written, "the extreme cognitive relativism which is implicit in Quine (though Quine himself might not wish to draw such conclusions) is reinforced in Kuhn's thought by the linguistic relativism of Benjamin Whorf, which seems to have played a big part in shaping Kuhn's ideas about science."[65]

Ian Hacking notes that before Kuhn and Feyerabend, incommensurability had a clear sense in mathematical terminology, namely, that which had "no common measure," as the length of the hypotenuse of a right triangle has no common measure with the length of the sides. Kuhn and Feyerabend made *metaphorical use* of this original sense, generating at least three distinct considerations.[66] First, Hacking identifies topic incommensurability—incommensurability of problems and perhaps of standards. The point is simply that the logical positivist notion of the reductive cumulation of successive theories does not fit all (if any) cases of historical scientific theories. Successive theories change topic; in other words, they identify or prioritize different problems, giving related observational data new and different salience. This was part of the historicization of science central to both Kuhn and Feyerabend and discernible especially in comparison with the contemporary assertion of the orthodox view in Ernest Nagel's *The Structure of Science* (1961).[67] The notion of theory succession on which the Received View proceeded required term-for-term reduction of the confirmed portion of each surpassed theory to its successor, because only this secured the logical comparison of the two theories as axiomatized systems required by the "context of justification."[68] Any failure of commensurability, any variance in meaning, would destroy the logical possibilities of reduction.[69] Here, clearly, the lure of logical precision carried philosophers of science beyond the pale of actuality, and this was the central and compelling thrust of the "historicizing" revolution launched by Kuhn, Feyerabend, and others.

But there was another sense of historicism lurking in the idea of incommensurability: not simply the sound claim that theories changed what they talked about, but that having changed it, the possibility of understanding across the difference could be fatally impaired. There is nothing in the mere fact that the subject has changed that would prevent scientists of one theory from understanding what was claimed by scientists proposing another theory. But if the very language in which the second theory was formulated referred to entities alien to the first scientist, then the prospect of understanding became severely compromised. This sense of literal incomprehension Hacking calls "dissociation," and he locates it aptly in historical distance.[70] The point cannot be stressed too heavily, for it was the inaugural experience for Kuhn himself. Kuhn has attested that he began his theorizing about the historical problem of understanding science when he approached Aristotelian theory of motion from the vantage of his own training in modern physics. He found Aristotle absurd for some time, but then he experienced a "Gestalt switch." Suddenly things snapped into place when he grasped that Aristotle was not undertaking at all the same project.[71] Reconceiving what Aristotle's project had been, Kuhn was able to make sense of Aristotle's passages in ways that were no longer absurd, though they remained utterly irreconcilable with modern physics.

This sense of historical alienation, of "dissociation" in Hacking's terms, is a historicization not so much of the object of inquiry as of the inquirer her/himself: the realization that we are always situated in a given set of categories through which we order the world, and that these are neither universal nor timeless.[72] In *Structure*, Kuhn attached enormous importance to this subjective suddenness of historical discovery. He therefore tended to exaggerate the discontinuity of scientific development, as he admitted in later writings: "In recent years I have increasingly recognized that my conception of the process by which scientists move forward has been too closely modelled on my experience with the process by which the historian moves into the past."[73] Given sufficient distance in time (or often, space) in our inquiry, Hacking summarizes, we come upon "ordering of thought that we cannot grasp . . . for it is based on a whole system of categories that is hardly intelligible to us."[74] Even when we have putative "translations" of the words and sentences of a figure like Paracelsus or Aristotle, they literally make no sense to us, *cannot* make sense to us in our own categories of thought. We are pressed either to abandon appraisal, since "we cannot attach truth or falsehood to a great many of [these alien] sentences," or to "rethink the works of our predecessors in their way, not ours," whereby the project of translation is supplanted by a substantially different idea of understanding.[75] While Hacking notes that this is a rather elementary methodological observation for historians, it represented a profoundly revisionist idea for philosophers of science committed to the timeless logicism of the Received View. Still, the issue remained precisely *historical* rather than *philosophical,* Hacking pointed out, since it still concerned facts of the matter about the world, in other words, what historical figures "were talking about" and not whether what they said "is true."

It is the third aspect of incommensurability that belongs strictly to the philosophers—specifically, the philosophers of language. That third aspect is "meaning-incommensurability." As Hacking has noted, "at the root of meaning-incommensurability is a question about how terms denoting theoretical entities get their meaning."[76] The meaningfulness of theoretical entities represented one of the classic conundra within the philosophy of science, dividing realists from antirealists, perhaps the most pervasive division in that discipline in modern times.[77] What is at stake can be illustrated by a single example. The antirealist Bas van Fraassen challenged his realist rivals by asking in what sense the referent of the term *electron* with which Thompson or Rutherford or Bohr or Schrödinger theorized could be "the same."[78]

The Philosophical Reception of Incommensurability

Kuhn had a variety of things in mind with his term *incommensurability,* of which language was only one, and not necessarily the most important. To most philosophers, however, incommensurability promised to plunge philosophy and science into subjectivism, relativism, and all the other horrors of dread night. What kept Quine and the other legatees of the Vienna Circle from total dissolution of epistemology was confidence in "observation sentences," or reliance upon the constancy of sensory input as the basis for linguistic efficacy as well as empirical knowledge. But what Kuhn and Feyerabend seemed to be proposing was the dissolution of that one last piece of solid ground in the epistemological sphere. Or, at least, when philosophers read Feyerabend and especially Kuhn, this is what they fastened upon as the important philosophical issue. Not only rationality seemed imperiled; so, too, did progress. These were, understandably, highly valuable commodities for philosophy of science, and a rescue effort was immediately mounted. A part of that rescue effort took up the issue where it emerged, in terms of the interface of competing scientific theories as a matter of historical fact.[79] But the part that occasioned the most attention went in another direction, a transposition of the problem of philosophy of *science* into the philosophy of *language.* What seems to have characterized the reaction to

Kuhn's historicist challenge was a striking "linguistic turn" in the philosophy of science, the view that problems specifically of philosophy of science
could be resolved (only) by semantic ascent into philosophy of language.[80]
Of the many "marriages" proposed for philosophy of science in that decade,
I surmise this was the most treacherous, perhaps because it seemed the most
natural.

It is crucial to discern how the philosophical reception shifted everything
Kuhn had been saying into the key of philosophy of language and focused all
its energy on the dicey notion of meaning and its (in)variance. Ironically, one
of the culprits of just this maneuver was Dudley Shapere, who came later to
rue this whole strategy in philosophy of science. Shapere's influential early
review of Kuhn's *Structure of Scientific Revolutions* largely executed a "semantic ascent" upon Kuhn's arguments, reading Kuhn as a (bad) philosopher of language to explain away his insupportable claims in the philosophy
of science.[81] Shapere linked Kuhn systematically with Feyerabend, arguing
that their positions were to all intents identical in these areas: "neither author gives us a criterion for determining what counts as a part of the meaning of a term, or what counts as a change of meaning."[82] Shapere found
them both guilty of "extreme relativism" in their invocation of meaning
variance: "the root of Kuhn's and Feyerabend's relativism . . . lies in their
rigid conception of what a difference of meaning amounts to."[83] "[I]nas-
much as all meanings are theory-dependent, and inasmuch as theories can be
shaped at will, and inasmuch, finally, as all observational data (in [Feyer-
abend's] sense) can be reinterpreted to support any given theoretical framework, it follows that the role of experience and experiment in science
becomes a farce."[84] Shapere claimed that "this relativism, and the doctrines
which eventuate in it, is not the result of an investigation of actual science
and its history; rather, it is the purely logical consequence of a narrow preconception about what 'meaning' is."[85] Clearly, Shapere assigned that responsibility for introducing "meaning" deleteriously into philosophy of
science to Kuhn and Feyerabend, not himself. He insisted, as he would later,
that "it seems unnecessary to talk about meaning" in order to "understand
the workings of scientific concepts and theories, and the relations between
different scientific concepts and theories."[86]

The frustrating thing about Kuhn's work is the equivocal formulation of
all its key ideas. Shapere could find passages in Kuhn's work which warranted his critique. Claiming to discern "the logical tendency of his position," Shapere argued that Kuhn's notion of incommensurability was predicated on the idea that "problems, facts, and standards are all defined by the
paradigm, and are different—*radically*, incommensurably different—for
different paradigms." That is, paradigms "disagree as to what the facts are,
and even as to the real problems to be faced and the standards which a suc-

cessful theory must meet." Rather than measure Kuhn's claim against the historical record, Shapere insisted it was a consequence of Kuhn's misguided notion of *meaning*, from which followed the dire consequence "that the decisions of a scientific group to adopt a new paradigm cannot be based on good reasons of any kind."[87] Concentrating his fire there, Shapere set the course of the entire reception of Kuhn off on a "linguistic turn."[88]

But Shapere is not the main perpetrator of the linguistic makeover of Kuhn. That honor, such as it is, belongs to Israel Scheffler. Like Shapere, Scheffler was philosophically attuned enough to realize the parlous state of the idea of "meaning" within philosophy of language. While he recognized that "*Both* sense and reference depend on the human context of word use," Scheffler was sensitive to the vulnerability of meaning; splitting it into *sense* and *reference*, he opted to try to salvage the latter. He argued that reference was "surely the clearer of the two."[89] Indeed, Scheffler invoked Quine to imply that the intensional sense of meaning verged on worthlessness. He reacted to the Kuhnian assault on meaning invariance by simply abandoning the idea altogether and resorting to reference alone to secure scientific rationality. Scheffler put the case as follows: "We have here another paradox, the *paradox of common language,* and its upshot is that there can be no real community of science in any sense approximating that of the standard view, no comparison of theories with respect to their observational content, no reduction of one theory to another, and no cumulative growth of knowledge, at least in the standard sense. The scientist is now effectively isolated within his own system of meanings as well as within his own universe of observed things."[90] Scheffler challenged the idea that simply because terms were theory-laden any test of theory against evidence must be compromised. "Observation may be considered as shot through with categorization, while yet supporting a particular assignment which conflicts with our most cherished current hypothesis."[91] Saving reference got Scheffler what he needed: "alterations of meaning in a valid deduction that leaves the referential values of constants intact are irrelevant to its truth-preserving character."[92] Scheffler went further, though, to rescue at least some element of meaning: "Meanings, that is, may be considered as *relative to* language systems without compelling the further assumption that they cannot be *shared by* different language systems."[93] "The fundamental subjectivist argument from meaning thus collapses. If meaning is not a mystery of sounds and shapes, neither is it a subjective prison of the mind. It is perfectly possible to share common meanings with those who disagree with us in belief. Since, indeed, disagreement, in the sense of explicit contradiction, requires common meanings in order to be differentiated from a mere changing of the subject, the collapse of the subjectivist argument reinstates the possibility of such disagreement."[94] For science, Scheffler proposed, recourse to reference escaped most of the paradoxes of meaning, since "the denotation or reference of [scientific] terms in specific cases can . . . be determined independently of a characterization of their respective senses."[95]

Taking up the question as Scheffler formulated it, Carl Kordig offered a critique of the position on theory-ladenness which argued that the dissolution of the theory/observation distinction bore relation to Quine's dissolution of the analytic/synthetic distinction, but that the difference was more important than the similarity: "[F]or Feyerabend, given his radical meaning variance position, this denial should take the form of a claim that nothing is synthetically true, that all truth or falsity is linguistic."[96] Kordig took Kuhn to be a radical idealist, for whom paradigms were incommensurable in the drastic sense that they *could not* be rivals because they could have nothing in common. But Kordig did recognize that Kuhn was striving to articulate a theory about the succession of theories in science which was far more faithful to historical actuality than that of the Received View. Kordig understood Kuhn to be claiming that problems changed, and with them the salience of data and perhaps even the standards of evidence.[97]

Michael Martin took up from Scheffler, too, in his reaction against the position advanced by Kuhn. He noted that Scheffler had argued that meaning variance alone did not make rational theory choice impossible, because Scheffler held out for "sameness of reference" according to which "objective theory testing [could] proceed."[98] Martin proposed to extend the argument by claiming that even reference variance did not impede theoretical comparison. Martin ascribed to Kuhn the notion of *total* incommensurability: "that all terms change meaning in theory change . . . [and] this is a necessary consequence of theory change."[99] Martin argued that referential variance needed to be established empirically and not by a blanket, a priori assertion. He claimed that history of science would not offer much support to this claim.

Jarrett Leplin argued that while the "meaning variance" interpretation of the historical relativism claims of Kuhn and Feyerabend was taken by both sides to be central, subsequent debate concentrated on reference variance. But, like Martin, Leplin argued "identity of referential expressions does not guarantee coreference among theories; neither does diversity of referential expressions preclude it."[100] More centrally, theoretical terms typical of scientific theories pose particular problems for reference. " 'Electron,' 'positron,' and 'photon' are terms whose reference was initially fixed by a certain theoretical context on the basis of properties descriptive of a certain theoretical role their referents were to assume in that context. This is not to say that their reference was fixed by theories. . . . It is to say that what initially fixes the reference of these terms is not any experience but certain properties whose ascription to their referents carries theoretical commitments underdetermined by experience."[101] Leplin concluded: "My basic argument is that available theories of reference do not meet the challenge [historical relativism] poses for [scientific realism] because, apart from internal weaknesses, they are inapplicable to the [theoretical-context]-terms which are the important cases for [scientific realism]."[102] That meant that the issue of incommensurability and historical relativism was still very much alive.

Summarizing the controversy, Robert Nola contended that Kuhn sought to come to grips with the problem of incommensurability "by seeking an epistemological resolution of a problem which is at bottom one in semantics, especially the theory of reference."[103] Nola noted explicitly that Kuhn was not well-versed in philosophy of language: "Kuhn is not clear in the way he employs the terms 'concept,' 'meaning,' 'definition' or 'interpretation': nor is his use of 'meaning' sensitive to Frege's distinction between sense and reference."[104] Offering a more sophisticated formulation of what Kuhn's incommensurability thesis might mean within the philosophy of language, Nola suggested that recent achievements in the "causal theory of reference" might offer resolution of the issue.

In his contribution to the debate, Michael Devitt offered a cogent representation of the course of philosophical discussion through approximately 1979, noting that philosophical rebuttal to Kuhn and Feyerabend followed the line set by Israel Scheffler. Devitt offered a helpful distinction between semantic and epistemological considerations of the incommensurability thesis. That is, he argued that Kuhn and Feyerabend were claiming not only that we could not be certain that there were grounds for theory comparison (the *epistemological* thesis) but also that *as a matter of fact* no sufficient relationship existed between theories to permit comparison (the *semantic* thesis). While Devitt was altogether willing to accept the former case, he found the latter unjustified. Devitt recognized that theories were special languages, that they isolated a particular domain of terms and relations (within the totality of a natural language). As Devitt put it, "talk of meaning is notoriously vague," and the real issue was "what sort of entities we are interested in comparing logically." He suggested that the appropriate conception was *tokens*. And he insisted that the incommensurability thesis "*does not follow from any thesis about semantic change that Kuhn and Feyerabend might reasonably offer.*"[105] Noting that logical relationships always required the conformity of an instance with the logical form, Devitt argued, "*even the existence of a logical relationship between statements does not depend on logic alone; it depends also on certain empirical facts about the contained items.*" The empirical facts conformed to the conditions of the logical form. Devitt concluded: "logical relationships depend on empirical facts about reference." He added, "judgments of reference are, in fact, heavily theory-laden." Thus, the question of incommensurability rested upon a question of the theory of reference, a theory which "will tell us the nature of the link between terms of each grammatical category and the world." Devitt took up the causal theory of Hilary Putnam as such a theory. It postulated the actual existence of natural kinds such that tokens of this type causally prompted the use of the term. To make any such connection required the employment of a variety of theories.

Devitt wished to establish that judgments of incommensurability, like every other form of judgment, were contingent, but this contingency by no means implied impossibility. Invoking Hartry Field's argument for "partial reference," Devitt argued that there was no form of the incommensurability thesis that could plausibly reject the possibility of shared partial referents, and hence comparability was always possible, especially when theories were taken to be "in the same domain." Devitt's ultimate point was to return the issue of commensurability to empirical inquiry rather than a priori reasoning: "The truth about semantic comparison of theories is not that it is impossible (as [the incommensurability thesis] claims) but that it is epistemically like all other attempts to understand the world."[106]

The one significant early philosophical defense of Kuhn's epistemology was an essay by Gerald Doppelt in 1978.[107] Doppelt offered a careful critique of the Shapere-Scheffler reading of Kuhn, and also of the Lakatos rejoinder, demonstrating that they had largely missed Kuhn's point by swiftly and falsely assimilating it to a position in the philosophy of language which Kuhn did not intend or require for his argument. Doppelt maintained that Kuhn's argument for discontinuity, incommensurability, and relativism between scientific paradigms had four dimensions: the issues of language or concepts, of observational data, of problems, and finally of standards of evaluation. Shapere and Scheffler presumed that the driving force in Kuhn was the first of these issues. They took his doctrine of incommensurability (assimilating it to that of Feyerabend, of whom perhaps it was more true) to be that of total and absolute linguistic difference, such that comparability between theories became impossible. Even within theories, they maintained, there was such a high degree of autoconfirmation that the very idea of anomaly became unjustified and incoherent. Doppelt argued that this missed Kuhn's point virtually entirely. According to Doppelt, the driving force for Kuhn lay in the choice of *problems*, in the redefinition of the disciplinary inquiry, which had, as its consequence, a shift in the *relevance of data* and also, perhaps even more importantly, a *shift in standards of appraisal*. Only as a consequence of these shifts did there emerge (logically and temporally) a full-fledged linguistic reformulation of the theory.[108]

Given this construal of Kuhn's theory, there were no drastic consequences for comparability of theories, as Kuhn had all along insisted. For Doppelt, the conflicts between paradigms "have an irreducible normative dimension."

This made them more like moral or political conflicts than questions of methodological procedure.[109] Kuhn's point, as Doppelt reconstructed it, was that a simple sense of cumulative progress, in which *all* an old theory's "genuine data" were preserved and all its problems retained the same importance, simply could not be warranted from a consideration of the history of science. First, some data were lost, some problems were either demoted or dismissed. A shift in the purpose or definition of the disciplinary inquiry entailed shifts in the relevance or centrality of data in answering the newly critical problems. It was in terms of this shift in priority that the crucial question of standards arose, and with it the question of rationality. "On [Kuhn's] view the most basic level of disagreement concerns the normative question of how the subject matter as a whole ought to be defined."[110] On the basis of such differing value premises, there would "always [be] 'good reasons' for adhering or switching to a paradigm."[111] That was why Kuhn resorted to sociological and psychological dimensions and used such phrases as "conversion." "Given that the balance of reasons is necessary but not by itself sufficient to explain when and why some but not other scientists embrace a new paradigm, what additional conditions are required to explain this process?"[112] Doppelt replied, "we can make sense of Kuhn's introduction of various sociological, psychological, biographical, and historical factors . . . without seeing it as a lapse into irrationalism."[113]

Doppelt then appraised whether Kuhn had in fact proven his case. The issue was not only what has come to be called (after Roy Bhaskar) "Kuhn loss"—genuine data that are not preserved in shifting paradigms—but even more, shifts in standards. Doppelt argued that in the long-term view Kuhn could not establish that there really *must* be data loss, since later theories had recovered, and future theories always might recover, data lost in interim theories. More problematic was the question of shifting standards. "If a new theory can . . . define and satisfy standards which incorporate the different (and 'incompatible' in the weak case) standards of preceding paradigms, the possibility of a 'cumulative' shift of standards obtains and the present relativism argument is weakened."[114] Doppelt argued that Kuhn could not prove this impossible. "The shift-of-standards argument did not rule out 'cumulative' shift in standards; furthermore it depends on a relativistic criterion of scientific knowledge which receives no systematic philosophical defence in Kuhn's work, in the context of an examination of other epistemological models (non-positivist and non-relativist)."[115] Nevertheless, to demonstrate the *short-term failures* of cumulation in data or standards was sufficient, in Doppelt's view, to shatter the grip of the positivist notion of progress and to put in question the Received View of philosophy of science. This called for the discipline to turn from questions of the appraisal of theories to questions of the explanation of process in order to understand actual science.

Incommensurability and the Causal Theory of Reference

One of the decisive philosophical responses to Kuhn was to bring philosophy of science under the "protection" of philosophy of language, to rescue theoretical terms of science from the conundrum of meaning variance via the causal theory of reference, a major philosophical undertaking of the 1970s developed by Saul Kripke, Hilary Putnam, and Keith Donnellan.[116] While it would be unwise to conjecture that the Kuhnian eruption within the philosophy of science alone precipitated this departure in the philosophy of language, the linkage is direct enough to suggest that it played an important role in promoting interest in the causal theory of reference. The clearest linkage between the crisis in philosophy of science and the emergence of this new theory is Israel Scheffler's *Science and Subjectivity,* which sought to evade the dilemmas posed by meaning variance by resorting to persistence of reference. Shortly thereafter, Kripke, Putnam, and Donnellan worked out the full-fledged causal theory of reference, and it became the basic recourse of philosophers of science in the 1970s seeking to overcome the incommensurability thesis.

The tradition from Gottlob Frege and Bertrand Russell in semantics conceived meaning in two aspects, *sense* and *reference.*[117] For some time after Frege, the theory of meaning held that the *sense* of a term designated its *referent* via a definitional criterion or "cluster" of criteria. But by 1962 this Fregean idea of *meaning* was in desperate straits.[118] The problem with this definitional approach to meaning was that sometimes the criterion would designate something that was *not* an instance of the term, and sometimes something that *was* an instance of the term would not satisfy the criterion. The elaboration of a cluster theory sought to mitigate the damage by claiming that instances need only satisfy some of the criteria to serve as tokens of a given type. Thus, in Wittgensteinian terms, they would share a "family resemblance" if not an "essence." But what haunted all such efforts at intensional meaning was the implication that meaning existed as some "mental entity" apart from its linguistic use. A great deal of what Willard Quine had been up to for some time already in his linguistic holism can be seen as a sustained effort to render the mental entity notion of meaning worthless, to externalize epistemology.[119]

In philosophy of science the main recourse to escape the problem of meaning variance was to abandon the intensional idea of *sense* and to hitch all the hopes of conceptual continuity to the idea of *reference.* The new ideas concerning a causal theory of reference seemed tailor-made for this project. The approach began with problems already virulent within traditional Fregean meaning theory, in particular the status of proper names. In a key essay, Keith Donnellan suggested that in the "definite description," which was critical to the notion of meaning, there were two aspects that could always be teased out: fixing reference and attributing some characterization.[120] Saul Kripke elaborated this into a theory of the "rigid designation" of proper names.[121] In the words of Stephen Schwartz, "Kripke distinguishes between fixing the reference of a term and giving its definition. . . . When I fix the reference of a term, I give a description that is to be taken as giving the referent of the term, not the meaning in the traditional sense."[122] That is because all the description is intended to do is discriminate some referent in the perceptual environment, not give a definitive characterization of its nature. This allowed for a slack in the definitional description; so long as the referent was fixed, as long as the linguistic community in question fixed upon the same token, it did not matter whether the description was accurate and certainly not whether it was exhaustively discriminating. The distinction allowed Donnellan and Kripke to solve outstanding problems in the philosophy of language revolving around identity and proper names. The crucial step for the philosophy of science and its problems with meaning variance came when Hilary Putnam extended the causal theory of reference to the domain of natural kind terms.[123] Reference that involved nonunique (or proper) designation inescapably required a categorization of things into various types, a taxonomy. How these types are instantiated (extension, reference) is one issue; another is how each is characterized (denotation). Reference to natural kinds had two elements, the notion of a *type* or *kind* and the notion of its *tokens* or *extension*, its instantiations.

Putnam's approach to these questions has seemed to many the best response to the problems posed by meaning-incommensurability. Putnam asserted that the causal theory of reference worked for natural kinds. These kinds preserved reference from an original dubbing, either by ostension or by a fixing description, through the transmission of this term by historical transfer. The point for Putnam is that the transmission no longer required defining criteria to be constant. It was "common for the first stab at defining a species to flounder," as Hacking puts it, but that did not matter, as long as *something* had been distinguished ("dubbed") as a type whose tokens could then be discerned.[124] Thus, "gold" remains a referent despite the theoretical revisions which alter the criteria of its stereotype. This becomes important for those realists who wish to hold that none of our theories is strictly true, that all theory is "born refuted," and yet that there is nonetheless something real that they are all about. Entities, the *extensions* of referring theoretical terms, could be "real" by persisting across drastically differing descriptions, or changes in theory.[125]

But two important reservations surfaced with regard to the causal theory and its rescue of philosophy of science. First, it was not altogether clear how one could be epistemologically comfortable that the "historical explanation theory" of reference—reference-fixing through a type or the tokens of a type—would pick out the *same* referent simply because the type or term was transmitted historically from its original dubbing to its current use. That was a stipulation which seemed to beg the core question of incommensurability. Second, and even more grave, it was problematic whether the key terms in philosophy of science, *theoretical terms,* fell within the scope of Putnam's "natural kinds." Indeed, even Keith Donnellan expressed discomfort concerning whether the original causal theory worked for natural kind terms themselves.[126] As Schwartz summarizes: "The error of traditional theory was in extending its analysis, based on nominal kind terms, to natural kind terms. The analogous error on the part of the new theory of reference is to extend its analysis, based on natural kind terms, to nominal kind terms."[127] But if theoretical terms—and most of the artificial designations which throng natural language—are *not* natural kind terms, the bulk of language and especially the language of scientific theory appears to fall outside the scope of the new theory, and that leaves the problem of incommensurability exactly where it had been before.[128] Berent Enç posed this challenge to the causal theory of reference.[129] His objection was that, while the arguments of Kripke, Putnam, and others seemed compelling for those things for which ostensive indication was possible, it proved inadequate for nonostensive, or theoretical, terms. Extension was insufficient to secure the referent, for that extension could not be construed without an account. These terms required, in addition to the baptismal designation, a *definite description.* This meant that for all nonostensible terms, Putnam & Co. had not really escaped the conundrum of theory-ladenness.[130] Jarrett Leplin took a similar stance on the inapplicability of the Kripke-Putnam theory to theoretical terms.[131]

We are still, in short, in an quandary over the continuity of conceptual content in such terms as *electron* and, hence, over its ontological aptness. With regard to any theoretical term ("T"), Robert Nola writes, "theoretical beliefs must come to play a substantial role in providing an analysis of the non-ostendable event which is to yield an object to which the term 'T' can refer."[132] Yet there remained a sense, Nola conceded, in which "[Benjamin] Franklin, a quantum physicist and an ordinary electricity bill payer use the term 'electricity' with the same reference despite differing, or no, theoretical beliefs."[133] Dudley Shapere took a much dimmer view of the viability of the causal theory of reference to resolve the issues in the history of scientific theories and therefore in the theory of reference in philosophy of science.[134] And he was not alone.[135]

Arthur Fine wrote in 1975: "Work at a decent level of generality is at an impasse today in the philosophy of science as well as in the history of science."[136] The source of the impasse lay "in the theory of meaning, more especially in the theory of reference."[137] Fine took the view that *neither*

Kuhn's incommensurability argument *nor* Putnam's causal theory of reference worked, and that their mutual interference constituted the impasse. His proposal for extrication from the impasse was twofold. First, like Martin, he insisted that reference was an empirical matter. But unlike Martin, he found this no consolation, for the facts of the matter could be construed in terms of either view and therefore would not resolve the matter. Nonetheless, Fine's second element was to simply "suppose that the worlds of the theories [in question] overlap in such a way that there is a correlation between the terms of the theories that makes correlated terms co-referential in each world in the region of overlap."[138] This tactic was not guaranteed to work, but "since later theories develop out of earlier ones, and since contemporaneous rival theories develop against one another, the supposition of overlap is likely to hold in all cases of interest."[139] In short, Fine gave the issue over entirely to historical-empirical investigation with the prospect that *given* overlap in reference, comparability would be possible.

One of the most explicit efforts to employ the causal theory of reference to address the issue of Kuhn's claims about incommensurability of historical scientific theories was Philip Kitcher's important essay of 1978, "Theories, Theorists and Theoretical Change." Kitcher developed a version of Putnam's causal theory to address the problem of theory succession in science, taking issue specifically with Kuhn's representation of the "chemical revolution" and the term *phlogiston*. For Kitcher, Kuhn represents the paradox of both insisting upon the relevance of history of science to philosophy of science and maintaining "that the task of the historian of science cannot be successfully completed."[140] Kuhn's notion of incommensurability signifies for Kitcher that "the content of past theories resists expression in modern terms" and, hence, candidacy for truth. Kitcher terms this doctrine conceptual relativism, and its key feature is "that the old language and the new language are not intertranslatable."[141]

Kitcher was attuned to the widest resonance of issues of interpretation in the history of philosophy.[142] He set out from the proposition that "Historians of science are interested in discovering what Priestley was talking about, and how much of what he said is true."[143] There is a tension in this proposition which is caught in the tense difference it employs. Truth, Kitcher implicitly affirms, is always only a present-tense matter. What Priestley was talking about, however, belongs in the past: precisely what gets eliminated is the possibility that truth could have been different in Priestley's time, for Kitcher is clear that truth must be tenseless. To get at the truth in what Priestley was talking about, we need to understand his language, and that brings Kitcher to incommensurability.

Kitcher identifies incommensurability with untranslatability. He claims that the difficulties conceptual relativism brings in its train are a legacy of the crisis of the Frege theory of "sense" as a component of meaning. The natural recourse to escape these difficulties, Kitcher avers, is "to turn to the notion of reference," and he cites Israel Scheffler as having pioneered this course. Against Scheffler, however, Kitcher argues that even reference change does not result in incommensurability, in conceptual relativism. Only reference change that results in "mutual inability to specify the referents of terms used in presenting the rival position" would "block understanding."[144] To circumvent such a possibility, Kitcher proposes we need a more sophisticated approach to translation. Instead of a one-to-one mapping of the tokens of the two conceptual systems, which Kitcher terms a "context insensitive theory of reference," he proposes an alternative which shows the sensitivity to context which is requisite for the functioning of natural languages and also relevant to the analysis of scientific theories.

Here he has recourse to the causal theory of reference, or what he calls, following Donnellan, the "'historical explanation' theory."[145] That is, an entity is fixed referentially in terms of the history of an expression's transmission from an original dubbing to its current use. Of course, "the evidence available to us may not enable us to construct explanations of the productions of all tokens . . . in sufficient detail," and "in some cases, the facts of reference may be genuinely indeterminate," yet "our predicament is not hopeless."[146] Rather than an all-or-nothing conception, Kitcher offers a spectrum of determinacy and suggests that "neither conceptual relativism nor the traditional approaches [term-term translatability] offer a satisfactory account of the languages of past science," while positions between the two extremes do allow for "the scientists in question [to] be able to *formulate* their disagreements."[147] To attain his goal, Kitcher resorts to something like Hartry Field's notion of "partial reference," namely that some tokens of a type of past science, for example, "dephlogisticated air," might refer while others failed to refer. Were we to "attempt to combine uniform semantic treatment for all tokens of the term with the demands of a legitimate constraint on translation," the result would be incommensurability.[148] But we have alternatives, such as Richard Grandy's principle of humanity, which allow us to interpret past scientific discourse in a manner which makes it more rational (by our lights, of course).[149]

To discriminate when Priestley's tokens refer to oxygen and when they fail to refer, Kitcher suggests, we must "construct an explanation of [the referent of a token's] production."[150] Thus, Kitcher proposes that the type be abandoned for the sake of the discrimination of referring and nonreferring tokens. Not only is this a violation of strict translation, it also, of course, renders opaque the connections which Priestley saw among his tokens as instances of a type: we rescue "truth" for some expressions at the cost of the coherence of Priestley's way of thinking. Our struggle to get at what Priestley was talking about is ultimately not to understand Priestley but to be able to discriminate the truth for us of his various expressions. Thus, Kitcher concludes, "I propose that we should abandon the traditional assumption of the philosophy of science, the assumption that we can reconstruct the language of a theorist by reconstructing the language of his theory. . . . there are some crucial expressions in the languages of theorists, separated by a large change in theory, whose reference potential cannot be matched by any expression of the rival theory. Nevertheless, successful communication can continue, even when reference potential has changed, because each theorist can specify the referents of his rival's individual tokens."[151]

While much of this suggests the fruitfulness of the causal theory of reference for the interpretation of scientific change, it also left Kuhn less than convinced that what resulted provided satisfaction either as history or as philosophy of science. But before we consider Kuhn's debate with Kitcher, we need to widen consideration to perhaps the most famous and probably the more fruitful rejoinder to the radical idea of incommensurability: Donald Davidson's articulation of the "principle of charity."

Incommensurability, Conceptual Schemes, and the Third Dogma of Empiricism

Both Hilary Putnam and Donald Davidson argued forcefully that Kuhn's notion of incommensurability, taken in its strong form, was literally incoherent. Each critic noted a decisive performative self-contradiction in the whole approach, namely that Kuhn, as before him Whorf, set about offering characterizations of the ostensibly incommensurable language in the home language.[152] But Davidson used this as the springboard for an even more dramatic argument, aimed as much at Quine as at Kuhn, in which he discerned and set about refuting what he termed a third and final dogma of empiricism, the scheme-content distinction. Davidson's essay, "On the Very Idea of a Conceptual Scheme," is one of the most important essays on interpretation in the whole era of post-positivism, with reverberations across many theoretical divides.

Davidson wrote his essay to rebuff a "conceptual relativism" that, drawing on Kuhn but also on Quine, seemed to be sweeping the intellectual scene by the mid 1970s. He undertook to deflate the enterprise which traded on the idea that "reality is relative to a scheme" to achieve "dramatic incomparability." That was hyperbolic to the point of unintelligibility, Davidson claimed, but "moderate examples we have no trouble understanding."[153] Obviously, Davidson's undertaking evokes the keenest sympathy on the part of the present author, even if some disagreements emerge in the detail. Total incommensurability and, with it, the sort of dramatic incomparability which came to be the postmodern uptake of Kuhn (and Quine) represent exactly the sorts of obstacles to sober empirical inquiry this study seeks to explode. Davidson's essay represents one of the landmark debunkings of postmodern excess.

Crucially, Davidson discriminated total incommensurability from partial incommensurability. He was not taking the stance that there were not differences among conceptual schemes; he was after the claim that the differences could be so extensive as to render them mutually incomprehensible. The crucial passage in the essay on this score goes: "We can be clear about breakdowns in translation when they are local enough, for the background of generally successful translation provides what is needed to make the failure intelligible."[154] In fact, local incommensurability was really all that Kuhn had any claim to, as he himself would acknowledge later in his career, but total incommensurability appeared to be what he was claiming in *Structure,* and it was assuredly what Whorf and Feyerabend were claiming. It is this "larger game" that Davidson set off after.

To come to grips with the issue, Davidson reconfigured it in his own terms. First, he identified conceptual schemes with languages. He felt entitled to do so, since Quine's dissolution of the analytical/synthetic distinction appeared to him (and to those he was criticizing) to have also dissolved the distinction between theory and language. The issue thus came to be whether languages could share conceptual schemes, and the test of such sharing was translatability. The claim of conceptual relativism was that conceptual schemes (or languages) were untranslatable. Davidson could only conceive this as the assertion that conceptual schemes differed in their relation to some primordial, uninterpreted "empirical content." The governing metaphors he discerned had the schemes "organizing" or "fitting" either the "world" or "experience" as an undifferentiated given.[155] That is, "it is essential to this idea that there be something neutral and common that lies outside all schemes."[156] But just this radical discrimination between conceptual schemes and some indeterminate given was a third dogma of empiricism, which "cannot be made intelligible and defensible."[157] Indeed, with its demolition, empiricism loses all warrant and vanishes as a meaningful philosophical position, in Davidson's view. Not only is Davidson in accord with Wilfred Sellars that there is no sense to be had in the idea of an experiential, primordial "given" like the old Kantian notion of intuition or the empiricist notion of sense-datum, but Davidson appeared to find no sense in the notion of a noumenal "world" prior to its organization in human schemes.[158] Here, Quine was as much Davidson's target as Kuhn and the relativists, or even more.

Two reservations need to be entered. First, it is not at all clear that Quine succeeded by his demolition of the analytic/synthetic distinction in totally liquidating the difference between theory and language. That issue has come to the fore in our treatment of Quine: semantic holism does not jibe at all with Quine's prudent formulations of underdetermination of theories. If theory dissolves entirely into language, it becomes impossible to make sense of the Duhem thesis as holding out any prospect of evidential warrant, which even Quine acknowledged. And similarly, if theory dissolves entirely into language, it becomes inconceivable how *two* conflicting theories could be formulated in the same language, which is the indispensable basis for Quine's claim about the underdetermination of theories. More generally, even Davidson has recognized that natural languages behave in ways that cannot be assimilated to the formal languages in which—at least conventionally within the analytic school of philosophy—theories have been conceived. All in all, it is simply not the case that Quine or Davidson have succeeded in erasing the distinction between theory and (natural) language, though they have certainly made it harder to define *theory*.[159] One way indeed to begin a conceptual discrimination of theory within language might be via the idea of a conceptual scheme. That is perfectly compatible with the "partial untranslatability" that Davidson finds both plausible and epistemologically innocuous.

Second, it is not at all clear that Davidson's crusade against the indeterminate given really gets at the core of Kuhn's concern with incommensurability because Kuhn, Feyerabend, and their sympathizers, were operating always with the notion that "meaning," as Davidson himself puts it, "is contaminated by theory." In other words, a conceptual scheme in Kuhn's sense does not work with theory-neutral observations but rather with already richly structured evidence which is "theory-impregnated," in Popper's phrase.[160] Refuting the third dogma seems a separate enterprise from refuting incommensurability. Davidson appears to be conflating targets, to the disservice of his main objective.

Davidson is on surest footing in the claim that "languages" that are totally untranslatable would offer no ground to be considered languages at all.[161] There could be no rational basis for recognizing them as rival formulations about the world because we could not recognize them as formulations.[162] The line of argument is central to his whole theory of interpretation. The notion that conceptual schemes work to "fit" the world is a notion of epistemic adequacy of sentences to their referents. But Davidson claims this "fit" of conceptual scheme to world is just a cumbersome way of saying that a theory is true, and the claim of incommensurability reduces to the idea that schemes are "largely true but not translatable."[163] Formulated that way, Davidson has the resources to demolish the claim, for he holds that the very idea of truth requires translation. That is the whole thrust of Alfred Tarski's famous "Convention T," which Davidson argues is the only viable conception of truth available.[164] If truth requires translation for its very intelligibility as a category, the construction "largely true but not translatable" becomes an absurdity.[165]

Total untranslatability is, in a word, unintelligible. But that opens the way for Davidson to go further. He wishes to question even the viability of any strong sense of partial untranslatability. How is it, Davidson asks, that we come to understand someone speaking an alien language? The condition of "radical interpretation" confronts the fact that without a knowledge of beliefs, one cannot grasp a speaker's meaning, and without a grasp of a speaker's meaning, one cannot fathom his beliefs.[166] The only recourse, Davidson insists, is to assume the speaker's beliefs largely parallel one's own—"subject to considerations of simplicity, hunches about the effects of social conditioning, and of course our common-sense, or scientific, knowledge of explicable error" (i.e., "differences of opinion").[167] That is Davidson's famous "principle of charity," and he claims "charity is not an option, but a condition of having a workable theory. . . . Charity is forced on us; whether we like it or not, if we want to understand others, we must consider them right in most matters."[168] That is, "to make meaningful disagreement possible . . . depends on a foundation—*some* foundation—in agreement."[169] That leads Davidson to the view that we cannot conceive even substantial differences in belief. This, as numerous commentators have urged, goes too far.[170] But the crucial argument holds, and it grants Davidson a substantial advance over the positivist-empiricist legacy of Quine, dispersing the skeptical veil of an intermediary between language and world. "We do not give up the world, but re-establish unmediated touch with the familiar objects whose antics make our sentences and opinions true or false."[171]

The question that remains in this context is whether Davidson has dispatched the whole basis for conceptual schemes, specifically in reference to less-than-holistic versions of incommensurability, and whether he has done justice to the epistemological and historical issues that Kuhn was seeking to highlight in formulating the idea of incommensurability in the first place. While there are few who would rescue total untranslatability from Davidson's critique, the ensuing debate has seen fit to quarrel with his handling of partial untranslatability. Nicholas Rescher, for one, maintains that "the rejection of the notion of conceptual schemes is not warranted."[172] He points to its prominence in the accounts of descriptive sociology, intellectual history, history of science, and philosophical epistemology. Rescher insists that "taxonomic and explanatory mechanisms . . . may differ so radically that intellectual contact with them becomes difficult or impossible."[173] While I would concur that difference renders matters difficult, with Davidson I resist the idea that it is impossible. The task of the hermeneutical disciplines that Rescher identifies with conceptual schemes is to find access, to bring the "horizons" of distinct discourse into fusion (in the terms of Hans Georg Gadamer). In the empirical human sciences such endeavors have been going on for centuries, and thus the empirical fact is that hermeneutic interpretation works robustly, despite difficulties.

Rescher is on sounder footing in raising what has become a central distinction in the whole debate about incommensurability, namely, the distinction of "translation" from "interpretation." If translation is taken in its full rigor, along the lines which Quine specified for his translation manual, then quite the contrary of Quine's own postulation emerges: there are not *many* equally viable translation manuals; there are *none*. "There is and can be no genuine translation where the descriptive, taxonomic, or explanatory mechanisms—the whole empirically laden paraphernalia of empirical reportage—are substantially different."[174] Yet what cannot be literally *translated* can be rendered *intelligible*, or interpreted via "circumlocution and paraphrase, . . . explication and approximation."[175] That is a crucial qualification of the problem of incommensurability, and one that Kuhn himself would invoke in his later writings against Quine (and implicitly Davidson). But it does not directly address Davidson's argument that translation is essential to any philosophical notion of truth.[176]

Rescher seeks to disentangle the issue of conceptual schemes from Davidson's attack on the scheme/content "dogma" by denying Davidson's claim that it is "predicated on the model of a common, preexistent raw material input which is processed differently by different schemes."[177] Rescher argues cogently that discriminating the input of a preschematically given from the contribution of the human scheme is utterly impossible, as Davidson himself would agree. But then "the idea of a preexisting 'thought-independent' and scheme-invariant reality that is seen differently from different perceptual perspectives *just is not* a presupposition of the idea of different conceptual schemes."[178] In my view, this is a fair protest against Davidson. Conceptual schemes present epistemological issues in a different vein from the problem of the raw given.

Rescher tries to illuminate that issue from another vantage: "Different conceptual schemes embody different theories, and not just different theories about 'the same things' (so that divergence inevitably reflects disagreements as to the truth or falsity of propositions), but different theories about different things."[179] In this sense, theories can be incommensurable simply because they are deployed in different domains. Thus, a theory of particle physics is obviously incommensurable with a theory of poetic sonnets. Rescher argues that the "key to scheme-differentiation lies in the nonoverlap of theses."[180] That is, "some truth-determinations from the angle of one scheme are simply *indeterminate* from that of the other in that it has *nothing whatsoever* to say on the matter."[181] Hence, "certain perfectly ordinary assertions from the perspective of one scheme are altogether *ineffable* from that of the other."[182] Of course, the question is, so what? Rescher is quite right that cultures find different projects worth pursuing and elaborate theories appropriately. We have all heard of the Eskimo categories of snow, of "calibrating" tense-structure between Hopi and English, or of color spectra between Navajo and English. Here is precisely where incommensurability is epistemologically innocuous. While there might well be interesting practical issues about the selection or even the amalgamation of such schemes, there is no cognitive-relativist implication at all in what Rescher has presented. The practical (methodological) problem would lie in the imaginative accession to the other scheme, and here difficulty is not to be underestimated, yet with a will such cultural learning seems at the very least possible. An a priori argument against it, certainly, would be ill advised.

Rescher, in my view, retrieves the notion of conceptual scheme for its hermeneutic usefulness, but he does not advance us much in the question of Davidson's disposition of the skeptical-relativist claims of incommensurability. The issue, as Robert Kraut notes, is with the "ontological relativity" of languages, especially since their "expressive resources" are by no means identical. It is by rendering such ontological relativity in the form of a drastic juxtaposition of scheme to an unschematized given that, Kraut suggests, "Davidson has placed an unreasonably strong demand on the scheme idea."[183] As Kraut has it, "scheming is the natural manifestation of a willingness to acknowledge alternative ontologies, as spawned by alternative expressive resources." Hence, "the distinction between scheme and content corresponds to the distinction between someone's individuative/discriminative resources and those resources we employ—at least initially—in providing an account of their language."[184] That is, "what counts as content, and what counts as scheme, is determined by the standpoint of the interpreter. But this does not invalidate the scheme idea; rather, it *relativizes* it, in much the way that the notion of *observationality* ('the given') is relativized to the linguistic practices of a community."[185] Kraut ends up with what I take to be the decisive rejoinder to Davidson on this issue: "The possibility of alternative schemes corresponds to the possibility of non-trivial expressive and discriminative disparities between theories. Such disparities need not entail global translational breakdowns; but they nonetheless have fairly interesting ontological consequences, consequences of the sort that schemers everywhere are trying to capture with their metaphors."[186] Unless we are prepared to dispense with the idea of theory rivalry, unless we are prepared to do away with the idea of inscrutability of reference—and these are Quinean legacies Davidson is quite loath to abandon—conceptual schemes have more claim than Davidson has acknowledged. Where Davidson prevails is in the rejection of *total* incommensurability. And we may also grant his point against the empiricist dogma of the given, but with the proviso that post-positivist theory of the relation of theory to observation hardly seems beholden to that notion. As Marie McGinn notes, for post-positivism "what constitutes the evidence for science makes no claim to be prior to theory, or prior to inference, but unashamedly makes use of descriptions involving, sometimes, highly sophisticated, scientific concepts." That is the positive epistemological consequence of the Duhem thesis; evidence "comprises what is not, in the given context, currently in doubt or under test."[187] But that is to see Davidson's argument in an entirely epistemological light. The possibility exists, in contrast, that it is an ontological issue that is ultimately in play here, the contest between scientific realism and antirealism. That is the direction, in any event, in which Rorty proposes to take Davidson's arguments.[188] Davidson will not have it that an epistemological claim be made that the world, fact, or experience justify belief. Only other beliefs justify belief.[189] But then the connection between language and the world seems to become attenuated.[190] Davidson believes he can conceive "reality without reference," but he also believes that there is a causal relation between the world and the conscious subject. That was indicated in his concluding sentence to the essay "On the Very Idea of a Conceptual Scheme." The question of Davidson's allegiance in the realism/antirealism dispute is a vexed one.[191] What the current discussion has established, at least provisionally, is that incommensurability needs to be taken strictly epistemologically, and when one does so, the idea of *total* incommensurability has been shown to be incoherent.

Kuhn's Continued Work on Incommensurability in His Later Writings

In Kuhn's own subsequent work, the problem of incommensurability loomed large. In reflecting on the thirty years since he published *Structure*, Kuhn wrote in 1990: "I emerge from those years feeling more strongly than ever that incommensurability has to be an essential component of any historical, developmental or evolutionary view of scientific knowledge."[192] Kuhn was all too prepared to assimilate his arguments about discontinuity in the practice of science to linguistic characteristics: "My original discussion described nonlinguistic as well as linguistic forms of incommensurability. That I now take to have been an overextension resulting from my failure to recognize how large a part of the apparently nonlinguistic component was acquired with language during the learning process."[193] Stated more clearly, whereas Feyerabend consistently "restricted incommensurability to language[,] I [Kuhn] spoke also of differences in 'methods, problem-field, and standards of solution' . . . , something I would no longer do except to the considerable extent that the latter differences are necessary consequences of the language-learning process."[194] Thus, "If I were now rewriting *The Structure of Scientific Revolutions*, I would emphasize language change more and the normal/revolutionary distinction less."[195] Frankly, more's the pity; history and philosophy of science need a richer sense of developmental process, not more wrangles over philosophy of language. Kuhn contributed to his own marginalization out of his hunger to belong to the community of (linguistic) philosophers.

In 1982 Kuhn presented a major address on incommensurability, and Kitcher was invited, along with Mary Hesse, to comment on his presentation. The resulting session is certainly one of the most illuminating in the extended debate on incommensurability. In "Commensurability, Comparability, Communicability," Kuhn sought at once to deal with the decades of "misunderstanding" which had greeted his old ideas of incommensurability and to present new and more cogent formulations of what he really meant. In particular, Kuhn addressed the criticism recently mounted by Davidson, Putnam, and Kitcher. Two overarching responses to his view drew his attention. The first claimed that Kuhn identified incommensurability with total untranslatability and radical meaning variance. It asked how such totally insulated theories could be compared. The second claimed that Kuhn practiced a bald performative contradiction, claiming that languages were untranslatable and then proceeding to translate them. In his new view, Kuhn radically reconstructed his stance. He denied that incommensurability meant incomparability and he insisted that he had always upheld a "more modest" theory of "local incommensurability." That is, "Most of the terms common to the two theories function the same way in both," which made it possible to perform comparisons and establish rational grounds for theory choice.[196] Yet he maintained there was "no language . . . into which both theories, conceived as sets of sentences, can be translated without residue or loss."[197] That was the basis for *local* incommensurability. This was Kuhn's (preliminary) response to the first vein of criticism. To the second, Kuhn responded by discriminating sharply between *translation* and *interpretation*. His critics misunderstood his practice as performative contradiction, because they failed to discern that when he discussed the incommensurable features of another language he did so interpretatively, not literally via translation.

Kuhn insisted that translation needed to be conceived rigorously along the philosophical lines that Quine had spelled out in *Word and Object*. That is, translation "systematically substitutes words or strings of words in other languages for words or strings of words in the text in such a way as to produce equivalent text in the other language."[198] Kuhn recognized that "equivalent text" was a very vague formulation, but he wanted it that way, for the moment, because it was close enough for the current purpose, namely, the *literalism* demanded of translation. In 1976 Kuhn wrote: "the problem of comparing theories becomes in part a problem of translation, and my attitude towards it may be briefly indicated by reference to the related position developed by Quine. . . . Unlike Quine, I do not believe that reference in natural or in scientific language is ultimately inscrutable, only that it is very difficult to discover and that one may never be absolutely certain one has succeeded. But identifying reference in a foreign language is not equivalent to producing a systematic translation manual for that language. Reference and translation are two problems, not one. . . . Comparing theories, however, demands only the identification of reference."[199]

Kuhn argued that it was undeniable that between distant scientific theories there could be no such literal, term-for-term translation which would not generate passages that would be unintelligible. Hence untranslatability. But that did not mean that the linguistically distant theory was itself unintelligible. The historian could access it, but only by learning the original language—by becoming bilingual. That was what interpretation was about. "Acquiring a new language is not the same as translation from it into one's own. Success with the first does not imply success with the second."[200] The first task entailed coming to terms with "an interrelated or interdefined set [of terms] that must be acquired together, as a whole, before any of them can be used."[201] In just this way, natural languages demonstrated their unique taxonomies of reality; these clusters or taxonomies "show . . . how the other language structures the world." Kuhn insists that intensions, the connotations of words, play as central a role here as extension or reference. Hence, translation that seeks to preserve reference will fail, as will any translation that works with an alternative taxonomy, for some part of the connotational cluster of terms will be lost in the selection of a term in the home language into which to render it. That is the point of the Italian adage "to translate is to betray."[202]

Even interpretation is strained, Kuhn claims. What we learn when we learn a different language is precisely its "taxonomic categories of the world." While its speakers will have achieved mastery of that taxonomy via different "reference-determining criteria," in order to become speakers in the same language community, their "taxonomic structures must match."[203] The linguist can eventually break into this circle. The problem is replicating it for the home language. Kuhn insisted that Quine and other philosophers of language mistook the capacities of natural languages to render any term or sentence from another language, even if only by extending the lexicon. For Kuhn, reporting in the home language about the meaning of sentences in the object language would not only involve circumlocutions and paraphrases, it would entail teaching at least a part of the object language in its own terms, the part that was incommensurable. Attaching these terms to the home language would make the sentences of the object language intelligible, but they would not make these neologisms full citizens of the home language; their eligibility to truth value would conflict irreconcilably with sentences still current from the unrevised home language.

Thus, the historical interpreter was faced with an unpleasant choice. The interpreter could save the truth value of as many sentences from the alien language as possible, at the cost of rupturing the coherence of the taxonomy around which it constituted itself. Or the interpreter could interpret the taxonomic coherence, but by introducing it as an irretrievably alien body into the home language. Its organization of the world could not but contest that of the home language. Kuhn charged that Kitcher was all too willing to opt for the first course, and Kitcher admitted it without regret. He expressed no interest in retrieving the holistic structure of former theories. Rather, he sought referential determination of the tokens of its terms, such that he could retrieve them into sentences in his home language and appraise their truth or falsity. "It seems to me that Priestley was partly right, and partly wrong—and that we can say *where* he was right and *where* he was wrong."[204] Since this measure of determinacy was possible across the historical distance, Kitcher found Kuhn's new formulation of local incommensurability epistemologically innocuous, and the theory of language learning in terms of holistic clusters uninteresting. All in all, he found Kuhn's reformed modesty "much less of a threat to orthodox empiricism."[205] From the vantage of the causal theory of reference, Kitcher felt he could both assimilate Kuhn's point and uphold a rigorous comparison of theory in terms of truth values.

Interestingly, Mary Hesse resisted Kuhn's conciliatory moves, insisting that there were no grounds for backtracking from the position that there could be no theory-neutral language. Therefore, incommensurability of theories represented a more drastic challenge to the Received View than Kuhn or Kitcher pretended.[206] Against Kitcher, she insisted on the centrality of the idea of holistic language learning, of the place of connotation and intension in language, and of a holistic approach to theory appraisal.[207] "In other words, we want to preserve Priestley's 'rationality,' not necessarily the truth-value of his theory."[208] Hesse, in short, holds out for a far more resolutely "historicist" approach, a more radical reading of Kuhn. It is no accident that she was a decisive mediator between Kuhn and the Strong Program.

In later writings, Kuhn claimed that incommensurability equaled untranslatability in the Quinean sense, in other words, the absence of any "string in one language [that] may, *salva veritate,* be substituted for a given string in the other."[209] Kuhn held that Quine's arguments for the indeterminacy of translation could be employed "with equal force" to establish that

no translation was possible. But Kuhn hastened to add that this did not signify that such untranslatable sentences were unintelligible. Rather, "anything that can be said in one language can, with sufficient imagination and
effort, be *understood* by a speaker of another."[210] That entailed *language
learning*, not translation, however. Thus, Kuhn claimed, historians become
in effect *bilingual* in gaining access to a temporally remote language. They
understand it, as would any other learner of that language, literally *in its
own terms*. Whether they can *translate* it remains quite undetermined; indeed, in these last texts, Kuhn is even chary of admitting that they can *interpret* it in the home language using glosses, circumlocutions, and paraphrases
that exceed strict translation.

Another key feature of Kuhn's later defense of incommensurability was
his insistence upon its *local* character. The phenomenon characterized "a restricted class of terms" which he specified as "taxonomic terms or kind
terms." Incommensurability referred strictly to "a sort of untranslatability,
localized to one or another area in which two lexical taxonomies differ."[211]
In defiance of Davidson, Kuhn insisted that the proper term for such local
structures was "conceptual scheme," and he elaborated extensive arguments
to describe how the learning of such "taxonomic modules" was holistic.
Learning such taxonomic modules was a form of initiation into a community (of discourse): "terms are introduced by exposure to examples of their
use, examples provided by someone who already belongs to the speech community in which they are current."[212] Kuhn argued that "transmitting a lexicon requires repeated recourse to concrete examples," picking up the
emphasis of exemplars he had long ago highlighted in *Structure*.[213] But he
now stressed the communal and the indeterminate elements of this language
learning. "What characterizes members of the group is possession not of
identical lexicons, but of mutually congruent ones, of lexicons with the same
structure."[214] While the initiation, the language learning, is an iterative circle of adequation to this abstract structure, "individuals can in principle
communicate fully even though they acquired the terms with which they do
so along very different routes."[215] The parallel with Quine's description of
the public nature of a language is patent.

The important issue is how the historian can escape the language of his
objects of study in order to represent his discoveries to his own home culture. Somehow he must find the resources for this in the home language.
Quine, Davidson, and most theorists of natural languages are convinced
that the resources *must* be available to the home language, if only by lexical
assimilation. Kuhn disagrees. To be sure, the historian can add such alien
terms to the lexicon of the home language, but only by sequestering them in
a special status, for "the two sorts of knowledge are incompatible," and "the
price of combining them is incoherence in the description of phenomena to

which either one might alone have been applied."[216] That is, even as assimilated into the home language, the sentences thus introduced would not be eligible for the same status with regard to truth value as the sentences composed of terms current in the home language. This makes the situation of historical reconstruction problematic once more. If Kuhn situated his whole discussion in the "sympathetic" endeavor of the historian not to take the scientist of the past to have been uttering nonsense, in other words, "passages that make no sense," this cannot mean simply that the sentences of that past scientist are taken to be true. They only become intelligible, whereupon it would still remain to determine whether "to ignore such passages or to dismiss them as the products of error, ignorance or superstition."[217] But can the reformulated sentences of the object language assume a form in the home language in which they are *eligible* for truth or falsity? Can the standards of truth and falsity of the home language ever be put in question by the introduction of the alien sentences or are these sentences inevitably dismissible? Kuhn argues that the judgment could only be made at the global level of the two entire language systems: "comparing the whole older system (a lexicon plus the science developed with it) to the system in current use."[218] The standard by which such a comparison could be undertaken would be "the relative success of two whole systems in pursuing an almost stable set of scientific goals."[219] Kuhn makes clear his belief in such continuous standards: "There are shared and justifiable, though not necessarily permanent, standards that scientific communities use when choosing between theories."[220]

Conclusion

Dale Moberg suggests that the idea that theories might be incommensurable arose "because the theory of theory comparison has not yet been sufficiently developed to deal in a satisfactory way with the problems of comparison that the history of science presents."[221] Ironically, Kuhn's misadventure with incommensurability demands a closer consideration of the relation of the history of science to the philosophy of science than even he achieved. One of the decisive questions to bring to bear upon the incommensurability controversy is its empirical historical applicability. As Moberg has argued, "neither Feyerabend nor Kuhn, nor anyone else has definitely established that any theories which historically have been rivals have also been radically meaning variant."[222] Nancy Nersessian in fact has suggested that a more thoroughly historical approach to scientific change would establish the concrete transitions through which meaning change occurs and yet allows for theoretical commensurability.[223]

As with Quine's deflated ideas of indeterminacy, Kuhn's ideas regarding incommensurability in their ultimate, pared-back form, remain significant enough that they figure as methodological and epistemological parameters of historical interpretation. What they underscore, however, is the contingency and fallibility of empirical inquiry, not its utter arbitrariness or impossibility. Moreover, the formulation of these issues in a strictly linguistic key impoverishes the resources available for the constitution and adjudication of empirical accounts. Practice, experiment, the complicated negotiations with the constraint or resistance of the world under investigation, and through all this a robust recognition of process, learning, and change: these are elements that Kuhn indeed helped thrust into the center of science studies. These need to be affirmed as a more empirically evidential and theoretically coherent basis for interpretation.

Kuhn played a substantial role in the proliferation of science studies over the balance of the century, but it was always both an unwitting and an ambiguous role. Steve Fuller put it very well in the passage cited as an epigraph to this chapter: "It might be said that *Structure* has a philosopher's sense of sociology, a historian's sense of philosophy, and a sociologist's sense of history."[224] In that work, Kuhn adopted a philosopher's sense of sociology: that is, he neglected virtually everything concrete about both its theoretical and its empirical practice, but invoked the *idea* of it, as it suited him. Kuhn adopted a historian's sense of philosophy: that is, he was far better at tracing what got said and why than at grasping what makes it true. Kuhn adopted a sociologist's sense of history: that is, he aimed always to glean from the historical record some generalization that governed it all. If, in the aftermath of Kuhn, history, philosophy, and sociology have interfused, it is alas all too apparent that they risk bringing out the worst in the interdisciplinary mutuality.

CHAPTER 4

Philosophy of science without history of science is empty; history of science without philosophy of science is blind.
— IMRE LAKATOS, "HISTORY OF SCIENCE AND ITS RATIONAL RECONSTRUCTION"

Why do historians and philosophers of science expect to profit from each others' inquiries: and why are their expectations so apt to end in disillusionment?
— NICHOLAS JARDINE, "PHILOSOPHY OF SCIENCE AND THE ART OF HISTORICAL INTERPRETATION"

Doing Kuhn One Better? The (Failed) Marriage of History and Philosophy of Science

Incommensurability, specifically as a matter of meaning variance, has been a dead end. The two decisive components of Kuhn's revisionism in philosophy of science now appear to have been the *historicization of reason* and the *social construction of knowledge*. In this chapter I will consider the first of these, and in subsequent chapters the second.

Kuhn claimed that history of science could be constructive for philosophy of science with reference to three core issues of the latter: "the structure of scientific theories, the status of theoretical entities, or the conditions under which scientists may properly claim to have produced sound knowledge."[1] Kuhn urged philosophy of science to move from a timeless sense of the "logic of justification" to a diachronic array of varying norms of scientific validity, what within logical positivism had been called a "relativization of the Kantian a priori," and more recently has come to be one sense of "evolutionary epistemology."[2] Gerald Doppelt claimed that Kuhn's true message was that philosophy of science should "turn to a more historical analysis and develop a different image of scientific rationality."[3] Indeed, Kuhn's ultimate importance to post-positivism, I believe, lies in the impetus he lent to revisionism along these lines.[4]

The historicization of philosophy of science, attunement to its diachronic dimension, became a decisive concern in the 1970s. Ironically, however, Kuhn appeared more a provocation than a model in the emergence of more complex and sophisticated programs to wed history of science with philosophy of science.[5] Though they disdained Kuhn, a historicist school of philosophy of science, led by Imre Lakatos, Stephen Toulmin, and Larry Laudan, turned to the interpenetration of history with philosophy of science over the 1970s and early 1980s.[6] By the time he gathered his papers on the dialogue of history of science with philosophy of science, in *The Essential Tension* (1977), Kuhn was no longer the guiding intellectual force in these fields. It was only in the early 1990s that philosophers of science became willing to credit Kuhn with a decisive role in bringing history into philosophy of science.[7]

Ronald Giere has argued that Kuhn was from the outset misplaced in this group of historicizers (Toulmin, Dudley Shapere, Ernan McMullin, Lakatos, and Laudan): "they wanted to restore to science the *rationality* Kuhn seemed to deny. The difference was that they appealed to a historical notion of rational progress rather than a logical notion of rational inference."[8] While his characterization of the historicists is quite apt, I question Giere's view that Kuhn aimed to undermine the idea of rationality in science.[9] I think, rather, that he sought to relativize rationality to history, not abolish it, and in this sense, the historicizing philosophers of science were authentically pursuing the enterprise Kuhn set in motion, even if they, like Giere, shunned the association.

Logicism versus Historicism—the Relativist Crisis in Epistemology

Philosophy of science, if not an ancient branch of philosophy, has a discernible tradition that goes back at least to the nineteenth century and, arguably, to the seventeenth. Especially starting with Kant, philosophy of science played an important role in the self-conception of philosophy as a discipline: if it could no longer be the "queen of the faculties," it could at least be the rational arbiter of scientific achievement. By contrast, the *history* of science was still a very young field even when Kuhn published *Structure,* and it had not fully established its disciplinary home. Some disciplines take a proprietary interest in their history; others leave it to general historians. Since natural scientists take up the history of their disciplines only occasionally in autobiographical reminiscences—or worse, in textbook vignettes— history of science has had no serious place within the existing natural science disciplines. At the same time, it seemed quite esoteric for history as it was conventionally conceived.[10] The decisive breakthrough of sophisticated history of science, spearheaded by Alexandre Koyré, drew the history of science very close to the history of philosophy, or a version of intellectual history hardly to be discriminated from it.[11]

Roughly contemporaneously with Kuhn's *Structure* and perhaps inspired by it, a new prospect began to develop in the Anglo-American academic context, namely, the establishment of combined programs in the history and philosophy of science (HPS).[12] A decisive force in this development was Gerd Buchdahl at Cambridge University.[13] As John Passmore cogently observed, "the organizational conjunction may have no stronger ground than that history of science, when not conducted on a scale sufficient to constitute a department by itself, is hard to 'place' in a normal Faculty organization."[14] At a symposium on "The Mutual Relevance of the History and the Philosophy of Science" held at the annual meeting of the American Philosophical Association in 1962, Norwood Russell Hanson dismissed the idea that philosophy had anything to learn from history.[15] Kuhn himself expressed ambivalence over placing the two research fields in a common institutional domicile. He wrote of it as a problematic "marriage."[16] Under just that metaphor, a debate developed over the relation of the fields for the next two decades.

The distinction of the context of discovery from the context of justification had not only conceptual but also disciplinary implications. First, it distinguished conceptually the *normative* enterprise of appraisal from the *descriptive* enterprise of accounts.[17] But second, it allocated these enterprises to separate *disciplinary contingents* in the study of science. Appraisal was the exclusive purview of philosophy, whereas accounts could be left to psychology or sociology or history. The matter they would take up was ad hoc and their methodology could scarcely transcend their matter.[18] By contrast, in the context of justification, logical rules would authorize appraisal of the validity of the most polished products of human empirical insight. This implied a clear hierarchy. Natural science supervened over the social sciences, but philosophy took the highest place of all.[19] Even history of science acknowledged this hierarchy in its distinction of internal from external history—of the immanent, rational evolution of scientific knowledge from the intervention of psychological, political, economic, or other "extraneous" influences.[20] Indeed, in the Received View, *any* historical or sociological factor "which is necessarily external for a given period is also necessarily irrational for that period."[21] As Robert McGlaughlin has put it, "To *define* the discovery/justification distinction in these categorical terms is merely to make logical reconstructions of discovery *guilty by definition* of psychologism!"[22]

To conceive the relation of history of science to philosophy of science in a purely disciplinary sense, however, is to miss the saliency of a crisis in fact entirely *within* philosophy of science.[23] This crisis—specifically of an authoritative epistemology to preside over the context of justification—got expressed as a philosophical contest between *logicism* and *historicism*.[24] "*Logicism*," Ernan McMullin explains, "is the metamethodology which would maintain that the *methods* of theory-appraisal have the status of logical rules, and that methodology reduces therefore to applied logic."[25] Gerard Radnitzky put the controversy clearly in 1976: "If '*logicism*' is taken as a label for the view that in theory-appraisal it is *only* the *logical* aspect which matter [*sic*], the opposite pole of *totalization* could be labelled '*historicism*.' The view that a purely descriptive approach can account for what scientists have done, and that the *past successful praxis of scientists*, history of science as a success story, *can serve as arbiter in appraising competing methodologies*."[26] Some philosophers of science proved unabashed in their affirmation of logicism: "The task of evaluation of the products of scientific inquiry, of establishing the epistemological import of such products . . . falls squarely on the shoulders of the philosopher of science," wrote Harvey Siegel.[27] Not even the scientists themselves were privy to this highest sanctum. "Scientific decisions are one thing, the justification of scientific decisions another. . . . For what we are after, in the context of justification, is an account of the justificatory force of the scientist's reasons for adopting [a given theory]; an account that enlightens us as to why those reasons are *good* reasons."[28] Siegel did hedge: "I am *not* suggesting that philosophers, rather than scientists, make (or ought to make) scientific judgments which need to be justified. Scientists, not philosophers, assess the adequacy of . . . competing accounts. To this extent I agree with Quine that there is no 'first philosophy.'"[29] But Siegel hardly clarified how "accounting for justificatory force" is different from "assessing the adequacy." And certainly his presumption that philosophers *could* discern "*good* reasons" reflected unmitigated confidence in epistemology's adjudicatory criteria. That belies his profession of agreement with Quine concerning "first philosophy."[30]

Such logicism—and with it first philosophy—cried out for a good bit of deflation. As Harold Brown has noted, logicism perpetuated a long tradition of absolutism in Western philosophy, while "historicism is a form of relativism."[31] Perhaps even more pertinent from our current philosophical and cultural vantage would be the discrimination between foundationalism and antifoundationalism. From at least Plato onward, Western philosophy has professed a radical disjunction between opinion and true knowledge, *doxa* and *episteme*, and it has struggled to make good on that claim by seeking "to identify those epistemic features which mark off science from other sorts of belief and activity," as Laudan puts it. But, he continues, "it seems pretty clear that philosophy has largely failed to deliver the relevant goods."[32] As Laudan summarizes his overview of the recent discussion, "neither verificationism nor falsificationism offers much promise of drawing a useful distinction between the scientific and the non-scientific."[33] Even the arch-Popperian W. W. Bartley gave up on the demarcation project, to the considerable ire of his colleagues.[34]

The crisis in the philosophy of science arose with the admission of the bankruptcy of justificationism (foundationalism) and with the growing realization that Popper's ingenious proposal of a falsificationist alternative was not proving adequate to the task.[35] Popper had been instrumental in constructing an alternative to the increasingly futile justificationist view of knowledge, opening up the possibility that there could be knowledge without a necessary claim to proof. This entailed embracing a thoroughly fallibilist theory of knowledge.[36] But the problem remained how to uphold *rationality* in those terms. Popper was, despite his repudiation of induction, a logicist. While the verificationists made use of *modus ponens,* Popper sought to make *modus tollens* the logical vehicle for a viable falsificationism. But the falsificationist account did not tally well with actual scientific practice. The Duhem thesis drove a decisive hole through any simple notion of empirical refutation. Moreover, the negativity of the method seemed out of sync with science both as an inveterately inductive practice and as an ideal of progress in knowledge. Thus, Popper flirted with verisimilitude to accommodate these recalcitrant actualities, leading Imre Lakatos famously to urge that Popper smuggle at least "a whiff of inductivism" into his theory of science.[37]

As Lakatos phrased the essential question, "Can we save scientific criticism from fallibilism?"[38] That is, "the methodological falsificationist realizes that if we want to reconcile fallibilism with (non-justificationist) rationality, we *must* find a way to eliminate *some* theories."[39] "Naive" falsificationism was not adequate to the task; Duhem had made clear how robust theory could endure anomaly. Kuhn made that central to his theory of "normal science." What Lakatos sought was a "sophisticated" falsificationism that could admit the fallible nature of all knowledge-claims and yet specify what rationality should signify, in this way rescuing the philosophy of science from relativism or, as he identified it, historicism.[40] Lakatos sincerely believed himself to be antihistoricist because he loathed the *radical* historicism that seemed to pervade Kuhn's *Structure.* This appeared to be a crippling relativism for philosophy of science. It threatened, in his view, "to surrender science to the rule of unreason."[41] In his contribution to the volume *Criticism and the Growth of Knowledge,* which was largely a confrontation of Popperians with Kuhn, Lakatos lashed out against Kuhn in the most intemperate terms: "*in Kuhn's view scientific revolution is irrational, a matter of mob psychology.*"[42]

What troubled the Popperians so much? "Kuhn's stress on science as a community enterprise, Polanyi's thesis of personal knowledge, are so many ways of bringing out that decisional elements play a far larger role in scientific appraisal than either logicists or scientists themselves would ever have been inclined to admit."[43] Popperians could accommodate some logic *in,* if no algorithmic "logic of" discovery.[44] What they could not endure was sociology or psychology in the context of justification. Kuhn had carried his challenge directly into the den of the Popperians in 1965 with his talk "Logic of Discovery or Psychology of Research?" He proclaimed: "Sir Karl's view of science and my own are very nearly identical. We are both concerned with the dynamic process by which scientific knowledge is acquired rather than with the logical structure of the products of scientific research."[45] That stretched truth by more than a bit. Popper responded less than enthusiastically to his recruitment into the ranks of historicists: "to me the idea of turning for enlightenment concerning the aims of science, and its possible progress, to sociology or to psychology (or . . . to the history of science) is surprising and disappointing." He lamented, "all I have said before against sociologistic and psychologistic tendencies and ways, especially in history, was in vain."[46] Popper minced no words in 1965: "while the Logic of Discovery has little to learn from the Psychology of Research, the latter has much to learn from the former."[47]

While Kuhn had made points against the "naive" form of falsificationism, Lakatos pronounced him inept against a properly sophisticated falsificationism such as Lakatos believed could be fashioned out of Popper (or various Popper-stages).[48] Lakatos insisted that "a sophisticated falsificationist theory is 'acceptable' or 'scientific' only if it has corroborated excess empirical content over its predecessor (or rival), that is, only if it leads to the discovery of novel facts."[49] Thus, crucially, no theory is ever replaced unless there is a superior rival to challenge it. This point strikes at the justificationist or skeptic who would demand absolute warrant for any knowledge-claim as well as at the (naive) falsificationist who can make no sense of the robust tolerance for anomaly in actual science.

Ironically, third parties ranged Lakatos *among* the historicizers. Lakatos proved willing to bring historical development into the context of justification, thus breaking decisively with Popper, even if he never admitted it.[50] He felt he had no choice. As McMullin observes, "Lakatos is correct in perceiving that if history cannot count as warrant in metamethodology, the only alternatives are absolute *a priori* ones of a logical or metaphysical sort or else one or other form of conventionalism."[51] Lakatos ranks among the most prominent of the historicizing philosophers of science in the 1970s, but before we can address his particular approach we need to set the backdrop: the debate over the "marriage" prospect.

The "Marriage" Controversy

The metaphor of a "marriage" between philosophy of science and history of science, though launched by Kuhn, really came to prominence in the aftermath of a conference held at the University of Minnesota in September 1969, whose proceedings were published as *Historical and Philosophical*

Perspectives of Science (1970).[52] The disciplinary boundary issues between history and philosophy of science had occasioned interest starting in the early 1960s, and they played a role in the 1969 Urbana discussions of scientific theories, especially in the exchange between Bernard Cohen and Peter Achinstein.[53] But the decisive discussion of the proposed "marriage of philosophy of science with history of science" came with the publication of *Historical and Philosophical Perspectives of Science*.[54] Ronald Giere's review of the Minnesota volume launched a lively controversy over the appropriateness of disciplinary integration in terms of the metaphor of a "marriage of convenience."[55] Ernan McMullin took up Giere's challenge on those terms, and Richard Burian added to the debate.[56] The irony is that when, eventually, some philosophers of science began to see the charm of such domesticity, many historians decided to walk out.[57] That makes the metaphor of marriage between the history and the philosophy of science a biting metaphor indeed, for there proved to be little love lost either way. Their "marriage of convenience" appeared only too inevitably headed for recriminations and divorce.[58]

For the Minnesota conference, McMullin composed an elaborate "taxonomy" of the relation between history and philosophy of science. He distinguished two senses of science—as *product,* a set of propositions with supporting evidence, and as *process,* which included as well the path (experimental, instrumental, and social-historical) by which the evidence was accumulated and the propositions developed. He argued that philosophy of science in the Received View cared exclusively for product, like the scientific disciplines in their own self-reconstruction (textbook history).[59] Philosophy of science saw its project as the appraisal of these scientific products in terms of validity, with no regard to the *process* of their emergence. In short, it followed Reichenbach's distinction of the context of discovery from the context of justification.[60] Only the latter was a properly philosophical subject.

The issue, McMullin contended, was whether the criteria for judgment of scientific validity were to be found *within* scientific practice or established *externally* by philosophy. He argued that most philosophy of science had adopted an external standard, whether metaphysical or logical.[61] He argued, further, that a philosophy of science which sought to derive standards for the context of justification *internally* from scientific practice faced severe problems, since it would rely upon an empirical inference from the process to appraise the product of that process, a (vicious) circularity.[62] In that light, McMullin turned to appraise the role that history of science could play in the philosophy of science. Clearly, philosophies of science that employed an external principle had no need for it; at most they used historical case-studies to illustrate models of scientific theory formation and justification. Philosophy of science seeking an internal or immanent warrant in actual scientific practice, on the other hand, became subject to history in two distinct ways. First, historical cases set the standard for validity. But second, history of science introduced *shifts* in the standards of scientific rationality itself. Changing scientific theories appeared capable of feeding back upon the criteria for the appraisal of such theories. That made normativity exceedingly difficult to extract from the process.

Still, McMullin acknowledged that it had become increasingly difficult to maintain—in the face of the challenges of skepticism, relativism, and historicism—the posture that there existed timeless and universal logical standards under which every scientific claim could be appraised.[63] In short, McMullin questioned whether any *external* (a priori) philosophy of science could be *actualized,* and whether any *internal* (empirical) philosophy of science could be *normative.* This has been, indeed, the essential dilemma, and it remains to be confronted and dealt with adequately if empirical inquiry is to establish a viable methodological standard.

In his 1970 review Giere took exception to any historical turn in philosophy of science. "Past success is not the same as truth," he asserted, and philosophy of science appropriately concerned only truth.[64] Giere was sensitive to the problem that the standard for truth could not simply be logic; he acknowledged that "there is such a thing as empirical validation." Somehow out of the negotiation among scientists and philosophers of science, Giere envisioned the emergence of "a unified method of validation to be applied in current scientific inquiry."[65] For that, Giere admitted, philosophers of science needed to be more intimately acquainted with "real science." But he denied that this meant *history* of science.[66] Knowing about actual science "requires no peculiarly historical techniques." Giere denied that these had "anything essential to contribute to the context of contemporary philosophy of science."[67] His concern was exclusively with *current* scientific inquiry. "Concern with the process of inquiry does not automatically make one an historian. The needed connection between process accounts of rationality and real science may be made solely within the context of contemporary science."[68]

Giere specifically rejected taking theories as evolving "historical entities," insisting that a scientific theory was a determinate set of claims invoking a certain body of evidence (McMullin's "product" conception) always open to appraisal in any given present. Whether the criteria of scientific validity might have evolved seemed irrelevant to Giere in terms of his presentist orientation: he was concerned strictly with rationality in current science.

McMullin responded by identifying the "question of scientific rationality and how it is to be determined" as the decisive question for philosophy of science. He indicated three contexts in which it could be appraised: theory assessment, scientific growth, and the ontology of theoretical entities. He urged that on all three counts the role of history of science in philosophy of science was more ample than Giere would allow.[69] McMullin distinguished *logicism* from *historicism* along lines drawn in a recent essay by Alan Musgrave.[70] He challenged the "intuitive" certainty of purely formal-logical standards, while at the same time acknowledging that contingent and fallible scientific practice lacked normative authority. " 'Pure' logicism or 'pure' historicism . . . are . . . plainly untenable," he concluded.[71] As he put it in another context: "The historicist is hardly any better able than the positivist was to see a special significance in the ability of a good theory to predict novel and unexpected results nor to help us understand what it is that makes it possible to separate *ad hoc* theories from good theories."[72]

Above all, McMullin sought to exclude historicism from the context of justification. "To make the *rationality* of the acceptance of a proposed new theory in science depend upon sociological, psychological or historical factors, runs the risk of either separating rationality and justification, or confusing causes and reasons."[73] He sharply attacked Stephen Toulmin for his effort to historicize rationality: "Toulmin . . . withdraws rationality from the context of justification altogether, and limits it to the context of discovery."[74] By contrast, McMullin claimed the Received View "talked not only about empirical content, but about predictive success over a period, about fertility in guiding research, about unexpected confirmations, about ability to absorb anomaly by internal modification, and so on. These are *historical* questions."[75] Yet this is hard to reconcile with John Passmore's apt observation: "the 'Young Popperians'—Bartley, Agassi, Feyerabend, Lakatos, Laudan—have all, even when they criticized Kuhn, attacked Popper on historical grounds."[76] McMullin's ambivalence about the "historicist turn" comes clear in his retrospect of the controversy, which laments that "we seem to have entered a period in which even the real virtues of formalism are often forgotten and sometimes denied."[77]

McMullin did accept from Toulmin the idea that scientific theory was itself evolutionary. While he quarreled with Toulmin's unit of change, the individual concept, he was very intrigued with the idea that theories should be seen as developing entities.[78] That brought into focus a criterion of theory strength not accessible in the logical assessment of science as a product, namely *resilience,* an "ability to meet anomaly in a creative and fruitful way," which could "manifest itself only gradually."[79] Thus, philosophy of science would do well to take up "theory considered in terms of its entire career."[80] McMullin thought this was best exemplified by Lakatos's idea of a "research program."

Richard Burian took up the matter there, acknowledging that by 1977 philosophers of science had been presented with considerable evidence that "real science" did not adhere to the standards for the "context of justification" that the Received View had postulated.[81] Of course, Burian noted, philosophers of science could shrug all that off as mere actuality and insist their domain was the ideal. He represented this logicist posture as that of the sovereign *judge* of scientific practice: "it is the philosopher who first reveals theoretical structures perspicuously, who clarifies theories and cognitive standards by which they are judged, and who assesses the logical consequences of theoretical claims."[82] Logicists could presume to do this because they upheld "universally valid methodological and epistemological standards."[83] Burian raised three objections to this grand self-conception. First, it made real scientific practice totally irrelevant, triggering the question why such logicism should be considered a philosophy *of science*. Second, it could not prove that these standards held constant for all scientific practice over the course of history. Third, it could not grasp the distinct "career of a theory" and its relevance to scientific rationality.

Nevertheless, Burian endorsed all the suspicions about the normativity of historicism which Giere and McMullin had articulated. The solution, he suggested, could only come by shifting to the question of the growth of science. This was a more dynamic inquiry, in which Burian was convinced that "specifically historical techniques and considerations," contra Giere, were indispensable.[84] He urged that in place of static "product" models of theory, philosophy of science should construe "theory-versions" in sequence, in other words, the "career of a theory" along the lines of Toulmin, Lakatos, or McMullin (and one might well add Kuhn, with his theory of normal science, whose symbolic generalizations were more law-sketches than laws).[85] Burian not only argued that it was necessary thus to widen the object of appraisal but added that further historicization was inevitable, since more sophisticated theory appraisals required the incorporation of "background knowledge" (Popper, John Watkins, Musgrave) and the criteria of appraisal themselves proved historically mutable (Toulmin, Shapere, McMullin).[86] Accordingly, Burian concluded, philosophy of science had to admit its dependence on "the best available interpretation of the historical record."[87]

One common ingredient in all of the essays in this exchange is a rather dogmatic sense of what *history* "as such" *is*, what "historiographical techniques" *are*. Foremost, all three philosophers had no doubt that history is *exclusively* particularist. Thus, McMullin, in his taxonomy, wrote that history "terminates in the historical singular." Since it will "attempt to reconstruct the cultural and intellectual milieu, . . . [t]his sort of historiography focuses on the historical singular in all its contingency."[88] In his most extended discussion, McMullin concluded: "The historian is concerned with what happened just because it *did* happen. . . . [H]is goal is not the assertion of a universal, a pattern, or the interlinking of such patterns."[89] Giere, as well, believed this: the historian was concerned strictly "to explain the occurrence of particular occurrences."[90] Burian, finally, saw history as "concrete and descriptive . . . seeking to understand the concrete particularity of complex and complexly related individuals and events."[91] While all three had conceptions of "rational reconstruction," they pronounced this (with different valences) to be a species of philosophy, not real history.

These philosophers, so attentive to methodology and epistemology for *natural* science, simply ignore parallel issues in historical research. What Wilhelm Windelband long ago termed the "idiographic" conception of history drastically underestimates the cognitive endeavor of many historians.[92] Such historians compose narratives that make substantive claims about the nature of science and the causal elements in its developmental sequence. These are, in short, *explanations* with their own theoretical burden.[93] Kuhn tried to make that point early on, in 1968, when he drew attention to what amounted to a *bias* among natural scientists and philosophers who regarded history as "*mere* description." He insisted that history was "an explanatory enterprise," though one with "almost no recourse to explicit generalizations."[94]

Yet even Kuhn wrote, "One can explain, as the historian characteristically does, why particular men made particular choices at particular times."[95] This assertion hardly makes it transparent how historians "can . . . explain . . . why"—and *particularity* is in itself not a sufficient condition. The historian of science Bernard Cohen observed in 1969 that "a greater sensitivity to the canons of history could readily produce a dramatic rise in the historical level of accuracy and authenticity of philosophical discourse."[96] But all the "canons of history" seemed to mean to Cohen was avoidance of anachronism.[97] "The historian's job is . . . to immerse himself in the writings of scientists of previous ages, to immerse himself so totally that he becomes familiar with the atmosphere and problems of that past age. Only in this way, and not by anachronistic logical or philosophical analysis, can the historian become fully aware of the nature of the scientific thought of that past age and can he really feel secure in his interpretation of what that scientist may have thought he was doing."[98] But *can* a historian "become fully aware"? *Can* a historian "really feel secure"?[99] Are there such ubiquitous "canons" or "historiographical techniques" whereby anachronism is avoided or accuracy assured?[100] This is not the place to undertake sustained discussion of philosophy of history, but what seems to be lacking here is recognition that the problems of validity the philosophers stress in their theories about natural science apply with equal force to the utterly fallible, ineluctably empirical endeavor of history.

Given his strong historicism, Larry Laudan disputed Giere sharply on the relation of history to philosophy of science. He found equivocation in Giere's position that while "actual science" is indispensable for the philosopher of science, history of science is not. Even contemporary science can be construed only under a description, and "a resolution of the normative/descriptive paradox is as crucial for those philosophies which are grounded in contemporary science as it is for those philosophies which look back beyond our own time."[101] Laudan also stressed that historical method is (or ought to be) more than merely idiographic. He urged historians to move beyond "exegetical" history to "explanatory" history; one should "not merely . . . rehearse what 'great minds' have said but also . . . explain *why* they have said it."[102] Laudan charged that intellectual history in its merely exegetical mode exhibited "explanatory bankruptcy."[103] Part of the problem, in his view, was that intellectual historians needed to make proper selection of units of analysis, on a line parallel to his own selection of research traditions rather than individual theory statements.[104] The historian should "explain the evolving vicissitudes of belief" in terms of (synchronic) *configurations of concepts* as elaborations of larger (diachronic) *research traditions.*[105]

Yet Laudan insisted in addition that such explanation could arise only if the historian were possessed of the appropriate theoretical conception of rationality: "It is the historian's intellectual—even moral—obligation not only to be self-conscious about the kinds of norms he is applying, but also to see to it that *he is utilizing the best available set of norms*" (original emphasis).[106] The problem with this mandate to historians of science (and intellectual historians generally) is that it begs a host of essential questions. Barry Barnes has made the case eloquently.

> If the individual historian cannot write other than selectively, and with prior tendencies of interpretation, how can we articulate an ideal which stresses the primacy of evidence, documentation and "what actually happened" as though the problem of selection and bias did not exist or could simply be overcome? . . .
>
> . . . The constraints upon the historian arise from his status as a member of a culture with a pre-existing common vocabulary, and a shared set of patterns, structures and forms of representation. This large and diverse set of cultural resources does define what the historian is able to represent the past as constituting, and to a large extent it restricts how the past can be perceived. . . . But it implies nothing substantive. . . . This involves no necessary "bias" in the old-fashioned sense. One *finds out* what best describes the past, by empirical study and documentary method. This is why it is possible to *learn* from history.
>
> By retaining a broadly empiricist ideal of history, and insisting that "what actually happened" is to be decided by concrete investigation, we make our available cultural resources and representations compete as a means of understanding the past. As a consequence, the results of historical enquiry can surprise us, challenge us, and educate us.[107]

This simultaneous articulation of the "situatedness of the historical subject" and insistence upon the possibility of learning, of disrupting presupposition, is the essence of a robust hermeneutical historicism, a key to the possibility of a naturalized epistemology, as I will argue at the conclusion of this chapter.

No historian of science would contest the first element in the squib Lakatos (and before him Hanson) lifted from Kant: indeed, "history of science without philosophy of science is blind."[108] The reasons why historians of science need philosophy of science, Kuhn observed in the late 1960s, "are, at once, apparent and well known."[109] His ally Norwood Hanson made much the same case even earlier.[110] They meant not only that historians borrow concrete *terms* from the philosophy of science, but that it supplies them with methodological and epistemological *standards,* and perhaps even with the *problems* that need investigating.[111] However, no *single* philosophy of science—and certainly not the Received View—proved adequate to historical purposes. If there *were* an unequivocally warranted philosophy of science upon which historians could rely in every context, the whole matter would long since have been closed. No such philosophy of science has withstood fundamental criticism with much of its edifice left standing. That historians have been skeptical and selective with such a volatile fund of critical resources seems eminently sane in that light.

Because there has been no *definitive* philosophy of science, historians have made the best of eclectic choices among plausible terms, programs of investigations, and problems from the available philosophies. That pragmatic eclecticism has its limitations, but it has been robust enough to generate a body of work. What remains to be clarified is the other element in the Lakatos squib: "philosophy of science without history of science is empty." What the "marriage" debate signals is that avoiding emptiness did not entail crude submission to *facticity.* Philosophy of science insists upon *normativity.* But then what exactly would a historically informed philosophy of science be? "Just *how* does history serve as evidence for normative methodology; how do the *facts* of the past connect with what *ought* to be done in the future?"[112] A logical place to start would be with the pundit Lakatos himself.

Lakatos, Toulmin, and Laudan: The Historicization of Philosophy of Science

It seemed to many that Lakatos extracted what was rational from Kuhn's reckless formulations and achieved the most promising incorporation of history into philosophy of science. Commentators preferred Lakatos to Kuhn because Lakatos's "main point is that the idea of progress can be rationally justified."[113] He offered a "rational reconstruction" of the history of science grounded in Popper's invocation of a "third world" of autonomous logical development.[114] He believed he could establish a continuous, logically warranted narrative of the "*products* of the cognitive activity of scientists, rather than the activity itself," in Tomas Kulka's words.[115] Lakatos replaced Kuhn's notion of paradigm with the idea of *research programs*. His unit of investigation was a series of theories in their chronological development: "theory, taken not as a timeless ensemble of propositions, but as a set of propositions and an accompanying model which are developed and modified over a period of time, with successes and failures of various sorts to take into account."[116]

As Lakatos saw it, "*my concept of a 'research programme' may be construed as an objective, 'third world' reconstruction of Kuhn's socio-psychological concept of 'paradigm'*" (original emphasis).[117] Lakatos developed an entire vocabulary to characterize his new rational reconstruction of historical science—a "hard core," a "protective belt" (Hacking calls these "pornographic metaphors"), and negative and positive heuristics associated with each—on the basis of which one could estimate whether the research program entailed progressive or degenerating problem shifts. Development was progressive if theoretical and empirical content of the theories grew in each successive iteration, and degenerating if these shrank. Theoretical content had to do with the problems for which the theory offered interpretation, and empirical content had to do with the confirmation of the solutions the theory proposed for these problems. Research programs struggled to incorporate anomalous findings while holding their hard core of theory roughly constant. "Duhemian adjustments" in the protective belt of auxiliary hypotheses allowed research programs to tolerate anomalies alongside new empirical findings and hence remain progressive. When, however, too many such adjustments took place, empirical findings dwindled and a research program entered upon a degenerating phase.

Inquiry into the history of science required a "rational reconstruction" of the methodology upon which research programs *in principle* relied and an appraisal of the degree to which actual ("external") history deviated from this ideal "internal" history. For Lakatos, the sets of sentences in the sequence of theories which make up a research program are the only objects of investigation, sufficient for the sort of historical inquiry appropriate for a philosopher of science. What mattered was this ideal history of the research program, and Lakatos relegated accounts of how actual history "misbehaved" to footnotes. He was prepared to do substantial violence to the actual history of science in order to accommodate his model of how it "ought to have happened."[118]

Lakatos proposed a new strategy for philosophy of science. Instead of asking how knowledge could be possible, the vein of epistemology since at least Descartes, Lakatos chose a different starting point, namely, "that knowledge has grown."[119] Hence, the premise "is not that there is knowledge but that there is growth." Lakatos believed one could establish this by "direct inspection," and so he could turn directly to "an analysis that will say in what this growth consists." "One does not defend the claim that certain cases exhibit the growth of knowledge, but uses the examples to define a new canon of 'rationality.'" Thus, Lakatos could pursue "internal considerations about the history of knowledge" without having to develop "any theory of truth."[120] Moreover, scientific growth supplied the criterion (*demarcation*—a word dear to Popperians) between rationality and irrationalism.[121] As Alan Musgrave put it, Lakatos's "central idea is that the history of science is the history of *competing* research programmes, and his central problem is how one research programme is *eliminated* or *rejected* in favour of another one."[122]

Typically *methodology* concerns "choice between competing procedures and courses of action for the future." However, Lakatos used it "as the name of his philosophy of science"—"something backward-looking," "a theory for characterising real cases of the growth of knowledge."[123] "His philosophy provides no forward-looking assessments of present competing scientific theories."[124] John Jamieson Smart made the same complaint: "His methodology is not meant as advice to scientists now and in the future but as an *ex post facto* appraisal of past science." Smart continues, "if Lakatos's methodological principles are not meant as heuristics, what is the point of them?"[125] Tacitly, Smart claimed, "we *are* saying that the methods which worked in the past are likely to work in the future," and so "even he can not quite evade the charge of historicism (despite his own disavowals of it)."[126] Lakatos meant to use his retrospective criterion of rationality in science as a prospective epistemological standard. That is, Lakatos can be construed as contributing to what his fellow Popperian Joseph Agassi christened as the "bootstrap theory" of knowledge.[127]

McMullin grasps the complexity of Lakatos's theory in this light: the fertility of a research program "is estimated by the *actual* success of the theory in opening up new areas, in meeting anomalies, and so forth. Thus it is *past*, not future, oriented; it is *proven* 'fertility' . . . that confirms the truth-value of a theory not its as-yet-untested promise."[128] But "what carries most weight is the potential the theory has shown to allow creative scientists to go *beyond* the original statement."[129] One of its vital resources is the range of *metaphor* it offers, in other words, "a tentative conceptual juxtaposition of elements capable of suggesting to the creative mind a whole range of possible further developments of the theory itself."[130] Here one has a right to think of Kuhn's conception of exemplars and analogies at the core of "paradigms" as a legitimate precursor.

Ewa Chmielecka has contended that Lakatos was attempting, through a "rational reconstruction" of the heuristics of actual science, to introduce into the context of justification elements that Popper had relegated to the context of discovery. But she concluded, "[A]lthough Lakatos has constructed the framework of his methodology so as to include the issues ignored by Popper in *The Logic of Scientific Discovery,* he has neither developed them nor analyzed the subjective factors stimulating the growth of knowledge nor—as a matter of fact—has he been able to show what role they play within his methodology of scientific research programmes."[131] For those accustomed to the terminology of history of science, Lakatos deployed a highly idiosyncratic distinction between internal and external history.[132] When Lakatos proposed a "test" of his ideal history against actual history, every actual deviation became by definition "irrational." Any consideration of actual history was, by Lakatos's own admission, always informed by the ideal history of science, introducing a "very strong circular element."[133] Actual history, in short, could never contest the theory, could never really matter, and so it *deserved* to be relegated to the footnotes.[134]

Kuhn, in his comments on Lakatos, claimed that what the latter was doing should not be considered history, but rather philosophy by example.[135] There is nonetheless a way to conceive what Lakatos endeavored as a radical innovation in historical interpretation rather than a repudiation of it. Noretta Koertge writes: "Agassi [in *Towards an Historiography of Science* (1962)] suggested that historians emulate scientists by using *bold* theories. Lakatos suggested that they go even further and use *idealizations*—theories which deliberately neglected factors known to be operative."[136] Smart puts it analogously: "Lakatos's historiographical idea is that we can judge a narrative in the history of science by the extent to which the narrative makes the actual history approximate to the internal history."[137] Hacking is more reserved: "As for the question, whether normative reconstructions are histories, the answer is a cautious 'yes.' But they are only applied history: the past applied to the solution of a philosophical problem."[138] These eccentricities made Lakatos's approach seem as problematic as Kuhn's to many philosophers of science (to say nothing of historians), though the impulse to relate theory to history remained powerful.

Lakatos died suddenly in 1974, leaving a substantial body of unpublished material in which he had hoped to develop a full version of his theory of the logic of discovery in the history of science, and in particular to refute the intolerable "historicism" not only of Kuhn but of a more immediate rival, Stephen Toulmin. In his final writings, Lakatos waxed intemperately hostile toward Toulmin's *Human Understanding* (1972). Toulmin found it hard to comprehend Lakatos's "violent expressions of disagreement . . . at a number of public meetings, right up to November, 1973."[139] He wondered why Lakatos insisted on seeing him, Polanyi, and Kuhn as a unitary "school" of "dangerous ideological tendency." It seemed particularly baffling to Toulmin since, in the view of many, Lakatos himself belonged among the "historically-minded philosophers of science" because of the "crucial roles that he gave to historical change."[140] Toulmin concluded that Lakatos was driven to this extreme hostility by an "oversimplification" which "led him to believe that all those positions in philosophy of science which attach central importance to the *praxis* of scientists are open to the charge of 'historical relativism,' in the same way as (say) the first edition of Thomas Kuhn's *The Structure of Scientific Revolutions.*"[141] Toulmin expressed equal hostility to this extreme position of Kuhn, which he characterized as relativist.[142] But he felt that his own position should not have been lumped together with that, and he implied that even the later Kuhn should not fall under Lakatos's blanket condemnations.

Toulmin wished to discriminate two senses of historicism, the radical one presented in Kuhn's first edition of *Structure,* and a more modest and justified version which in fact Lakatos *shared* with Kuhn, Polanyi, and Toulmin himself. The problem for the philosophy of science appeared, in Toulmin's formulation, to be finding a more viable, moderate formulation of historicism.[143] The stress on praxis was crucial for Toulmin, as it has proven crucial for the course of the subsequent debate. Toulmin noted: "If the intellectual content of any genuine natural science embraces, after all, not only *propositions* but *praxis*—not only its theoretical statements, but also practical procedures for their empirical application—then neither scientists nor philosophers can afford to confine their 'rational' or 'critical' attention to a *formal idealization* of its theories, i.e., to a representation of those theories as comprising merely systems of propositions and inferences structured into logico-mathematical form."[144] Toulmin insisted that, in this vein, he and Polanyi (and, I would add, Kuhn) represented an important advance over Popper and Lakatos, in developing "rational criticism" not only of the *words* but of the *works* of scientists.[145] Lakatos (and presumably Popper) had made no room for such matters in the "third world" of objective knowledge.

Toulmin's *Human Understanding,* volume 1, *The Collective Use and Evolution of Concepts* was an unequivocal failure.[146] Its length was a token of its inconclusiveness.[147] His aim was clear: "in science and philosophy alike, an exclusive preoccupation with logical systematicity has been destructive of both historical understanding and rational criticism."[148] But because he was unable to clarify his argument, he had to reformulate his book's essential claims more pointedly in a separate article.[149] He proposed that scientific theories needed to be considered in genealogical terms, on a par with Darwinian evolution, such that theoretical "populations" of scientific theory could be compared with "organic speciation."[150] That is: "concepts for Toulmin are analogues of individual organisms, with the collections they form—namely, intellectual disciplines—being counterparts of organic species. Disciplines instantiate . . . 'populations.'"[151] That was the core of his idea of evolutionary epistemology. But his strictures against logic and systematicity were too extreme, and his effort to come up with a better theory of rational change never achieved the appropriate concreteness. I. C. Jarvie put it with cruelty: Toulmin's conception of rationality "makes it possible for astrologers and witch-hunters to claim rational equality with science."[152] Perhaps just as cruel is the judgment of a number of reviewers that despite his earnest efforts, Toulmin is not at all that far from Kuhnian relativism.[153]

Larry Laudan endeavored dramatically to historicize the philosophy of science in his monograph *Progress and Its Problems* (1977).[154] He sought, indeed, to derive rationality from advancing success in the solution of problems ("progress"), reversing the traditional derivation of progress from rationality. He drew upon and refined the ideas of "paradigm" from Kuhn and "research program" from Lakatos to construct his own idea of "research traditions" as the appropriate units of consideration in the historical philosophy of science. Laudan elaborated his notion of a "research tradition" to garner the best of both approaches, recognizing that research traditions, like paradigms or research programs, "are *historical* creatures."[155] What distinguished his view from that of his predecessors was the notion that elements even in the "hard core" or the exemplar could be modified and the research tradition still retain its historical continuity. Hence, instead of essential attributes persisting in the descent of theory-versions within a given research tradition, Laudan conceived something closer to the variations of family resemblance in a genetic tree.

For Laudan, Lakatos's scheme was, "in many respects, a decided improvement on Kuhn's."[156] Laudan was particularly unhappy with what he saw as the "rigidity" of Kuhn's notion of *paradigm,* which did not allow for "the historical fact that many maxi-theories have evolved through time."[157] He was also unhappy that Kuhn did not allow for the competition of theories at any one point in time, though historical research clearly indicated such competition was typical. In addition, he criticized Kuhn's insistence that a paradigm could not be fully articulated theoretically. Most important, Laudan quarreled with the global character of change as Kuhn conceived it: "The core ontology of a world view or paradigm, along with its methodology and axiology, come on a take-it-or-leave-it basis."[158] Laudan was convinced that historical transitions between theories and larger units like "paradigms," "research programs," or "research traditions" came in more piecemeal fashion, with critical consequences for their comparability and the rationality of theory choice.[159] Laudan criticized Lakatos as well, primarily for his logicism, with "its dependence upon the Tarski-Popper notions of 'empirical and logical content.'"[160]

Laudan meant to historicize rationality not only for the *heuristic* or discovery contexts, but also for the *evaluative* or justification contexts of theories and research traditions: "any appraisal of the rationality of accepting a particular theory is trebly relative: it is relative to its contemporaneous competitors, it is relative to prevailing doctrines of theory assessment, and it is relative to the previous theories within the research tradition."[161] While Laudan stressed that rationality in problem-solving needed always to be set in a comparative context, McMullin questioned whether alternative theories were *always* available. "Kuhn may have exaggerated the normality of 'normal science,' but it would seem equally unjustified to take the opposite extreme and assert that theories may progress *only* where alternative theories are also being propounded."[162]

A more general question is whether there is not a fourth level of relativity—one relative to *our* (presentist) criteria. The problem, which Laudan recognizes, is that without some ultimate criterion, the historian or philosopher of science seems to lack a standard to distinguish *good* science from actual science. That is the logicism that remained in Lakatos (derived from Popper). There was, as Ernan McMullin detected, a significant equivocation between presentism and historicism in Laudan: "Laudan rejects the maxim he attributes to Popper and Lakatos that 'we should evaluate historical episodes using *our* standards.' . . . But if the historian is also supposed to apply the best model of rationality he can discover today, how in practice is he to 'attend to' the actual beliefs and canons of rationality of past scientists?"[163]

Laudan, like Kuhn in his revisionist stance, acknowledged that one way out was to admit that "there are certain very general characteristics of a theory of rationality which are *trans-temporal* and *trans-cultural*."[164] But he sought to displace this logicist option with a naturalist or historicist one. Philosophy of science could find its warrant in episodes of scientific progress whose success was unquestioned. Believing that there were certain such episodes that no one could reasonably dispute, Laudan invoked "pre-analytic intuitions" to identify this criterial class of cases. Using such "intuitions," he suggested, these episodes of the "actual past" could be discriminated directly without contamination by accounts of that past.[165] A model of scientific rationality drawn from these cases would then supply the basis for historical interpretations. "*The task of the historian of science . . . is to write an account . . . of episodes in the history of science . . . utilizing as his criteria of narrative selection and weighting those norms contained in that philosophical model which is most nearly adequate to representing* [our preferred pre-analytic intuitions about scientific rationality]" (original emphasis).[166] The element of circularity in this endeavor was not lost on Laudan, but he believed he could weather it.

His critics have not agreed with his judgment. First, is there *any* methodological possibility of discriminating the actual past from its representations?[167] Second, what *is* a "pre-analytic intuition"?[168] Third, how can one *select* the privileged instances in the history of science without already *utilizing the criteria* they are supposed to warrant? Finally, *whose* rationality governs the appraisal of historical actors?[169]

McMullin expressed discontent with Laudan's very idea that problem-solving could serve as a characterization of the essential aim of science. He preferred a more traditional sense of understanding, of "explanation-giving" in the sense of "charting of regularities."[170] McMullin's review developed a more pervasive concern: he found Laudan's recourse to problem-solving an effort to create an all-too-positivistic quantification calculus—*counting* problem-solutions as between rival theories or research traditions—in order to achieve adequately "objective" evidence.

This suspicion of an unacknowledged positivism was underscored in the review by the Popperian I. C. Jarvie, who alleged Laudan's "ideas are Kuhnian, his language Popperian, and his deepest presuppositions inductivist."[171] Indeed, Jarvie's review developed the thesis that "underlying [Laudan's] view is an unacknowledged positivism" which became clear especially in his prescriptions to history and sociology.[172] Above all, Laudan's turn to history for resources for his model showed painfully little sense for history or social theory. It was, Jarvie averred, "a reversion to those abstract or philosophical theories of scientific method that Polanyi and Kuhn broke away from. . . . Laudan's characterization is wholly intellectualist: traditions are treated simply as sets of assumptions. . . . The intellectualism clearly stems from a preoccupation with attempting to improve on Lakatos' methodology of scientific research programmes."[173] The plausibility of these allegations is heightened in light of Laudan's subsequent leading role in the VPI Program for Testing Philosophies of Science.[174] There, in the terms of Steve Fuller, "Laudan has abstracted a lowest common denominator from the views of various philosophers of science and then reified it as the essence of the scientific method."[175]

Gerald Doppelt, by contrast, saw Laudan sliding into the all-out relativism associated with Kuhn. "Laudan's position, while more clearly developed and philosophically rigorous than Kuhn's, is much closer in spirit and substance to the latter than Laudan recognizes."[176] Doppelt emphasizes as Laudan's "most striking and controversial aspect" his dramatic widening of the contextual or background elements in theory-appraisal. "Laudan has in effect expanded the conception of scientific rationality and progress to encompass the broader epistemic goals of a global, mutually supporting system including all of our fundamental beliefs."[177] Hence his "bold claim that genuine conceptual problems can arise *for science* (and not merely for philosophy, society, culture, or religion) from an incongruity between a scientific theory and an extra-scientific system of beliefs or 'world-view.'"[178] But the result, for Doppelt, was a "most serous epistemological problem," namely, how to "overcome historical relativism": "does Laudan's criterion of problem-solving effectiveness slide into a Kuhnian relativism for which every theory is free to define its own standards?"[179] That is, Laudan failed to show that "all research traditions agree on and share some neutral, external standard of problem-solving effectiveness (of the sort Laudan's avoidance of relativism requires!)."[180] Finally, Doppelt saw a tension between Laudan's theory of progress and his rejection of convergent realism, which expressed itself in a hyperbolic impulse in Laudan's argument: "Laudan flies from the fire of 'total cumulativity' into the frying pan of 'zero cumulativity.'"[181] But Doppelt reasonably contended that Laudan need not and should not have made that move: "Some significant degree of overlap, 'partial cumulativity,' at the level of empirical problems (and the concepts required to commonly identify them) seems to provide the criterion for individuating research traditions which belong to the 'same' scientific discipline and can thus exhibit progress with respect to its domain of problems and concerns (however much these change)."[182] Such partial cumulativity is an essential component of the "bootstrap" theory of rationality to be defended as a core element in naturalized epistemology at the conclusion of this chapter.

James R. Brown suggested that Laudan, whose general orientation was antifoundationalist, had smuggled a form of foundationalism back into his theory with the privileging of certain historical moments, creating "spurious entities, *normative historical facts.*"[183] He found little merit in Laudan's idea that "preferred pre-analytic intuitions" about key episodes in the history of science offered an escape from the "vicious circularity" that Laudan himself indicated would threaten any enterprise to derive norms from actuality. Brown queried Laudan's very recourse to "intuition," as McMullin had queried it in the logicist philosophies of science. But Laudan's confidence in selected episodes from science also troubled Brown. He claimed Lakatos was more consistent than Laudan in holding that an adequate theory of science should fit the broadest spectrum of actual history of science. Still, he could not find a way out of circularity for Lakatos either. "Any written historical episode must be a theoretical reconstruction; that is, it must invoke . . . concepts which owe their being to some methodology of science. This is because our histories are explanatory and they appeal to reasons as the causes of events."[184] The elaboration of a "normative naturalism" seemed doomed to "vicious circularity," according to Brown and many others, if it presupposed such "theoretical reconstruction" as its basis.

In the context of justification Anthony Murphy and R. E. Hendrick demand a foundation which needs no further warrant. On that basis, they charge Lakatos and Laudan (and by implication *any* historicist approach) with ad hoc abortion of the question of justification. The authors invoke two arguments: the "Scylla of circularity" and the "Charybdis of the normative/descriptive paradox," in other words, first Laudan's own articulation of the "vicious circularity" problem and then the "logical impossibility" of deriving norms from facts.[185] Where the authors are persuasive is in their sustained critique of the presumption in Laudan and Lakatos that the philosopher of science can somehow access actual history unmediated by historical interpretations. This is indeed a flagrant problem in their work, but it need not be so for all historicists. To seize the other horn of the dilemma, while Murphy and Hendrick take as the best effort at foundational analysis the "comprehensive critical rationalism" of W. W. Bartley, they demonstrate that even this cannot achieve the requisite closure.[186] By contrast, a severe problem with Murphy and Hendrick's evaluation of Lakatos and Laudan is that they do not recognize that there could even be a difference between the vicious and the hermeneutic circle; all circles seem vicious and that is what they take the hermeneutic circle to signify.[187]

The Failure of the "Marriage"?

Laudan's work failed to win over most philosophers of science. By the beginning of the 1990s, the link between history of science and mainstream philosophy of science became significantly attenuated. Thus, Steve Fuller observed in 1991, "the past fifteen years have witnessed a steady retreat behind disciplinary boundaries."[188] "Sad but true, . . . HPS has gradually lost its momentum. The younger generation of historians of science . . . [t]o a person . . . define themselves explicitly, if sometimes oppositionally, to the theoretical agendas of the sociology—not the philosophy of science."[189] Specifically, "no *historian* after Kuhn has tried her hand at divining philosophical lessons from the history of science—though some philosophers have: Lakatos, Laudan, and Shapere."[190]

The important point here is that it was the *historians* of science who were finding it increasingly unproductive to link up with the philosophy of science. This estrangement became apparent to those who had earlier sought to bring history and philosophy of science together. In 1989 Larry Laudan was invited to write a retrospective at the twentieth anniversary of the journal *Studies in History and Philosophy of Science*, which he cofounded with

Gerd Buchdahl in the late 1960s. In his retrospective, Laudan lamented: "Contemporary philosophers of science, whatever their persuasion, are now prepared to grant that historically-based philosophy of science is not only a viable but a valuable venture. By contrast, many (perhaps most) professional historians of science have refused to see the point. Indeed, the distance between mainstream history of science and the philosophy of science is probably greater now than it has ever been."[191] In 1992, the Philosophy of Science Association devoted a panel to this issue.[192] Robert Richards echoed Laudan in maintaining that "history of science remains sublimely indifferent, if not hostile, to philosophy of science."[193] Picking up the "marriage" metaphor, he observed that institutional cohabitation had "demonstrated . . . instability."[194] Rachel Laudan observed that among new historians of science, "philosophy of science is regarded as outmoded, trivial and irrelevant." That was because "history of science is undergoing a sea change," she explained; " 'high theory' has entered history of science, drawn from an eclectic mixture of continental philosophy, literary theory, cultural anthropology and the gurus of postmodernism."[195]

But there was another reason, elaborated cogently by Marga Vicedo, namely, that philosophy of science—even as practiced by figures like Larry Laudan—proved inconsistent with history of science. "By using the history of science only as a repository of cases, we are not dealing with the history of science, but only with past episodes in science," and "the evidential role that can be attributed to an isolated episode from the history of science is usually very low."[196] The very achievement in combined "history-and-philosophy-of-science" that Larry Laudan had held out as his instance of the intellectual "blossoming" of the field, the VPI project that resulted in the publication of *Scrutinizing Science,* came under withering fire from *both* historians of science and philosophers of science.[197] In the words of Steve Fuller, Laudan's "project has been resisted from all quarters."[198]

The grand-scale "normative" prepossession of philosophy of science seemed to be insufferably obstructive to any effective *descriptive* or *explanatory* investigations of concrete areas or problems. As Thomas Nickles puts it, "providing intelligible descriptions and explanations of conceptual discoveries" is extremely difficult both for philosophers and for historians.[199] He elaborates, "reasoning is so complex and content-specific that no simple logic of discovery could possibly do justice to it."[200] The implication is that localizing inquiry holds the best prospect for substantive outcomes. Much, if not all, that is interesting in philosophy of science is best pursued *locally* or *internally* (methodologically) within empirical scientific practice itself ("naturalized epistemology").[201] Many practitioners of philosophy of science were already tacitly acknowledging this by their actions. Grand theory of science as a whole has come increasingly to be displaced by disciplinary or problem-specific methodological and metamethodological study.[202] For the empirical disciplines, it seemed there remained precious little reason to think *external* philosophy of science had anything at all worth hearing. In the words of Mary Hesse, "What has in fact happened is that, far from philosophy providing criteria for history, all forms of historical investigation, internal as well as external, have led to radical questioning of all received philosophical views of science."[203]

Philosophers of science—starting with Kuhn himself—believed they could go to history to equip themselves for their own endeavor, to answer their questions about validity, rationality, truth. But a marriage could only prosper if there were some reciprocal value for the historian of science. Kuhn and Hanson made the obvious point that historians borrow terms, methods, standards, and sometimes even questions from philosophy. Yet philosophy of science has not proven all that fruitful for historians of science. It has scorned history's own concerns, or trivialized them. "Philosophical models of proper evaluation are irrelevant to the historian's task. Indeed, with their typical stress upon the formal, abstract properties of verbal argument, they can even impede an adequate naturalistic understanding of actual judgments."[204] Historians need no longer be seeking to answer the philosopher's questions. Without accepting the distinction, when philosophy of science abandoned the "context of discovery" to the empirical human sciences, it left a major *theoretical* matter to be clarified, namely, how it was that history (and the other empirical human sciences) were to proceed in such empirical inquiry and to appraise their own efficacy. As Nickles put it, "historians and philosophers *have* succeeded in making intelligible the routes to several important discoveries."[205] *How* did they do that? What does *intelligible* signify? If, indeed, historians can provide "intelligible descriptions and explanations," what do "description" and "explanation" signify? If historians go about "making intelligible to reason (insofar as possible)," by offering "accounts," how is that done, and with what claim to reasonableness? What *is* "historical *explanation* as opposed to mere chronicle or anecdote"?[206] It is not just condescending but obscurantist to write this off to mere ad hoc groping in the inductivist dark, even if so grand an authority as Karl Popper proposes it. On the other hand, no one is really trying to propose that there is some deep *algorithmic* theory of discovery to be worked out.[207] That is, historical practice presumes that "reasoning [in science] usually is so complex and context-specific" that it is pointless to seek a universal algorithm.[208] That does not make it pointless, however, to take methodological bearings, to theorize the endeavor, and to use, in our theorizing, the exempla of "intelligibility" which we appear to possess. What prospect is there that a naturalized approach might provide a more fruitful partnership between history and philosophy of science?

The Prospects of Naturalized Epistemology

What is it that we wish to understand about science? Traditional philosophy of science has been concerned to understand its truth-claim, its validity. But one could also be interested in the *success* of science and in its *process*, and in the link between that process and success. The traditional concern sets out from the premise that there is a standard *higher* than science for the ascertainment of knowledge or truth. That standard was traditionally associated with *logic*. But one of the ways to construe post-positivism in the study of science is to see it as the displacement of this ideal of logic by a concern with rationality.[209] The essential question then becomes: what *is* rationality? And above all, must it always have been the same? Is rationality independent of science (as process) and criterial for it? The key here is *change*. We have compelling historical evidence that change has taken place in science. That is, not simply has there been a change in the content of scientific knowledge or in scientific theories, but we have strong evidence that there have been changes in methodology and even in aims. The most compelling case has been made by Dudley Shapere.[210] As Thomas Nickles summarizes, it is "Shapere's main thesis that, historically, scientific goals, methods, criticism of appraisals, *etc.*, and not just empirical and theoretical content, have been learned."[211]

Shapere asks us to consider the "dynamics of rational change."[212] That is indeed the question, but its formulation could prompt a misguided rejoinder: what makes the change *rational*? That is, by making change and not rationality the object, we presuppose a criterial independence for rationality. The question should be reformulated: we are concerned with the *dynamics of changing rationality*, with the historicity of reason. Formulated that way, the account of change—the "dynamic"—*constitutes* the entity, rationality, which is changing. And that is the point of the decisive term in Shapere's formulation, namely *learning*. Shapere, as Nickles points out, is taking up an intervention from Charles Peirce, namely, that science is the essential story of *how we learned to learn*.[213] In short, instead of asking what makes change rational, we must ask how rationality changes. That is, we must consider it immanent in the growth of knowledge, as emergent in and through the process of science. This is what Nickles and others have termed "a broadly 'bootstrap' account of the growth of knowledge."[214] That is the core idea of a naturalist, or historicist, or evolutionary epistemology.[215]

The decisive starting point for post-positivism has been the recognition of a tacit collusion between positivism and relativism: first, that, in the absence of absolute certainty in cognitive assertion, "anything goes"; and second, more insidiously, that, in accordance with the fact/value dichotomy, *all value judgments are (equally) arbitrary*.[216] With the failure of the claims to absolute certainty (to foundationalism), it has been all too facile to infer a radical historicism or relativism. But this was wrong-headed from the out-set.[217] The "place" from which such absolutes could be pronounced—pro or con—was *no human place*. This was a "view from nowhere." Humans are always already situated. Therefore, with Nickles, I am "forced to reject [such radical historicism], historicist though I am, as an untenably strong form of historicism."[218] I agree further that we must "temper our histori-cism with a dose of (pragmatic) naturalism" and achieve thereby a "more Deweyan sort of balance."[219] As Nickles puts it, "a defensible historicism does not rule out a bootstrap account of the development of knowledge; on the contrary, it requires it!"[220] That is, "a moderate historicism itself implies that any adequate account of the growth of knowledge will be broadly cir-cular. The growth of knowledge will have the character of a self-transform-ing, ultimately self-supporting or 'bootstrap' process."[221] Nickles clearly discerns that this implies what he calls a "multi-pass conception of science," in other words, an iterative process of adjustment, for which the term *di-alectical* is not inappropriate.[222] "Human knowledge has grown by means of a self-transforming, dialectical or 'bootstrap' process, rooted in variation, selective retention, and triangulation of historically available resources."[223] Nickles affirms a Deweyan Hegelianism here: embracing "Hegel's 'method-ological' insights, his anti-dualism, and his historicist and sociological ten-dencies."[224] This pragmatist-naturalist reception is, he aptly affirms, only "*weakly* Hegelian because . . . it presupposes no transcendent Reason that shapes the overall developmental process."[225] That, in my view, is the "liv-ing" heritage of Hegelian dialectics. This moderate historicism is, I submit, highly *robust* and sustainable. In the terms of Dudley Shapere, "The view would imply that we learn *what* 'knowledge' is *as* we attain knowledge, that we learn *how* to learn in the process of learning."[226]

The key point for philosophy of science—and, indeed, history of sci-ence—is that "science is, after all, a paradigm case of the knowledge acquir-ing process."[227] Shapere draws a strong, but in my view correct, inference: "It is a condition of the adequacy of any philosophy of science that it show *how* rational change in science is possible, and a philosophy of science which, after asking *whether* scientific change can be rational, denies that it can be, must be rejected."[228] Shapere and others think that Kuhn denied such rationality. I think they are wrong. But there have been those who not only thought Kuhn did so, but that he was correct. With Shapere and Nick-les, I take the strongest exception to that view. Rationality may be imma-nent, but it is not a delusion. Absolute Reason, timeless and divine, and with it "first philosophy," have no place in a naturalist philosophy of science, but rationality clearly must. As Hilary Putnam put it, enough isn't everything, but enough is enough.[229] In the words of Larry Briskman, "we want an epis-temology which allows that we can *learn to be more rational*—that we can learn to pursue better (or more rational) aims and can equally learn to pursue these aims better (more rationally)."[230]

As Philip Kitcher has observed, "the failure of appeals to conceptual truth, to analyticity, is fully general."[231] That is, "virtually nothing is knowable *a priori*, and, in particular, no epistemological principle is knowable *a priori*."[232] Kitcher believes these results stem, as this study has sought to document, from "arguments from Quine and Kuhn."[233] And he concludes, "the denial of the a priori thus leads . . . to a position whose emphasis on the growth of knowledge invites the title 'historicism.' "[234] Since "knowledge is embedded in the history of human knowledge, and not detachable from it," we are "ineluctably dependent on the past."[235] The collapse of justificationist philosophy of science—of the claim that there exist a priori standards against which philosophers could hold scientific achievements to assay their worth—has left the question of normativity in limbo. Naturalism (and historicism) dealt severe damage to traditional logicism, but could naturalism offer an alternative normative epistemology?[236] In Kitcher's terms, "the central question is whether naturalism allows any way to save the traditional meliorative project of epistemology."[237]

The character of such an enterprise is clear.[238] John Passmore put it as follows: "To pick out from what scientists have done the strategies which make possible the emergence of science as a successful enterprise entails looking carefully at what has happened, contrasting successful and unsuccessful science and asking how far the failures were the result of the use of bad strategies as distinct from ignorance of crucial facts, experimental limitations and so on."[239] Of course, the response of the traditionalist would immediately be that success is not the same as validity, and the question of "good science"—or even what is to count as "science"—gets begged when history is taken unproblematically to constitute the relevant evidence. Far more troubling is the *radical relativist* response, which is that there is no place *left* for epistemological questions.[240] Naturalist philosophy of science "emerges from the attempt to understand the growth of scientific knowledge," but radical relativism disputes *any claim to growth,* to even partial cumulativity, drawing upon and radicalizing what Kuhn averred.[241] Thus, Kitcher writes: "We can imagine numerous possible histories of science, yielding divergent conceptions of nature and rival sets of epistemological principles . . . there is no basis for concluding that the actual evolution of science is self-correcting. . . . The history of science comes to resemble a random walk, not a unidirectional process."[242] This "dangerous form of skepticism threaten[s] to collapse traditional naturalism into a radical position."[243] But Kitcher also maintains that "the general arguments for that collapse are not cogent."[244] That is, the dispute boils down, ultimately, to an *empirical* question about the history of human knowledge. "Only a serious examination of the historical, sociological, and psychological material can resolve the issues that are raised by these versions of skepticism."[245]

The rise of evolutionary or naturalized epistemology, a philosophical approach to science which is *historical in essence,* finds its models and metaphors for theory in a science quite different from the Received View's physical science. Biology, and specifically evolutionary biology, is the inspiration of this new approach to philosophy of science.[246] As Kitcher has noted, there are "*two* ways in which evolutionary concepts can be introduced into epistemology: one can try to appeal to selection pressures on hominids . . . or one can try to frame epistemological theories that have formal analogies to the theory of evolution by natural selection."[247] Kitcher has little faith in the first of these ventures providing much reinforcement, and I concur. It is the second vein that holds promise. Richard Burian fixes upon the essential point: There are "very deep and difficult problems of establishing criteria of identity and individuation for growing, developing, changing theories," and the best available analogy is biological speciation.[248] That intuition grounds some of the most exciting current philosophy of science.[249]

Ronald Giere has summarized three widely invoked arguments against naturalized epistemology.[250] First is the familiar "vicious circle" argument.[251] Logicists claimed that, without some *prior idea* of good or successful science, selection could never even get off the ground.[252] Giere thrusts the traditionalist scruple about the circle back upon the critics themselves. Every one of their own formulations has led to infinite regress, so that it hardly seems that theirs is a superior epistemic stance.[253] In Harold Brown's apt words, "a priori epistemology faces problems that are at least as serious as those faced by naturalized epistemology."[254] The circle is vicious only if one privileges the foundationalist ideal of absolute certainty, which itself has proven perennially inept against skeptical pressures demonstrating internal inconsistency and infinite regress.[255]

The second argument brought forth against naturalized epistemology, according to Giere, is the distinction of description from prescription.[256] Naturalism in epistemology ostensibly seeks to use empirical success to warrant legitimacy, but this leap from "is" to "ought" violates the sacrosanct fact-value dichotomy upheld from Hume forward. However, it remains possible to conceive reason as an emergent—ontogenetically as well as phylogenetically—and seek to ascertain its structure from within the circle of its own emergence. As Harold Brown argues, "A consistent empiricism requires that even truths of logic have [a] kind of empirical prehistory."[257] That is, "norms, in the form of both ends for science and methodological imperatives, are introduced and evaluated in the same ways as theoretical hypotheses, experimental designs, new mathematics, and other features of the so-called content of science."[258] He explains: "To be sure, these truths are deeply embedded in our present conceptual system, but this need show only that such elementary principles as noncontradiction and *modus ponens* are so simple and so useful that they were built into our language in the distant past; that they have been passed down from generation to generation through the normal means of cultural transmission; and that no reasons have appeared to cast them out."[259]

The final objection to naturalized epistemology trades on dread of a slide into relativism.[260] That is, as soon as one abandons the security of a priori principles, there is no stopping on the slippery slope to total relativism, to the "death of epistemology" or indeed of philosophy. Once one begins to "de-transcendentalize," in Rorty's terminology, there is no purchase left.[261] As a historian of philosophy, I cannot help but be reminded of Friedrich Jacobi's vehement assertion, in the "*Pantheismus-Streit*" of the 1780s, that once the sound tenets of faith were surrendered and one ventured into philosophy, there could only be a slippery slide into atheism and nihilism.[262] Thinking in such an absolute "either/or" manner may have the virtue of purity, but it lacks a human measure. If philosophy has resolutely continued without the recourse to divine sanction, so must it continue without the surrogate divinity of the a priori, and without precipitate despair.

G. H. Merrill recognizes that the dilemma presented by "pure" logicism versus "pure" historicism has motivated the moderate position, but he believes that it cannot be sustained.[263] Merrill terms the moderate historicist strategy an "*empirical approach*" which takes for granted that most of what we *call* science has been that. The question that Merrill raises is whether moderate historicism can avoid slipping into the "radical historicist" position for which "there is no distinction between actual science and good science."[264] The moderate position "is intelligible only if it can accommodate the *possibility* that some actual science fails to be *good* science," but that "must presuppose a history-independent distinction between good and bad science."[265] Merrill presumes the availability of "more basic logical and philosophical" criteria, "appropriate canons of reasoning" that "may be identified and elucidated through philosophical analysis."[266] No such philosophical analysis has withstood criticism. Moreover, one of the most compelling complaints has been that such models have proven woefully incongruent with the actual processes and successes we discern in science.

The response to Merrill is that beyond an absolute skepticism, which is not intelligently to be debated, the emergent and immanent standards that moderate historicism proposes represent the best available accounts of the growth of knowledge and the constitution of empirical inquiry.[267] Merrill himself acknowledges that "science is thought of as a paradigm epistemic enterprise."[268] That "success" is always an immanent and emergent principle is just what it means to adopt an antifoundationalist, historicist approach.[269] *We have no higher warrant for rationality than the success of science.* As Nicholas Rescher puts it, "Science *as we have it*—the only 'science' that we ourselves know—is a specifically human artifact that must be expected to reflect in significant degree the particular characteristics of its makers."[270]

I agree with Ronald Giere that "attempting to draw a fundamental distinction between rational and irrational activities is itself not an effective way to understand science, or any other human activity," especially, I would add, if the idea of rationality that is in play here should be what Habermas long ago called the "positivistically halved idea of reason," reason stripped of its pragmatic and creative, its dynamic and dialectical character.[271] But there are other traditions of thought about rationality: most promising is a convergence of pragmatist naturalism in the spirit of Dewey with elements in dialectical historicism in the vein of Hegel. At the very least, we need to think about how history can be *rationally understood* to change reason.[272]

Giere launches another line of criticism, namely, that thus far naturalizers or historicists have sought to formulate theories of evolutionary or naturalized epistemology so general that their models are too thin and undifferentiated. Giere writes, "most historically oriented philosophers of science since Kuhn seem to be aiming at a similarly grand theory of development."[273] Giere questions whether any general theory or model of development that historians could muster would be adequate to account for all of real science. That seems to him especially true about claims concerning shifts in standards and aims. "Following Kuhn, historically minded philosophers of science have argued, using historical examples, that not only the content of science changes with time. Its aims and methods change as well."[274]

Larry Laudan is a primary proponent of such a normative naturalism. As Adam Grobler puts it, "Laudan's offer consists in fact of replacing the hierarchical model of scientific rationality by a reticulated one. The three levels—the factual, the methodological and the axiological—are mutually related there."[275] It is a crucial feature of Laudan's normative naturalism that methodological rules are strictly *hypothetical*, not *categorical* imperatives; they are always "directives for achieving scientific ends." Moreover, Laudan argues that "history shows that there are no fixed ends for science and that the available background knowledge also changes. Thus there can be no fixed, categorical methodological directives for the pursuit of science."[276]

Laudan has defended his normative naturalism against a substantial and subtle critical inquisition.[277] While it has hardly emerged unscarred, such a complex theory of rational-scientific development in a fully dialectical sense deserves further exploration, not rejection. While Gerald Doppelt, perhaps Laudan's most inveterate critic, acknowledges that Laudan has deployed arguments sufficient to hold off extreme or radical relativism, he urges that Kuhn's moderate relativism, as Doppelt himself reconstructed it, remains a better alternative despite Laudan's best efforts.[278] Their debate leaves a reader less than edified, because they often talk past one another. Laudan is arguing against the view that there are *never* rational grounds for adjudication at the level of the aims and methods of science, while Doppelt is arguing against the view that there are *always* rational grounds for adjudication. Both of them are right, and each of them misunderstands the other. In any event, whatever moderate relativism or moderate historicism might respectively be, each can accommodate a contingent, fallibilist notion of rationality. In short, naturalism can live with that much anomaly.

Not every circularity is necessarily vicious, in particular not the hermeneutic circle in interpretation.[279] James Brown argues that as soon as one recognizes that philosophy of science is "not foundational," the circularity argument against normative naturalism becomes only a "pseudo-problem."[280] By carefully distinguishing between the use and the accreditation of a theory in the construction of history, Brown shows that it is possible both to utilize theory in historical research and then to critique its result and work toward a more plausible theory. Two strengths of the naturalist circle are its embrace of success in science, however approximate we may have to be about both science and success, and the more general principle of growth of knowledge, of *learning itself,* as the most discernible basis for any idea of rationality we might hope to find. The very idea of the hermeneutic circle has been that, via a process of successive approximations against wider and wider resources of application, interpretation achieves an immanent and self-correcting rigor.[281]

There is reason at the very least for exploring the idea that epistemic norms are immanent in cognitive development, that there can be an *iterative* normativity, if not an ultimate one. On the other hand, a *general theory of scientific rationality* derived from history may be too ambitious, and perhaps we should eschew theories on a grand scale. We can be satisfied for some time with *local* and *comparative* elucidations of rationality.[282] Smaller endeavors at developmental accounts and medium-range theories hold considerable prospect for achievement.

Admittedly, normative naturalism remains empirical, but it does not follow that it cannot be critical. In the words of David Hull, "The stories that historians tell are theory-laden but not so theory-laden as to be useless."[283] Hull admits that "bringing history of science to bear on general claims about science is extremely difficult" in virtue of theory-ladenness, but he makes the essential rebuttal: "No matter how strongly one's general views color one's estimations of data, sometimes these data can challenge the very theories in which they are generated."[284] This idea of "resistance" or "constraint" operates to keep both substantive scientific inquiry *and* the metainquiry into scientific justification from the total arbitrariness associated with the doctrine of theory-ladenness. It is the key to *moderate historicism.*

"Bootstrap" rationality is immanent intelligibility. The want of totality does not betoken want of concreteness; again, enough is not everything, but enough can be enough. The whole idea of dialectic is that one never starts from nowhere. A problem is inconceivable without a context. It is always already mediated. If there is no ultimate *foundation*, there is always some *platform*. And that platform is far more elaborate than a bunch of mere "data." "Problem contexts include much more than empirical datum constraints on adequate problem solutions. Large, conceptual problem contexts contain constraints of many kinds, many of them previous *theoretical* results, which function as consistency conditions, limit conditions, derivational requirements, etc."[285] We can conceive of scientific problems as emergent from "a structure of constraints (on the problem solution) plus a general demand, goal, or explanatory ideal of the research program in question that certain types of gaps in those structures be filled."[286] These constraints "constitute a rich supply of premises and context-specific rules for reasoning toward a problem solution."[287] Thus every problem is an emergent, in a situation, and the constraints that situate the emergent problem also equip the inquiry with (some of) the terms for its solution. Moreover, this whole syndrome must be taken as a dynamic, not a static, process, resulting in "successively sharper reformulations of a problem."[288] Such reformulations can result in radical departures, and it has been empirically the case that "it frequently takes science a good deal of time and effort after a discovery to say what exactly has been discovered."[289]

If this contextualist, thickly descriptive approach to the genesis and pursuit of a scientific problem is one harvest of the naturalization of epistemology, another is the recognition that intelligibility entails more than overt logical operations. "Not all of our rational activities and capacities can be made intelligible in terms of a fully conscious deliberation, but that does not make them any less rational."[290] This was what Michael Polanyi was seeking to articulate by "tacit reasoning." It also embraces what Marx Wartofsky has invoked as the "heuristic tradition," namely, noninferential judgment, the "craftsmanlike skill . . . exemplified in legal, clinical, aesthetic, and historical judgment."[291] Two points emerge. First, *retrospectively* we can trace the routes of (some of) our insights. Second, making *intelligible* those successes can empower us, both psychologically and methodologically, to undertake new inquiries. Not only can we be confident that "scientific problems and constraints do not fall out of the sky," but we can be hopeful that we can reiterate at least some of the moves that led us through prior solutions and *learn* how to attempt new ones.[292] Rationality, then, becomes not the *cause* of our success, or even a supratemporal *standard* for it, but rather the *tentative harvest* of our history of discovery. That is simply what it means that "philosophers take scientific activity as somehow paradigmatic of rational behavior."[293]

Rationality is the concept we can articulate to affirm that science has indeed found a way of "bootstrapping stabilized past results and practices into the future."[294] Scientists *learn* and they *use* what they learn: "Once a scientific claim is reasonably settled, they do not hesitate to use the newly accepted entities and processes in new research."[295] There is nothing frightening or illegitimate about this, even if some interpreters have called such presumptions "black boxes." As Nickles rightly asserts, "Something already black boxed can later be used." Indeed, that is what "progress"—or, less complacently, cumulation—betokens: "Often the most successful steps [in human learning] have become 'black boxed' and taken for granted by later generations, who employ these capacities and their products as 'givens' rather than items that were once constructed. In these cases the apparatus of construction and maintenance has become invisible and remains so as long as things work well enough."[296] The shift of focus from science as product to science as process, from the context of justification to the context of discovery, from logic to rationality, from timelessness to historicity—all these betoken the robustness of a naturalist, historicist, evolutionary epistemology. This is the great harvest of post-positivism in science studies.

Logicist resistance has not vanished, of course, and anything that does not satisfy the logicist model continues to get labeled irrational. Thus Colin Howson writes of the "poverty of historicism" with as little reservation as had Karl Popper several generations earlier.[297] Not everyone will concede that foundationalism is a hopeless cause: "Thus the failure of ahistorical justificationism, or of any other proposed epistemological programme, can never *logically* force us into adopting a psychologistic, sociologistic, or historicist approach to knowledge—for we are always free to try again."[298] Indeed, but I would think we are entitled to argue that this is (in the language of Imre Lakatos) a paramount instance of a "degenerating problem shift." I suggest such recalcitrant logicists appear amazingly like the returning aristos of France after the Great Revolution: they have learned nothing and forgotten nothing.

CHAPTER 5

> The question is: can we rationally reconstruct the growth of scientific
> knowledge or do we have to be satisfied with a socio-psychological de-
> scription of paradigm changes? An authoritative answer would largely
> determine whether there remains room for a philosophy of science, or
> whether this discipline is to become a chapter of descriptive sociology.
> — TOMAS KULKA, "SOME PROBLEMS CONCERNING RATIONAL
> RECONSTRUCTION"

How Kuhn Became a Sociologist (and Why He Didn't Like It): The Strong Program and the Social Construction of Science

The philosophical community Kuhn sought to join continually rejected his
ideas. By contrast, the discipline he invoked somewhat cavalierly to illustrate his views, sociology, took his ideas up in their most drastic formulations and launched a research program in his name, a "Kuhnian" sociology
of scientific knowledge. In the early 1970s, Barry Barnes, David Bloor, and
Steven Shapin—all of the Edinburgh Science Studies Unit—developed the
Strong Program in the sociology of scientific knowledge with explicit reference to Kuhn's revisionist interpretation of science.[1] Their conception of a
social construction of scientific knowledge took up the very elements of his
work which had become an embarrassment to him in the context of the immediate philosophical reception of *Structure*. Kuhn (largely unwittingly) invited this sociological conscription by his emphasis upon the social in
Structure and especially in its defenses, but he swiftly became disenchanted
with this appropriation of his work and took great pains to dissociate himself from those who invoked his name.[2]

Despite Kuhn's scruples, these sociologists opened the way for the most
"radical" impulses in the post-positivist theory of science. They contended
that nothing meaningful was left to the philosophy of science which could
not better be construed by the sociology of scientific knowledge. They insisted that philosophy of science, with its *prescriptive* account of science, had
become thoroughly discredited by such critics as Kuhn and hence needed to
be supplanted by a naturalistic and *descriptive* sociological account. Thus,
the most fateful reception of Kuhn's theory has taken form "as an interdisciplinary polemic between philosophy and sociology" in which "sociologists

often flatter themselves that their discipline has displaced philosophy or is about to do so."[3]

There are two wider contexts into which this story of Kuhn's conscription into sociology must also be set. First is the political climate of the 1960s, with its rebellion against established authority and its questioning of the complacency of liberal ideologies and the complicity of liberal politicians in repressive regimes and global imperialism. One element in this political restiveness was suspicion of the pretense of science to be above the political fray. Scientific responsibility for the nuclear arms race and the all-too-patent connection of scientific research with what even Eisenhower called the "military-industrial complex" led to a strong suspicion of science. Jürgen Habermas's "Science and Technology as Ideology" can be taken as a representative of this suspicion at its intellectually most rigorous.[4] The ties of science with society, and the suspicion that scientists within their specialties were embroiled in "politics" incongruous with their ostensible norms, prompted sociologists to believe that there was work for them in this arena. In 1971 Kuhn wrote, "one reason [the turn to the external history of science] now flourishes is undoubtedly the increasingly virulent antiscientific climate of these times."[5] Yet the idea remained one of deviance from the ideal, a "sociology of error" or "ideology-critique." The last and most essential barrier to a "sociology of truth" had not yet been breached.

Beginning in the mid-1960s, a growing restiveness against the conservative implications of functionalist orthodoxy in sociology generally led to a proliferation of alternative approaches.[6] Among these was a revitalization of the sociology of knowledge. The sociology of knowledge originated in the Germany of the 1920s, especially in the work of Karl Mannheim.[7] Mannheim's approach came under tremendous criticism in its original German context and when it made the transatlantic passage.[8] Nonetheless, a vestige of sociology of knowledge gained a foothold in American sociology.[9] Robert Merton took it up and reorganized aspects of it into what came to be known as the sociology of science.[10] In the 1960s Peter Berger and Thomas Luckmann reintroduced the sociology of knowledge to American social science by assimilating it to the American pragmatism of George Herbert Mead. They titled their work *The Social Construction of Reality,* but the title's resonance with current usage makes it seem far more radical than it was.[11] Like Mannheim's original approach, it concentrated primarily on the political analysis of social stratification, and there was little impact on the sociology of science.[12] By the 1960s the sociology of knowledge reasserted itself in Europe as well, primarily in Germany, via the ideas of "critical theory" and "ideology-critique" sponsored by the neo-Marxist Frankfurt School.[13] The showdown between the functionalist and the critical viewpoints in German sociology came in the famous "Positivism Dispute" of 1969.[14] Karl Popper's participation in that dispute as the protagonist of orthodox social theory betokens the convergence of wider agendas in this moment of confrontation.

The other contextual backdrop for the emergence of a "Kuhnian sociology" draws together the thought of one philosopher with a global interpretive reorientation. The later philosophy of Ludwig Wittgenstein seemed to offer social scientists resources to come to grips with the increasingly troubling problem of Western condescension toward non-Western cultures.[15] The key locus was the so-called Rationality Dispute in social anthropology, spawned largely by the Wittgensteinian theories of Peter Winch.[16] What the Rationality Dispute explored was twofold: must rationality be narrowly identified with the orientation of the contemporary West, and specifically with Western scientific reasoning; and how, alternatively, could the ways of thinking of non-Western cultures be grasped as "rational" on a wider view? What Wittgenstein's philosophy offered, as Winch and others saw it, was a substantial loosening of the logical frame of Western rationality as well as a reinterpretation of the conventional character of practices of knowledge. Perhaps the clearest assertion of this is Winch's notorious line: "Criteria of logic are not a direct gift of God, but arise out of, and are only intelligible in the context of, ways of living or modes of social life."[17] This allowed not only a reorientation in the anthropology of non-Western societies along a more interpretive and less normative line but also a reflexive reappraisal of Western scientific practice and its ostensible norms.[18]

Within the social study of science, these impulses led to the search for a rational reconstruction of the so-called context of discovery. The ambition of that endeavor was to establish a bridgehead of continuity between discovery and justification in terms of the rationality found in the former. Thus, the most interesting work for the later emergence of "social construction" was the effort to find the terms of a *nonformal rationality*, what was termed "implicit" (Mary Douglas), "tacit" (Michael Polanyi), or "natural" (Mary Hesse) reasoning. Central here was the development of what came to be called the interaction theory of metaphor as a key ingredient in scientific creativity. Originating with Max Black's revindication of metaphor for philosophy, this interpretation led to a new conception of language learning and creativity.[19] As a theory of enculturation, it was extended into social anthropology by Mary Douglas and, as a theory of inference, into the philosophy of science by Mary Hesse. Both drew not only on Wittgenstein but upon the idea of a "web of belief" in the philosophy of Quine.[20] In a series of works, Douglas set about trying to reconstruct cultural patterns of knowledge organized in symbolic structures (myths, cosmologies) by relating such taxonomies or "grids" to "group" structures present in the social order.[21] Her first great breakthrough was in her theory of taboo, which had resisted most interpretive efforts until she found the key in taxonomical cosmologies.[22]

Her work in social anthropology highlighted problems of language acquisition and general enculturation which had direct philosophical implications.[23] Equally important was the interweaving of Quine's theories of a network of belief with Black's theories of metaphor by Mary Hesse in a series of works culminating in 1974 with *The Structure of Scientific Inference*.[24] Hesse was instrumental in aligning Quine and Kuhn together under the crucial rubric of a "new empiricism" and in transmitting this synthetic vision of a post-positivist history and philosophy of science directly to the founders of the sociology of scientific knowledge. The work on models, metaphors, and analogies in scientific discovery complemented the last linkage in this configuration of ideas: the concrete studies of scientific practice by Michael Polanyi in terms of "tacit knowledge" and by Norwood Russell Hanson in terms of "patterns of discovery."[25] A substantial body of theory lay waiting to be galvanized into a coherent research program. The spark for that synthesis came with Kuhn's *Structure of Scientific Revolutions*. Reflecting on the impulses that were reconfiguring science studies in 1977, Marx Wartofsky made that point clearly: "*The Structure of Scientific Revolutions*—an irritating, naive, confused and provocative work . . . brought it all together in one glorious explosion."[26] Twenty years later Ronald Giere acknowledged, "Without doubt, Kuhn's work was the single most influential force in creating the intersection of history, philosophy and sociology of science that became identified as 'science studies.'"[27]

The intuition of intersection is caught well in a little-known essay by Derek Phillips from 1975: "There are certain strong similarities in the writings of the philosopher Ludwig Wittgenstein; the historian Thomas Kuhn; and several contributions of two sociologists of knowledge, Karl Mannheim and C. Wright Mills."[28] Even earlier (1974) Phillips offered a penetrating characterization of the developments in the sociology of knowledge leading up to the Strong Program.[29] Another—backhanded—recognition of Kuhn's importance for sociology came from the irascible Popperian I. C. Jarvie in his withering review of Laudan's *Progress and Its Problems*. Jarvie contrasted Laudan with Kuhn at the expense of the former, despite his own deep, Popperian reservations against Kuhn's "irrationalism."[30] Jarvie's main contention was that the social sciences received Kuhn very favorably, in part because "he created a legitimate place within his theory for the work of the social scientist," and also because, in Jarvie's estimation, "Kuhn is a good sociologist: a scrupulous and sensitive ethnographer of the community he belongs to—science."[31] Jarvie could not resist adding that Kuhn's very errors made him attractive. The absence of rationality in theory choice "reduces the crucial changes in science to sociological, psychological or economic questions." And incommensurability suggested that there was nothing superior about scientific rationality. Hence, "Kuhn, without realizing it, demystified the natural sciences, when what he set out to do was to describe their enchanted status."[32] But with this last formulation, Jarvie accents the crucial possibility that the sociological synthesis represented a *mis*interpretation of Kuhn by sociologists, as Kuhn steadfastly maintained to his death.[33]

How did this misappropriation arise? In the postscript of 1969, Kuhn strove to clarify his idea of paradigm not only by linking up his usage of incommensurability to Quine's theories of indeterminacy of translation, but also by invoking what he called a more sociological approach to the idea of the scientific community. Thus, Kuhn raised the problem of a circularity between paradigm and scientific community and insisted that methodologically it would be best that scientific communities "be isolated without prior recourse to paradigms."[34] As he put it in "Reflection on My Critics" (1970), "Some of the principles deployed in my explanation of science are irreducibly sociological, at least at this time."[35] He avowed that such an "intrinsically sociological" position was "as such, a major retreat from the canons of explanation licensed by . . . justificationism and falsificationism, both dogmatic and naive."[36] Yet in the same essay he admitted that he shared with Popper the view that "the generalizations which constituted received theories in sociology and psychology . . . are weak reeds from which to weave a philosophy of science."[37]

While in the postscript Kuhn did gesture to sociological research into the community structure of science and claim that the "empirical techniques . . . for its exploration are non-trivial," that hardly constituted high praise or systematic incorporation into his theory.[38] Therefore it was perfectly reasonable for Keith Jones to query, "Is Kuhn a Sociologist?"[39] Jones stressed Kuhn's admission "that his work does not rely on received sociological or psychological theories."[40] Indeed, precious little social theory informs Kuhn's theories. Kuhn wished to draw sociological considerations *into* philosophy of science, not displace the latter by the former, since what he sought was to "explain the social process through which it becomes reasonable for a scientist to attempt the development of a new paradigm."[41] Only his intuition that shifting to the scientific *group* could not leave traditional epistemology unaffected led him to suspect a role for what is now called social epistemology. He transferred the locus of epistemological discrimination from the individual scientist to the scientific specialty as a *collective*.

What is striking about the elaboration of Kuhn's theory of the disciplinary matrix is how *conceptual* it remains, underscoring Jones's perplexity over Kuhn's assertion that his view was "irreducibly sociological."[42] What Kuhn offered is very much in the line of traditional internal interpretation. It is no easy matter to construe how Kuhn proposed to incorporate the external factors in scientific development conventionally consigned to the domain of the sociological. Most early sociological commentators found Kuhn *oblivious* to such questions. Thus, in addition to Jones, John Urry claimed that Kuhn offered "an account of change *within* a scientific community [which] abstracts it from the social, economic and value concerns of the encompassing society."[43] Lee Harvey claimed that Kuhn "fail[ed] to make any substantive links with the wider social context."[44] Douglas Eckberg and Lester Hill held that Kuhn's "sociological base" amounted really to the "fairly rigid, highly elaborated framework of beliefs" that a paradigm fastened upon its practitioners.[45] Herminio Martins made the same point: Kuhn's "account deliberately [held] constant 'external' societal factors" because they appeared "irrelevan[t] to *The Structure of Scientific Revolutions*."[46] Indeed, Martins grasped the profoundly problematic relationship that resulted from Kuhn's ostensibly sociological contribution: "There is something paradoxical in Kuhn's appeal to sociology, for Kuhn's work *combines sociologism with anti-sociology*, thus simultaneously satisfying two mutually incompatible idols of the age. It is sociological in that it presents something like a social theory of natural-scientific knowledge. . . . On the other hand, it can be used as an ideological tool of anti-sociology, in so far as sociology appears to be lacking in the diagnostic criteria of scientific maturity."[47] To put it bluntly, a *lot* of imputation would be required to muster Kuhn into the camp of sociology.[48] That it nonetheless transpired speaks to the pressing nature of the need for him in that quarter.

The Emergence of a "Kuhnian" Research Program in Sociology

The immediate origins of the sociology of scientific knowledge are to be found in the efforts of a few sociologists to break free from the constricting framework of Merton's functionalist sociology of science at the close of the 1960s. Perhaps there is no better way to grasp the "image of science" by which the times were possessed around 1960 than to retrieve to memory the orthodox formulation of Merton's sociology of science. In standard functionalist terms, Merton identified certain key norms (four in his original formulation, with several others added by his school) that worked to structure and stabilize the social system of science.[49] These norms were universalism, communism (renamed communality in token of the Cold War), disinterestedness, and organized skepticism. Elaboration of the model led to the inclusion of originality and individualism. Collectively, these norms constituted the "ethos of science," which governed the social interaction of scientists and distinguished their behavior from that of other groups in society. Merton insisted he sought "not the methods of science, but the mores with which they are hedged about," and he disowned any "adventure in polymathy" that would draw him into assessing the methods or substance of science.[50]

Merton and his school adopted an approach to science that conformed perfectly to what philosophers of science in the logical positivist/empiricist tradition had articulated to define their own pursuits. This *positivist collusion* revolved around Reichenbach's discrimination of the context of discovery from the context of justification, now reformulated as the discrimination of scientific practices from scientific outcomes. Merton and his colleagues were prepared to study only practices, consigned already by the philosophers of science to the murky actualities of psychology and happenstance, while outcomes were reserved for the pure logics of the philosophers of science and for the natural scientists themselves. Both groups deemed it inappropriate to extend sociological inquiry into justification, for that risked a "genetic fallacy." No account of the process whereby a scientific idea came to be formulated could ever govern its validity. Validity and justification were either matters of approximation to the real ("truth"), which only the natural scientists themselves could judge, or they were formalities best appraised by the elaborate logical apparatus of the philosophers of science.

The unquestioned authority of natural scientific knowledge, not only in its own domain but as the standard against which social science measured its own legitimacy, presented a formidable impediment to the reinvigoration of sociology of knowledge—and more particularly its extension to natural scientific practices. In the face of this positivist collusion, the emergent sociology of scientific knowledge had to challenge not only the established field of sociology of science as developed by Merton and his followers, but also the philosophy of science of the Received View. Once we recognize that positivism or scientism was perceived as the decisive barrier to the extension of sociology of knowledge to science itself, the importance of Kuhn becomes clear. Kuhn's revisionism was exactly what sociologists needed to free themselves from the oppression of positivist scientism.[51]

An early recognition of this resource was M. D. King's essay "Reason, Tradition, and the Progressiveness of Science" (1971). King pointed clearly to the self-constriction of sociology of science, the view that scientific thought "is distinguished from other modes of thought precisely by virtue of its immunity from social determination." Given this, sociologists accepted that "science as a system of knowledge is . . . none of their business."[52] Thus arose "the virtual bifurcation of scientific 'product' and the process of scientific 'production'" which I have termed the positivist collusion between Mertonian sociology of science and the Received View's philosophy of science.[53] For King, the importance of Kuhn was precisely to challenge these taboos. The question King posed was, "What kind of sociology of science might be developed from Kuhn's conception of research traditions?"[54]

King focused on the interpretation of scientific disputes in the competing frames of Mertonian and Kuhnian theory. While Merton saw disputes as inevitable but merely as a question of prestige and reward, Kuhn suggested something more fundamental: "recognition of priority is recognition of *intellectual authority*—of the right of the discoverer or his epigoni to set out research strategies for the new field of inquiry."[55] As King conceived the matter, "Kuhn is concerned with the socio-psychological processes through which specific authoritative traditions of scientific thought and practice are established, perpetuated, elaborated, and, in time, undermined and displaced." Far from presupposing a perennial normative rationality, "Kuhn adopts a more radical position. He questions what Merton, no less than the positivists, simply takes for granted: that the cognitive development of science is a rational process governed by timeless rules of procedure. In fact, he denies that such standards exist and maintains that the practice of science is monitored not by universal rules but by 'local' traditions of thought."[56] King highlighted the element of arbitrariness that Kuhn unearthed in scientific revolutions, their kinship to conversion experiences. Moreover, he challenged Kuhn's effort to retreat from his radicalism and reestablish "the *overall* progressiveness of science." Only the Kuhn of "paradigm incommensurability," of local order and discontinuity, seemed useful for sociology.

King drew a very fruitful analogy between Kuhn's model of paradigm shift and the idea of "living law" over against statute law in legal theory. The argument King derived was that practice constituted theory, not the converse. Drawing Kuhn's notion of a paradigm together with this idea of the living law, King concluded: "This surely is the essence of Kuhn's notion of a paradigm—that science is governed by tradition rather than reason."[57] For King, "This opens a whole new field for sociological inquiry. It invites us to study the *contingent* relationships between the rise and decline of such tradition-bound ways of doing science."[58] King summed up his stance: "What I would advocate is a kind of 'epistemological agnosticism' . . . which would give sociologists the opportunity of developing the kind of approach that serves more to illuminate actual historical processes of change in the patterns of thought, mode of practice, and social situation of scientists, than to meet the demands of epistemology."[59] Kuhn himself did "not succeed in developing a sociological theory of scientific change" because he remained too wedded to the idea of scientific progress, but sociologists could take off from him in developing this necessary theory.[60] King clearly grasped by 1971 how to take Kuhn at his most "radical" and therewith revamp the sociology of science.

The intellectual energies for the breakthrough gathered primarily in Britain.[61] They found in Kuhn exactly the resource necessary for the endeavor. Michael Mulkay, who became one of the leading figures in the new sociology of scientific knowledge, gave the earliest indication of the agenda of the gathering movement in an essay entitled "Some Aspects of Cultural Growth in the Natural Sciences" (1969).[62] Mulkay's essay set out as a critique of Merton's theory by stressing the power of *cognitive* as opposed to behavioral norms in the conduct of science. It was in this context that Mulkay invoked Kuhn, particularly Kuhn's essay "The Essential Tension: Tradition and Innovation in Scientific Research."[63] Instead of viewing science and its progress in terms of openness, as Merton's norms proposed, Mulkay claimed, "it is quite possible to view scientific growth as a product of intellectual and social closure," and he invoked Kuhn as his star witness.[64]

The decisive elaboration of a Kuhnian sociology of scientific knowledge came from the figures of the Edinburgh Science Studies Unit, Barry Barnes and David Bloor.[65] The team of Barnes and Bloor simultaneously assaulted the existing sociology of science of the Mertonians and the existing philosophy of science of the logical positivists and the critical rationalists, and with abandon. In the course of their various expositions, Barnes and Bloor emphasized that, in contrast with the philosophy of science's postulation of what science *should* have been (or *must* have been) doing (e.g., Lakatosian "rational reconstruction"), a sociological approach would simply get at what actually happened—*wie es eigentlich gewesen*. Empirically it promised to provide *causal explanations* of theory choice and change in science, and thus to provide an *actual* history of science vastly superior to the conjectural history of the Received View or even of its normatively oriented revisionists like Lakatos, Laudan, or Toulmin.

In the early writings of Barnes and Bloor we can trace the incorporation of all the elements outlined previously into a coherent theoretical perspective. The first publication along these lines, significantly enough, appeared in a journal of social anthropology and bore the title "Paradigms—Scientific and Social." In this essay, Barnes explicitly linked advancing the sociology of science to the problems arising from the Rationality Dispute in anthropology: "perhaps we shall begin to see science as so complex that anthropological interpretations of primitive beliefs will serve as models in understanding it."[66] For Barnes, there was a sociological research agenda to be found in holding these seemingly polar domains together: "it should prove profitable to look at how paradigms are articulated and developed in science and how they are overthrown, and enquire to what extent similar processes are possible in primitive societies."[67] This prospect of an alternative to the Received View of scientific rationality "very largely bases itself on views presented by Thomas Kuhn."[68]

Above all, what intrigued Barnes was that the scientists of Kuhn's normal science looked very much like the practitioners of non-Western modes of thought, constantly faulted by Western interpreters for seeking secondary elaborations in order to accommodate their theories to the anomalies presented by reality. Scientific paradigms seemed far closer to Azande witchcraft than even Robin Horton had surmised.[69] Working from Kuhn's idea of "normal science," Barnes argued in another essay, "Evans-Pritchard's celebrated account of how Azande can always interpret the behaviour of their poison oracle within the framework of their particular belief-system must not be regarded as an antithesis to good scientific practice; one way to become convinced of the value of secondary elaboration is to study the history of chemistry."[70] From such a viewpoint, Barnes would naturally have to dispute Merton's approach to the sociology of science. This he did, in collaboration with R. G. A. Dolby, in a key essay of 1970. The thrust of their attack was to suggest that Merton's model was irredeemably static. "It is our view that the general orientation which starts by identifying 'given' governing norms within science has resulted in neglect of the processes by which normative structures have changed."[71] Above all, Barnes and Dolby queried any timeless standard of scientific rationality, and they suggested "that the notion of 'rationality' as a specific scientific norm be abandoned."[72] Against Merton, Barnes and Dolby invoked Kuhn.

In "Sociology of Knowledge in Natural Science" (1971), which appeared in one of the inaugural issues of a new journal, *Science Studies,* emanating from the Edinburgh Science Studies Unit, Dolby diagnosed the positivist collusion between Merton and the Received View and its consequences for the sociology of scientific knowledge: "American sociology of knowledge has usually followed Merton in accepting the logical empiricist distinction between discovery and validation. Merton's influential 'Paradigm for the Sociology of Knowledge' systematically reviews the questions in the social origin of knowledge, setting aside the question of the validity of the knowledge-claims involved."[73] Against this, Dolby invoked Kuhn's emphasis on the historical contingency of scientific practice. "The approach introduces a new relativism into philosophy of science, one which opens up new vistas for the sociology of scientific knowledge."[74] Dolby emphatically welcomed this relativism. Invoking Kuhn's notion of "incommensurability," he claimed that such arguments "lead us to recognize a sociological and cultural relativity in *all* scientific knowledge-claims."[75]

In that same first volume of *Science Studies* in 1971, David Bloor published a very important review of *Criticism and the Growth of Knowledge,* which he clearly represented as a disciplinary controversy: "the clash that occurs when the sociological approach makes incursions into the field normally occupied by philosophers of science."[76] For Bloor, Kuhn needed to be understood as the entering wedge of sociology, and he justified this interpretation by exploring "the evolution of Kuhn's position since 1962."[77] Bloor seized upon all Kuhn's statements that "what is required is a sociological approach," that the "basic analytical unit . . . has become a grouping of scientific practitioners."[78] Bloor accentuated "Kuhn's growing emphasis on the primacy of the sociological approach."[79] At the same time, however, he pointed out that it remained merely a gesture: "Kuhn's sociological picture does not seem to have built into it any 'laws of motion' indicating what structural changes may have taken place historically in science . . . a very fruitful line of development for the Kuhnian research programme."[80] Thus Bloor appropriated Kuhn as a sociologist and announced the pursuit within sociology of science of a Kuhnian research program.

Putting together an anthology on the sociology of science for Penguin in 1972, Barnes was able to select the key developments and indicate the new prospects in the field. In his introduction, Barnes described the Mertonian view as "for a long period the only theoretical approach available," but he observed that in the 1960s "the monopoly of the approach weakened."[81] He identified the work of Kuhn as the source of this revisionism, and he claimed that "an entirely new theoretical approach" in sociology could be derived from it.[82] The question of science as an agent of social change could, Barnes suggested, be reversed: "Instead of analysing it as an agent of social change, or as an encapsulated sub-culture, the extent to which its own social structure and culture are derivative of those of the wider society can be investigated."[83]

Barnes firmly believed that the imposition of external, normative standards of rational conduct profoundly compromised the prospect of empirical adequacy in either social anthropology or sociology of science. What Barnes objected to was that "in practice at least, 'rationality' acts as an evaluative not an explanatory term."[84] The methodological principle that beliefs and actions should be approached via the actor's own situation, in terms of the social context in which the actor operated, constituted the homology Barnes discerned between social anthropology and the sociology of scientific knowledge. Barnes cited J. D. Y. Peel for the crucial argument: " 'True beliefs as well as false ones are the product of social forces and their origin is a perfectly legitimate concern for the sociologist; causal explanation is not to be restricted to what the sociologist's own society considers false.' "[85] This anticipated (and provided warrant for) the key idea of symmetry in the ultimate formulation of the Strong Program.

Barnes gathered these thoughts together into the first mature statement of what would become the Strong Program, his essay "Sociological Explanation and Natural Science: A Kuhnian Reappraisal" (1972).[86] Barnes now stressed that the cognitive privilege science had claimed could no longer be recognized: "*all* the beliefs and practices of *any* culture are, in principle, open to sociological explanation."[87] For Barnes, this signified a rejection of any foundationalist or timeless notion of truth or rationality. In proposing an alternative, "sociological explanation," Barnes distinguished between normative accounts and interpretive ones. The key to an interpretive sociological explanation was that it "does not depend upon . . . externally derived rationality criteria," but rather investigated the position and significance of beliefs and actions "within the actor's own overall perspective."[88] A second theme of the essay was to incorporate into the Kuhnian notions of paradigm and normal science the "recent work by philosophers on the importance of models, analogies and metaphors in science."[89] *"The crucial element in 'normal science' is the extension, alteration or re-interpretation of models and metaphors, or the concrete exemplars related to them, so that they indicate a new kind of concrete puzzle and the solution it is likely to have, or more commonly, they become congruent with an existing practical problem and indicate its solution"* (original emphasis).[90] Barnes proposed a theory of "natural" or "bootstrap" rationality along the lines of what had been developed by Mary Hesse and also by the theorists of naturalized epistemology.[91]

Barnes insisted that sociology had to assume, as its methodological premises, both skepticism and relativism. "It is sceptical since it suggests that no arguments will ever be available which could establish a particular epistemology or ontology as ultimately correct. It is relativistic because it suggests that belief systems cannot be objectively ranked in terms of their proximity to reality or their rationality."[92] Though a skeptic and a relativist, Barnes hastened to make clear that this did not imply that he questioned reality. "It is important not to lose sight of the connection which does exist between knowledge and the real world. . . . Knowledge arises out of encounters with reality and is continually subject to feedback-correction from these encounters."[93] Already, however, Barnes was aware that there were those who pressed for a more radical position. "Occasionally, existing work leaves the feeling that reality has *nothing* to do with what is socially constructed or negotiated to count as natural knowledge, but we may safely assume that this impression is an accidental by-product of over-enthusiastic sociological analysis, and that sociologists as a whole would acknowledge that the world in some way constrains what it is believed to be."[94] Barnes had explicitly criticized what he took as the metaphysical "idealism" of Peter Winch. Barnes objected to Winch's repudiation of causal analysis for the interpretive approach and held out for the plausibility of causal analysis "based on the regulatory concept of normal practice."[95]

In an important review of the new currents in sociology of science, John Law and David French tried to derive principles for the general direction of sociological inquiry. As they saw it, "the distinction between the 'normative and 'interpretive' modes in sociology is fundamental."[96] They found that the normative/interpretive distinction cut across the Mertonian/Kuhnian division. Mulkay, in their assessment, was a normative Kuhnian. By "normative" they meant he used an approach that perceived action as rule-governed (*deterministic*). They viewed Mulkay (and implicitly Kuhn) as claiming that within a given paradigm, such rule-governed behavior in fact prevailed. By contrast, "an interpretive model examines the way in which meanings are negotiated in interaction and assumes that such meanings may vary over time and between different interactions."[97] They stressed, accordingly, the openness even of a student's response to scientific enculturation against what they saw as Kuhnian determinism. Yet they also recognized in Kuhn elements "at least consistent with an interpretive analysis of action," especially in his notion of puzzle-solving in normal science, which did leave open a sphere of actor-independence.[98] To accentuate this potential in Kuhn, they urged its assimilation to what they called "the social pragmatism that is implicit in the writings of Wittgenstein."[99] Their point was that "learning cannot be seen as a simple internalisation of norms, but rather as an interactive process."[100] Thus, even within a paradigm, in contrast to Kuhn, "an interpretive analyst would not assume such a degree of consensus and uniformity."[101]

Richard Whitley made a survey of the new literature in sociology of science in 1972 which appraised much of the material we have been surveying. "Kuhn's *Structure of Scientific Revolutions* has had considerable impact on the British Sociology of Science. . . . By emphasizing the cognitive aspects of science and linking changes in cognitive structures to sociopsychological phenomena, Kuhn legitimated sociologists' revolt against Merton."[102] Historical criticism of assumptions about a single scientific method, Whitley argued, created the opening for a new anti-Mertonian sociology of science. In stressing the historical variance of knowledge-principles, however, Whitley contended that a normative element was inevitable for the new sociological approach, thus diverging sharply from the emergent Edinburgh school. "Kuhn has created the epistemological irrationalist revolution by asserting that knowledge governed by one paradigm is 'incommensurable' with that governed by its successor."[103] Whitley felt that "to deny *a priori* the possibility of scientists choosing between 'paradigms' on a rational basis . . . is to assert the primacy of sociological factors in scientific decision making" and thus to hold "an extreme relativist theory of knowledge."[104] Thus, Whitley was alarmed by the "radical" message of *Structure*. He felt that it represented a thoroughgoing relativism, but he recognized that Kuhn backtracked toward rationality in the postscript to *Structure*. Whitley welcomed this move and urged that "if some element of rational choice is introduced, sociologists would have to gain an understanding of substantive scientific issues," in other words, scientific truth and epistemological rationality.[105] Given his predilections, it is not surprising that Whitley turned to Popper and Lakatos to explore a more rationalist alternative to Kuhn. While he recognized problems in their positions, he concluded by insisting, "What is required for an understanding of the nature of scientific knowledge and its growth is a combination of sociological rationality and epistemological rationality."[106]

A few years later, Nico Stehr attempted an even wider integration of the controversy over scientific "ethos" into trends in the development of the sociology of knowledge, with special reference to European developments.[107] He began with a careful exposition of the Mertonian position, demonstrating how the normative approach was integrated with a notion of the goals of science to develop an account of the growth of scientific knowledge. Then he noted "the immense influence generated by Kuhn's *The Structure of Scientific Revolutions*," which made possible "an alternative theory of the social relations of science."[108] By raising questions about the timelessness of the norms and standards of science, the new approach created a new criterion for the sociology of science: "Any assessment of the degree to which the ethos of science claims to facilitate the production of scientific knowledge presupposes an explication of a historiography of scientific knowledge, and a specification of criteria for the growth of knowledge."[109]

What emerges from these essays is that the revisionists had succeeded by the mid-1970s in establishing an alternative research program deriving from Kuhn, a sociology of scientific knowledge.[110] It was one for which Kuhn had no fondness at all, and which embarrassed him by invoking his name. In 1977 he expressed his distaste: "In the literature of the sociology of science, the value system of science has been especially discussed by R. K. Merton and his followers. Recently that group has been repeatedly and sometimes stridently criticized by sociologists who, drawing on my work and sometimes informally describing themselves as 'Kuhnians,' emphasize that values vary from community to community and from time to time. In addition, these critics point out that, whatever the values of a given community may be, one or another of them is repeatedly violated by its members. Under these circumstances, they think it absurd to conceive the analysis of values as a significant means of illuminating scientific behavior."[111] Kuhn wished to make it plain "how seriously misdirected [he took] that line of criticism to be."[112] He referred explicitly to Barnes and Dolby and to the journal *Social Studies of Science,* the Strong Program. As Steve Fuller has correctly observed, "Kuhn disavowed every one of these appropriations of his work."[113] Indeed, "at every opportunity, Kuhn disowned all of the more exciting and radical theses imputed to him by friends and foes alike."[114] Fuller associates this with Kuhn's "utter lack of historical self-consciousness," his "acritical" mentality.[115] He even offers some leads for a psychoanalytic explanation.[116] This study will not go there.

The Strong Program as Rival to Philosophy of Science

It would appear that Kuhn was defending orthodox *sociology* of science from the new school, but in fact his hostility arose from the even more adversarial stance of the Strong Program toward the *philosophy* of science. From the outset the Strong Program pursued confrontation with *philosophy* of science even more than with sociology of science. The Strong Program undertook to displace philosophy by sociology: its tenor and its reception cannot otherwise be accounted for.[117] This disputation of the authority of philosophy of science emanated most forcefully from David Bloor. Bloor's book *Knowledge and Social Imagery* was aptly characterized as "a sustained tirade against philosophers." Bloor "sets out to re-define the disciplinary boundaries for the study of science, giving sociology pride of place . . . and dealing philosophers . . . largely out of the game."[118] From his earliest writings, Bloor directed himself to challenging philosophers; he had sublime confidence in the aptness of his own sociological orientation. Notably all of Bloor's earliest essays represent disputations with philosophers, assaulting the prevailing orthodoxy in British analytic philosophy with an ill-disguised disdain.[119] The title of one of his earliest essays puts things aptly: "Are Philosophers Averse to Science?"[120] In his major work of 1976, Bloor put it in provocative terms: "To ask questions of the sort which philosophers address to themselves is usually to paralyse the mind."[121] At the same time, Bloor's early essays demonstrate his culling of philosophical ideas and impulses to buttress his own orientation.[122] In these early essays, then, we can discern both the *animus* and the *ancestry* of Bloor's distinctive formulation of the Strong Program. Bloor minced no words on the perceptiveness of philosophers: "the methodological commitments common among contemporary British analytical philosophers may have prevented them from grasping the facts of the case [i.e., the relation of science to society] . . . [indicating] significant weakness in the conception of their own discipline."[123] Bloor detected in these philosophers an ideology similar to the sociological theory of functionalism, with its deeply conservative longing for stability. Immediately Bloor's extraordinarily speculative impulse showed itself: "We might further wonder whether believing a philosophical analogue of naive sociological functionalism does not perhaps derive from holding implicitly the sociological version as well . . . a picture of society being projected, as it were, onto the logical realm."[124] This drastically Durkheimian notion held the germ of Bloor's later Strong Program, namely, that knowledge is shaped after images of societal order.[125]

In Bloor's review essay of *Criticism and the Growth of Knowledge* for *Science Studies*, his agenda emerged clearly. He viewed the book as a confrontation between sociology and philosophy, with Kuhn as the vanguard of the sociological incursion. By forcing attention to "non-logical criteria," Kuhn opened the prospect for "the sociological investigation of the scientific community."[126] Bloor found in Lakatos "a wholesale absorption of the most characteristic Kuhnian positions," making Lakatos an unavowed apostate from the Popperian approach.[127] Bloor's Strong Program represented foremost an all-out relativism on Kuhnian premises, an assault upon the rationalist reconstruction of science in the vein of Lakatos or Toulmin.[128] In his review of Toulmin's *Human Understanding* under the provocative rubric "Rearguard Rationalism," Bloor accused Toulmin of "repeating the regrettable performance of Imre Lakatos who calls Kuhn an 'irrationalist' and then proceeds to advocate a theory of science which is structurally identical to Kuhn's in the respects criticized."[129] Bloor saw it fundamentally as a disciplinary conflict: "the question of how sociological and philosophical approaches toward scientific knowledge are related to one another."[130] In a 1973 essay, Bloor situated himself over against Peter Winch in similarly disciplinary terms: "Whereas Winch thinks that much sociology is misbegotten philosophy, the argument of this paper has been that much philosophy is misbegotten sociology. . . . Rather than philosophy illuminating the social sciences Winch unwittingly shows that the social sciences are required to illuminate philosophical problems."[131]

An essay by John Law, "Is Epistemology Redundant? A Sociological View" (1975), explicitly proclaimed the agenda of the Strong Program to be displacing philosophy.[132] For Law, the accommodation between Mertonian sociology of science and specifically Popperian philosophy of science could no longer be justified. The resource needed to spring the Strong Program free was the Wittgensteinian relativization of logic to social convention. Law, following Bloor explicitly, rejected Toulmin's endeavor to constitute an evolutionary epistemology, insisting that any move in this direction could not stop short of "a fully sociological account of rationality."[133] Law would have nothing of "success" or "progress" as criteria for rationality or objectivity, contending that these had been shown to be historically quite problematic. All that remained was to adopt an utterly interpretive, nonnormative approach to the practices of science. Consequently, Law suggested, epistemology was irrelevant; sociology should dispense with such considerations.

In his early essays, Bloor gathered all the elements for his own provocative synthesis. It appeared in the essay "Wittgenstein and Mannheim on the Sociology of Mathematics" in 1973, the formal announcement of the "strong programme of the sociology of knowledge."[134] In this seminal article, Bloor enunciated the four key principles of the Strong Program, stressing, as he always would, the fourth and most controversial, namely, "symmetry."[135] In *Knowledge and Social Imagery* (1976), Bloor articulated more fully these four principles for the Strong Program: causality, impartiality, reflexivity, and symmetry. He insisted that sociology, as science, had to offer *causal* explanations (though how these worked remained awkwardly vague).[136] Second, sociology needed to investigate social beliefs without imposing the standards of the investigator upon the subject of investigation. More traditional treatments called this "value-free" or "objective" inquiry, and the problems associated with that notion are not unfamiliar.[137] A third principle suggested that the theory of sociology should not be immune to its own argument; it must be possible to conduct a sociology of sociology.[138] The implications of this principle for the Strong Program become a bit less comfortable when one considers some of the corollaries of the fourth and most controversial principle, symmetry. This principle held that sociology should apply exactly the same causal methodology to true beliefs as to false ones.

Conceived in Bloor's preferred polemical way, the symmetry principle had the implication that epistemology was irrelevant. For Bloor, "truth" could serve neither as a *cause* (because it was an outcome, not a source) nor even as a *criterion* (because it was at best a vague ideal without determinate consequences).[139] Knowledge, not truth, was the object of inquiry, and knowledge was irretrievably a *social* construct, something "better equated with Culture than Experience."[140] Critics have suggested that it might have been more accurate still to suggest that what the Strong Program proposed was a sociology of *belief,* for it proposed to efface any distinction between what a group *took* for knowledge and what, by some standard, it was *justified* in taking for knowledge.[141] More broadly, Bloor rejected the idea that truth is self-explanatory and that only error requires sociological investigation.[142] That, he contended, was the weakness in Mannheim's program.[143]

Bloor suggested Mannheim's earlier program "faltered" before the application of sociology of knowledge to the sacrosanct domains of science, mathematics, and logic. Recent research suggests that Bloor's reading of Mannheim was heavily colored by Robert Merton's construction of Mannheim's thought. "Bloor has recalled . . . that his reading of Mannheim was indeed significantly shaped by his reading of Merton's review essays."[144] The suspicion arises that Bloor never fully realized the particular contextual and conceptual concerns that animated Mannheim's original form of the sociology of knowledge, and consequently misunderstood Mannheim's attitude toward natural science and mathematics. Via Merton he was "handed an interpretation of Mannheim's sociology which emphasized its failure to bring the knowledge of the physical sciences within the scope of its analysis."[145] Furthermore, because displacing Merton was the immediate agenda of Bloor's own theorizing, it appears Bloor suppressed a significant measure of his dependency on Merton in making Mannheim his interlocutor. That is, Bloor may well have got *both* Mannheim and Merton wrong, and precisely because of the contextual situation of his own theorizing—irony, indeed, for a sociologist of knowledge!

Bloor insisted that a strong sociology of scientific knowledge would not hesitate to seek causal accounts as well of "true" beliefs as of "false" ones. It would reject the proposition that truth sufficed as an explanation for a belief. In a word, the Strong Program proposed to *eliminate* the context of justification by *assimilating* its residue into the context of discovery, and it proposed to *explain* the context of discovery in terms of *social causes*. This causal account should attain "maximum attainable generality, given currently accepted scientific standards."[146] In his review of Popper's *Objective Knowledge*, Bloor proposed a systematic translation of Popper's talk of a "third world" into a theory of the "social world." That is, "the objectivity of knowledge resides in its being the set of accepted beliefs of a social group."[147] While, when challenged, the Strong Program conceded that there were *other* causes, and that other beliefs *could* feature in causal chains, and hence that *reasons* could be causes, the thrust of its theoretical and empirical endeavors was clearly to stress *social* causes to the virtual exclusion of all others.[148] "Social organization, then, is *the* crucial variable determining the perception of the truth and falsity of any given theory." Hence, objective knowledge "refers to something like the state of a discipline, or a part of a culture, at any given time." It is "the property of a collectivity."[149]

Bloor proposed a drastic theory of social determination: "theories of knowledge are, in effect, reflections of social ideologies."[150] Hence, the title of his book. The point is essentially Durkheimian: the structure of a society's knowledge emulates the structure of the society.[151] In his important essay "Durkheim and Mauss Revisited: Classification and the Sociology of Knowledge," Bloor attempted to vindicate the long-disputed claim that "the classification of things reproduces the classification of men."[152] Equally Durkheimian is Bloor's account of any resistance to his idea. Bloor imputed to his opponents the idea that science is sacred and, accordingly, the need to preserve it from desecration by prosaic accounts.[153] For Bloor, Frege's rebellion against "psychologism" on behalf of the autonomy of logic and mathematics "is a beautiful example of the purity rule in action."[154]

The essay on Durkheim and Mauss offers us a clear indication of Bloor's intellectual sources. He builds his effort to redeploy Durkheim's proposal primarily on the works of Mary Hesse and Mary Douglas. Hesse provides him with the "network model" of classification, grounded in the heritage of Wittgenstein, Duhem, and Quine.[155] Douglas provides Bloor with an updated mechanism for linking cognitive with social structures. The importance of Hesse and Douglas for the conceptualization of the Strong Program and for its rootedness in the entire post-positivist revision in philosophy cannot be too strongly underscored. Bloor addressed himself to Hesse's model in a number of contexts. His review of Hesse's *The Structure of Scientific Inference* tried to explain why it could be "so valuable to the sociologist." The network model worked out there demonstrated that "there is no opposition between causality and rationality." In other words, it grounded the symmetry thesis.[156] Moreover, Hesse's articulation of the theory of language from Quine underscored the "looseness of fit between language and the world," which allowed Bloor to raise the question of the "gap" between these and to offer social causes as the decisive intermediaries.[157] That is, "the similarity of structure between knowledge and society is itself the effect of the social use of nature. This is the real cause."[158] While Hesse offered two important parameters of linguistic-classificatory organization, which she termed "correspondence" and "coherence" conditions, in homage to traditional theories of truth, Bloor hastened to the idea of the indeterminacy of reference to minimize the first and concentrated his energies upon reformulating the second in a social direction that Hesse had not herself originally envisioned. Bloor was emphatic: "an important and ineradicable class of coherence conditions could be identified with social interests."[159] He termed this the "crucial formula" for the sociology of scientific knowledge (SSK).[160] It is noteworthy that Hesse welcomed this intervention as an enhancement of her sketchy endeavor.[161] Indeed, Hesse proved one of the staunchest defenders of the Strong Program.[162]

Mary Douglas proved an essential sponsor of the fledgling Strong Program, especially with her review of Bloor's first book.[163] The importance of Douglas for the entire group of the Strong Program emerges in a review of her *Implicit Meanings: Essays in Anthropology* (1975) by Barry Barnes and Steven Shapin, entitled "Where Is the Edge of Objectivity?"[164] They celebrated her as having a "range perhaps wider than that of any other living anthropologist" and for having developed "a fully general social epistemology ... with continuing reference to the archive of anthropological fieldwork."[165] Barnes and Shapin drew her contention that "knowledge is constitutively social" directly into connection with Kuhn: "Thomas Kuhn's account of the nature of modern natural science is recognizably a version of this thesis," with the implication that "the internal history of science would be correctly conceived as essentially social history."[166] That is, they elaborated, "a social epistemology erodes currently dominant legitimations of science and undermines the presently accepted way of distinguishing its internal history from the study of external pollutions."[167] In place of these sterile conventions, Douglas, following Durkheim and Wittgenstein, upheld the view that "there is no logical necessity, no 'logic of justification.'"[168] Given that knowledge "is constitutively practical," Douglas offered a more concrete theory of this practical relation: "The structure of social relation-ships between men in a society may be reflected as homologies and isomor-phisms in systems of natural classification."[169] Barnes and Shapin explicitly linked this theory with "the argument of Mary Hesse's *Structure of Scientific Inference,* which points out the necessity of culturally given, pre-existent networks of belief and general coherence conditions."[170] In short, Hesse and Douglas were the crucial mediators between Quine and Kuhn and the Strong Program.

Douglas offered Bloor the clearest "mechanisms for linking the social and the cognitive."[171] In "Polyhedra and the Abominations of Leviticus," his provocative juxtaposition of Imre Lakatos with Mary Douglas, Bloor sought to demonstrate how each developed a method for rendering the treat-ment of anomaly intelligible. It hinged upon recognizing that "in the inter-ests of overall coherence any particular achievement may be subverted," even in so "entrenched" a system as mathematics. "Lakatos . . . treats mathemat-ical kinds as being our creations." Furthermore, he "is saying that concept-stretching and the redrawing of classificatory boundaries is an integral part of mathematical reasoning."[172] Bloor drew a hyperbolic inference: mathe-matics "logically . . . is totally underdetermined . . . it is socially determined in the course of negotiations."[173] Douglas demonstrated how this kind of negotiation was itself structured. "Our thoughts are segmented in ways which mirror the segmentation of our social life. . . . Contingent social boundaries channel the flow of influence and interaction of our ideas."[174] Douglas was developing the "grid and group [model] of social structure" which would make it "possible to develop a typology of all possible struc-tures of interest."[175] But, as Edward Manier has contended, "Bloor's use of Douglas' theory . . . is insufficiently critical."[176] His "hope of resting the strong programme on Douglas' account of classification and control is based on the doubtful premise that it has sufficient theoretical coherence and em-pirical significance to warrant its translation into the sociology of sci-ence."[177]

The problems with Bloor's undertaking were severe. No sufficiently lucid theory of the concrete causal mechanism was offered—especially not by Bloor.[178] More problematically still, Bloor, like Quine, seems to have re-tained some of the behaviorist and reductionist elements from the earlier tra-dition of positivism which most post-positivists found disreputable.[179] Just like Quine, Bloor flaunted a scientism that appears incongruous with the otherwise revisionist tenor of his program.[180] The essential issue here was one of *reflexivity,* namely, how Bloor could adopt uncritically a model of sci-ence prevalent in the social sciences and presumed to be in alignment with the best scientific practices of the day, while undertaking a fundamental in-quiry into what it is that constitutes science. Laudan put the problem well:

"There is something profoundly paradoxical in saying that we are setting out scientifically to figure out what central features of science are."[181] A line of defense is Quine's similar posture, namely, that we must stand somewhere, and the most plausible place to stand is within the ambit of currently sanctioned scientific practices. The problem here is two-fold. First, it is not clear that there is *one* ubiquitously accepted current standard of scientific practice; indeed, there are strong grounds to suspect the "disunity of science."[182] Second, Bloor seemed to be operating on a frankly crude model of what science, causality, and method may all be taken to be under the rubric of best current practice. He labored, in a word, under what Habermas termed a "positivistically halved notion of rationalism."[183] And that is bitter irony, indeed.

One important contextual source for this regressive impulse lay in Bloor's training in experimental psychology in the early postwar era, when that discipline was dominated by behaviorism.[184] In "Remember the Strong Program?" Bloor pointed back to his training as the source of "an orientation I have never lost."[185] Slezak charges that Bloor never internalized the paradigm shift introduced by Noam Chomsky's devastating review of B. F. Skinner's *Verbal Behaviour*.[186] In Slezak's words, "Bloor is exactly forty years and a major scientific revolution too late."[187] His aversion to any psychological explanation of knowledge resulted in a failure to come to terms with modern cognitive psychology despite its direct bearing on his claims. While Bloor, when pressed, avowed "a multifactorial view of the causes of belief," including biology and psychology, in fact, Robert Nola charged, "only lip service is paid." That is, Bloor always operates on the view that "beliefs, and reasons, are the epiphenomenal froth on top of the flux of social and other causes and do not themselves enter, as active causal ingredients, in the mix of causes."[188]

Clearly, to offer a sociological account of mathematics or formal logic represented a dauntingly difficult program. But that was what Bloor sought to undertake.[189] He meant to "explain the logical necessity of a step in reasoning or why a proof is in fact a proof" in terms of social causation.[190] Of course it might seem absurd to ask after a causal account of the belief that twice two is four, Bloor averred. "The explanation of the belief is that it is true," seemed the only sensible response. But he insisted that this judgment relied upon an unexamined metaphysical hypostasis: "Mathematics and logic are seen as being about a body of truths which exist in their own right independently of whether anyone believes them or knows about them."[191] To break out of this debilitating stance, Bloor invoked Wittgenstein's *Remarks on the Foundations of Mathematics*, which, in his view, "offers what might be taken to be a forceful refutation of Realism."[192] Wittgenstein created a paradox about the obviousness of rule-following involved in mathematical logic, namely, that one must always ask for the rule that makes any given rule appropriate for the case at hand, precipitating a fatal regress. Bloor commented, "The trouble with Realism does not lie in the puzzling nature of its ontology but in the circular character of its epistemology."[193] He proposed that Wittgenstein offered what was in fact a sociological resolution. "Wittgenstein does not deny that logic compels. What he offers is an explanation of the content of that compulsion."[194] That is social practice, not metaphysical transcendence. As he wrote elsewhere, "the force of reason—where it is not animal instinct—is the force of society mislocated and mystified."[195] Mathematics must be seen as "invention rather than discovery," as a social institution, though as an institution "not subject to individual whim."[196] There could be a "sociological account" of "objectivity." Thus, "the sociologist is no longer excluded *a priori* from dealing with mathematical activity itself." For Bloor, "the great insight of the *Remarks* is that it treats the grip that logic has upon us as a fact to be explained rather than the revelation of a truth to be justified."[197] Accordingly, the bulk of Bloor's study concentrates on the social explication of the history of mathematics and logic, seeking to establish "that logical necessity is a species of moral obligation," a social imperative, not a transcendent one.[198]

That position has been challenged not only as a misrepresentation of Wittgenstein's stance, but as incoherent in its own right.[199] Michael Friedman contends that the problem in both senses lies in the "significant tension" between the aspirations of the Strong Program to generate empirical sociological explanations and its aspirations to "replace the philosophical tradition."[200] Friedman suggests that the Strong Program might well have its key thesis of symmetry if it were willing to abandon its quarrel with traditional philosophy as irrelevant. But it would not abandon this quarrel, because it believed that the history of science (and philosophy) had *refuted* the postures of traditional philosophy, of epistemological foundationalism, and made relativism the *only* possible cognitive position.[201] To negotiate this new terrain, only a social vantage could provide a naturalistic—in other words, a contingent and fallible—orientation. But Friedman argues that the Strong Program "acted precipitously in actively seeking to embrace and to advocate one particular side in a traditional philosophical debate."[202] Friedman's own reconstruction of the history of recent philosophy confirms that it has indeed come to the realization that "there is *no* over-arching notion of universal validity or 'correctness' independent of the particular, and diverse, rules of the particular, equally possible and legitimate, formally specifiable calculi."[203] That is, the "process of mutual scientific and philosophical evolution has, in the end, resulted in a thoroughly relativized and conventional conception of scientific rationality," what Rudolf Carnap called the Principle of Tolerance or the distinction of internal from external questions of language.[204] Friedman concludes: "Carnap plus Kuhn equals the philosophical agenda of SSK."[205]

All this would seem to *affirm* Bloor's agenda, but Friedman now makes his key move. In making *Wittgenstein* their "main philosophical hero," the proponents of the Strong Program missed something essential, both about him and about the situation of philosophy.[206] Wittgenstein, according to Friedman, insisted upon the autonomy of philosophy, an autonomy purchased precisely by renouncing once and for all the effort to be foundational. Philosophy at one and the same time acknowledged that "language in the end rests simply on the indefinite multiplicity of human forms of life," that "all signs, in the end, only unfold their meanings in and through their concrete applications at the hands of actual human beings," but that nonetheless human beings always already find themselves *within* a language.[207] As participants, "the question of the 'hardness of the logical *must*'" is a matter of first-person, normative responsibility.[208] Just this "non-empirical, essentially normative element in our thinking [of which] logic and mathematics continue to be paradigmatic" is the "province of a peculiarly philosophical investigation," according to Wittgenstein, *that cannot be reduced to external-sociological explanation or to transcendental-foundational justification.*[209] Friedman claims both that "there is no trace of socio-cultural relativism in Wittgenstein's own philosophizing" and that his *philosophical* project "is to wean us once and for all from the need for [foundational] justification."[210] In seeking to provide external, sociological explanations of the force of necessity in logic, or the causal reality of reasons, Bloor misunderstands philosophy *and*, more centrally, how normative rationality features in actual, contingent knowledge, both ordinary and scientific. Explaining normativity from the outside is failing to come to terms with its essential dignity and authority, which can only be experienced in the first person. Here is a critical sense of the phrase "there is no view from nowhere." Any view from somewhere is a view which acknowledges the necessity of the norms of that place. *Within* a language game, the rules *bind*. When Bloor claims that logic is a matter of *moral* necessity, he thinks that entitles him to step *outside* and *explain* the game itself on external (social) grounds. But that very move makes *moral* necessity as mysterious as the logical necessity Bloor takes it to ground. Another way to put this is to contend that insight into this experience of necessity requires a holistic human perspective, and the question of the role of specifically *social* causes in that perspective *needs independent warrant.*

Barnes took to heart the point that his Edinburgh colleague Steven Shapin stated so decisively: "The mere assertion that scientific knowledge 'has to do' with the social order or that it is 'not autonomous' is no longer interesting. We must now specify how, precisely, to treat scientific culture as social product."[211] The problem was how to establish a causal linkage. As Peter Slezak puts it, "the question of how . . . social contexts might conceivably act as causes of scientific or other beliefs has not even been addressed and is thoroughly obscure."[212] Barnes sought to take up this challenge. To succeed, he reflected, the interpreter "must make action intelligible through detailed and extensive insight into the nature of actors' perspectives, their categories and typifications, the assumptions which mediate their responses, the models which organize their cognition, the rules they normally follow. Then he must construct theories which treat this material systematically."[213] Barnes endeavored a hypothesis: "*all* knowledge, 'scientific,' 'hermeneutic' or otherwise, is primarily produced and evaluated in terms of an interest in prediction and control."[214] In his second monograph, *Interests and the Growth of Knowledge* (1977), the decisive theoretical resource upon which he drew was Jürgen Habermas's *Knowledge and Human Interests*.[215] While Barnes found Habermas too obsessed with the positivist model of science to grasp actual scientific practice, he took Habermas's formulation of theoretical interests together with Habermas's model of hermeneutic thinking to form the basis of a research agenda for sociology of scientific knowledge: "Knowledge grows under the impulse of two great interests, one overt interest in prediction, manipulation and control, and a covert interest in rationalization and persuasion."[216]

"Knowledge is not produced by passively perceiving individuals, but by interacting social groups engaged in particular activities," and "representations are actively manufactured renderings of their referents, produced from available cultural resources."[217] Barnes stressed that this interest could only work with the cultural resources at the disposal of the particular community. "Judgments were made with *particular* sought-after kinds of prediction and extensions of technique being taken into consideration. Distinct specific predictive and manipulative goals prestructured judgment and evaluation."[218] At the same time, the specificity of the claim got watered down: "The general point is not that the goal-oriented character of scientific judgment implies its relation to any particular contingency, or to external factors, or political interests; what is implied is that any such contingency *may* have a bearing on judgment and that contingent social factors of some kind *must* have. What these factors are is always a matter of concrete empirical investigation."[219] Barnes was candid enough to admit that grounding ideas in social structure had had "only very limited success."[220] "There simply is not, at the present time, any explicit, objective set of rules or procedures by which the influence of concealed interests upon thought and belief can be established."[221] As Bloor noted (conceded?) in his debate with Laudan, "The question of the kind or scope of the social factors at work in a system of knowledge is entirely contingent and can only be established by empirical study." Still, he insisted that "the 'interest' model has been shown to work convincingly and in detail in a large number of cases."[222] A decade later, in the afterword to the second edition of *Knowledge and Social Imagery* (1991), Bloor conceded that "the terminology of interest explanations is intuitive, and much about them awaits clarification."[223] That was—or should have been—a devastating admission.

Increasingly, the Strong Program appealed to a growing body of empirical historical work to corroborate its theoretical position. Bloor wrapped himself in case studies in his debate with Laudan. Later, he took the same tack: "There are numerous studies of how the divergent influences that scientists draw from the same body of data are explicable by reference to identifiable and divergent interests."[224] Barnes and Shapin maintained in their anthology of case studies, *Natural Order,* that "The best way to establish the possibility of doing something is to do it."[225] Shapin contended that historical case-studies substantiated a sociological approach to science. "One can either debate the possibility of the sociology of scientific knowledge or one can do it."[226] Actuality is a strong proof of possibility. But the question is: what exactly had these historical case-studies *done*? Empirical case-studies needed to establish the *codetermination* of interests, opening up inquiry into external, not just internal influences in the growth and change of science. As Karin Knorr-Cetina put it, "What is needed . . . is . . . a *genetic* model which demonstrates exactly *how* and in virtue of which *mechanisms* societal factors have indeed entered (and are hence reflected in) particular knowledge claims."[227] But critics found the case studies wanting precisely in providing the rigorous causal accounts the Strong Program promised. Paul Roth has attacked the causal warrant of case studies in history and sociology of science in a particularly vigorous manner.[228]

An internally mounted critique of "interest" explanation within SSK proliferated. Steven Yearley observed: "[M]any different empirically plausible interests can be ascribed to protagonists and their beliefs can be related to these interests in equally numerous ways. . . . [I]nterest theory provides no algorithm for deciding between them." The lack of rigor in accounting vitiated the whole program. "The search for a definitive sociological explanation of belief is frustrated by the essential flexibility of interest imputation."[229] Moreover, case studies could not exclude the possibility that "social circumstance merely accompanies belief rather than causes it," Robert Nola charged.[230] Peter Slezak developed this criticism extensively. "Sociologists of knowledge have avoided confronting such issues by their stratagem of developing elaborately descriptive accounts of the social and cultural circumstances surrounding various incidents in the history of science. However, the *causal* role of such circumstances is not thereby established. . . . Demonstrating the usefulness of some theory as a vehicle for promoting social interests, or even its role in determining patterns of acceptance and rejection, does not *ipso facto* establish these factors as causes of the theory's contents."[231]

Throughout, the Strong Program remained highly theoretical—even philosophical—in its articulation.[232] No sufficiently compelling account was offered for why *social* causes should have a paramount place.[233] Instead, the Strong Program appealed "in principle" to Quine's thesis on the underdetermination of theory by data, the Duhem-Quine thesis (semantic holism), and the theory-ladenness of data—in short, the whole panoply of post-positivist philosophy of science.[234] Bloor and Barnes believed that the Duhem-Quine thesis allowed communities to resist anomaly indefinitely and that the underdetermination thesis warranted the view that alternative theories adequate to all (possible) factual evidence were always available.[235] On these grounds they concluded that *only* social factors could account for the adherence to any given theory.[236] Yet, as we have seen, it is not a sound construction of the Duhem-Quine thesis (if, indeed, that is *one* thesis) to claim that anomalies can be resisted indefinitely.[237] Slezak questions, too, the use to which the Duhem-Quine thesis is taken as a warrant for such moves. "[E]ven if theories are radically unconstrained by cognitive factors beyond the 'raw' observational evidence, it by no means follows that it must be *social* factors which are the decisive ones in fixing their content."[238] As Jan Golinski writes, "the much-touted 'Duhem-Quine thesis' cuts both ways: phenomena are indeed interpreted in relation to a network of assumptions and beliefs; but it is precisely by becoming entrenched within those networks that they are experienced as passive constraints upon the production of subsequent knowledge."[239] Moreover, to stress infinite flexibility in the face of anomaly introduced significant tension into the Strong Program's reception of Kuhn, for whom mounting anomalies were the force for scientific revolutions. The Strong Program tended to minimize Kuhn's theory of scientific revolution and concentrate on his theory of normal science.[240]

Their view of the underdetermination thesis was also extreme.[241] "Barnes and Bloor tend to regard the existence of genuinely underdetermined theories as not only unproblematic but also readily available."[242] It is not the case—especially as a matter of empirical fact, to which they ought to be particularly responsive—that alternate theories are always available, or even, crucially, that there are not sufficient "correspondence" and "coherence" conditions, in the terminology of Mary Hesse, to discriminate one theory as superior to all its rivals on rational grounds.[243] Moreover, even if, counterfactually, the underdetermination thesis could be upheld, the project of a sociological theory of causation is still far from fulfilled.[244] In fact, the empirical research program of the sociology of scientific knowledge got freighted with an unbearable load of philosophical ambitions.[245] As Slezak writes, "stating the formal constraints on theory discovery is not the same as describing the processes actually involved."[246] After one has considered all the epistemological claims of the Strong Program, one is still entitled to raise a whole series of *methodological* queries about its empirical practice.[247]

As Roth puts it, "whatever 'gaps' are evident in orthodox explanations of scientific behavior, the strong programmers have been far too sanguine in their conclusion that only sociology remains to fill the breach."[248] The crucial issue is *efficacy* in explanation. Roth accused Bloor and SSK generally of "setting a standard for philosophy of science to meet which is so high that failure is inevitable, using this failure to license the need for (and hence the legitimacy of) sociological explanation, then forgetting all about the unrealistically high standard when it comes to *this* brand of explanation."[249] Bloor claimed to apply "the same causal idiom as any other scientist," namely, observation of regularities leading to "general laws"—in his particular case, "relating beliefs to conditions which are necessary and sufficient to determine them."[250] Roth queried: "What is the notion of causation assumed by the strong programmers? Basically, it is Humean concomitant coincidence."[251] But then, Roth pursued, "How is the move made from temporally associating to causally relating certain social movements or changes with concomitant changes in scientific outlook?"[252] His claim was that if the Strong Program could not produce a definitive answer here, it could not pretend to have displaced the rationalist account of philosophy of science. And, he maintained, "the causality principle of the strong programme . . . is unacceptably vague."[253] As far as Roth was concerned, "No argument yet made by advocates of the strong programme establishes that they have any monopoly on a general causal explanation of social beliefs." Furthermore, "no clear *mechanism* of causal relations is established."[254] That is, "the problem is a question of the *causal* mechanism at work; it is a problem of knowing where and at what to look in order to discern the engine driving scientific and social changes." Specifically, "Barnes's theoretical analysis gives no account of the interaction of interests, routines, and concept application."[255] Given "the complete lack of specification of a mechanism for belief formation and change," Roth argued, "we do not know what either the necessary or the sufficient conditions of such formation and change are."[256] The Strong Program offered "a completely unverifiable condition" as a "necessary condition."[257] Roth complained that "we find nothing approaching a proof."[258] "The use of terms such as 'necessary,' 'sufficient,' 'dependent upon,' and 'possibility' is merely rhetorical ruse. In context, the terms have no meaning; indeed, in the absence of the requisite theory of belief formation and change, they can have none."[259]

One might respond to Roth, however, that the standards of rigor he was applying to the causal claims of the sociology of knowledge would obviate virtually *any* empirical, contingent results of inquiry, and consequently that he, too, was holding up an impossibly strong standard, in the same manner for which he condemned the Strong Program. If all that Roth intended was to dispute the hyperbolic claim that the Strong Program had superseded philosophy of science by offering definitive causal accounts, he surely made his case. If he wanted to dispute the tacit *epistemology* of Bloor and SSK, likewise. If, on the other hand, what he intended was to dispute the very possibility of empirical causal accounts in the human sciences, he risked becoming yet another practitioner of hyperbole. An empirical human science such as history cannot be preemptively disqualified by an a priori argument of the sort that Roth appeared to be making. As Paul Thagard put it, undramatically but aptly: "All that needs to be abandoned is the hubris that claimed that social factors can do it all."[260]

To be sure, "the vulnerability of the strong programme derives from its extreme imperialism."[261] But it is also a result of rash appropriations from the discipline it set out to supersede. The Strong Program represents the first instance of an indiscriminate appropriation into science studies of philosophical dogmas in extreme and indefensible forms. It was, in fact, the uncritical uptake of Quine and Kuhn by the Strong Program which licensed ever more dubious postures in the "social constructionist" science studies agendas that followed. What was a *questionable* inference in the Strong Program became *unquestioned* convention for the generation that followed.[262] It is precisely the burden of the previous chapters of this study that such cavalier appropriation of the "dogmas" of post-positivism could have nefarious theoretical consequences.

The natural world in no way constrains what is believed to be.
— HARRY COLLINS, "SON OF SEVEN SEXES: THE SOCIAL DESTRUCTION
OF A PHYSICAL PHENOMENON"

As we come to recognize the conventional and artifactual status of our forms of knowing, we put ourselves in a position to realize that it is ourselves and not reality that is responsible for what we know.
— STEVEN SHAPIN AND SIMON SCHAFFER, *LEVIATHAN AND THE AIR PUMP: HOBBES, BOYLE, AND THE EXPERIMENTAL LIFE*

All the Way Down: Social Constructivism and the Turn to Microsociological Studies

Revisionism away from the original orientation of the Edinburgh school (the Strong Program) followed two distinct paths, laboratory studies and controversy studies. The pathbreaking work in laboratory studies was Bruno Latour and Steven Woolgar, *Laboratory Life* (1979).[1] Other major achievements include Karin Knorr-Cetina's *The Manufacture of Knowledge* (1981) and Michael Lynch's *Art and Artifact in Laboratory Science* (1985).[2] Bath sociologist Harry Collins was the pioneer in controversy studies. He consolidated and revised his many seminal papers into a book, *Changing Order: Replication and Induction in Scientific Practice*, in 1985.[3] Along with Collins, major work in this vein was achieved by Trevor Pinch and Andrew Pickering.[4]

Both paths shared a heightened empirical concreteness over against the theoretical generalities of the Strong Program. There was an insistent shift from the macrolevel (society and science as wholes) to the microlevel (a laboratory site or a "core-set" of disciplinary specialists) and an even more adamant insistence upon considering what scientists *do* (practice), as contrasted with what they *say* (discourse or, in a more traditional vein, theory). The distinctive move in recent social study of science has been the shift from conceiving of science as *knowledge* to conceiving of science as *practice*.[5] That is, "the precise meaning of a theory, or programmatic statement, is not contained in the statement itself, but in its exemplification in practice."[6] In particular, these new social studies of science recognized and documented the vast inculcation of "tacit knowledge" which constitutes actual membership in a scientific practice.[7] Finally, the theoretical hallmark of this shift in the social study of science was the radicalization of the relativism of the original Strong Program toward a *social constructivism* "all the way down."[8]

The most influential of all the laboratory studies was clearly that of Latour and Woolgar.[9] This joint publication made each of the authors a major figure in science studies and their divergent paths after this collaboration define key features of the subsequent history of the field. *Laboratory Life: The Social Construction of Scientific Facts* contained in its subtitle, as Ian Hacking noted, "a manifesto in its own right."[10] Besides this radical constructivism, of which more will be said shortly, what was noteworthy was the *ethnographic* approach. As Latour observed retrospectively, "putting to use the methods of anthropology in order to understand *our* sciences is only recent." He stressed that he and Woolgar used "the most outdated version of anthropology: the *outside* observer who does not know the language and the customs of the natives, who stays for a long time in one place and tries to make sense of what they do and think by using a metalanguage."[11] That is, the ethnography of laboratory studies "treats scientists as strangers, yet observes them in the midst of laboratory activity." This status of participant-observer preserved "simultaneous insiderness and outsiderness."[12] What was crucial, Latour insisted, was the agnostic posture of the observer.[13] Latour and Woolgar deliberately sought "anthropological strangeness," something like Bertolt Brecht's *Verfremdungseffekt,* an "attempt to make the activities of the laboratory seem as strange as possible in order not to take too much for granted."[14] Latour made the essential methodological argument for this practice: "No account qualifies as an explanation *if it simply restates the account it is supposed to explain*" (original emphasis).[15] Hence they resorted to systematic redescription (in a "metalanguage").

The structure of *Laboratory Life* is organized around what Latour calls the "three main elements" of "what science is"—*inscription devices,* a *body of scripture,* and an *agonistic field.* "Through the use of inscription devices the scientist might be able to modify the status of an assertion inside the body of scriptures (its modality), if he is able to win in an agonistic encounter."[16] To unpack the idiosyncratic terminology ("metalanguage") of this assertion is to grasp the thesis of the work. Latour and Woolgar wanted to know "what motivates scientists?"[17] The idea they developed is a variant of one worked out by Pierre Bourdieu. Bourdieu wrote of "credit" as "symbolic capital" which scientists could wield in the accrual of influence over the allocation of future resources in the scientific field.[18] Latour and Woolgar modified this slightly into "credibility," which had to do with the acceptance and use of the statements scientists entered into the discourse of their field.[19] This "credibility" also served as a form of symbolic capital, but Latour and Woolgar focused more on the discursive transformations of statements to achieve credibility than upon the outcomes.[20] Scientists lived in a world of "literary inscriptions." This was not only their most important product, in the form of scientific papers, but it was the structure of their material site with all its apparatus: "the laboratory began to take on the appearance of a system of literary inscriptions."[21] Latour and Woolgar characterized the laboratory apparatus as "inscription devices." In their terms, "an inscription device is any item of apparatus or particular configuration of such items which can transform a material substance into a figure or diagram which is directly usable."[22] This terminology was explicitly lifted from Derrida, and it partook of what Hacking terms "the now-outdated fascination with the sentence so characteristic of Paris intellectuals in the late sixties."[23] If the laboratory was redescribed into devices to "reduce (translate) complex processes to features that can be represented graphically in two dimensions," the crucial feature is that these became immediately "usable" in the generation and especially the modulation of scientific statements.[24] Latour and Woolgar conceived of five different forms of statement, which varied by "modality." The first type represented mere speculation. The second alluded to supporting evidence. The third advanced to the premise that the claim was "generally assumed." The fourth was a simple, unqualified assertion. The fifth was "taken for granted" or tacit: something "everybody knows."[25] For Latour and Woolgar, the ambition of scientists was to generate type-four statements and have them accepted in the wider public.[26] "The objective was to persuade colleagues that they should drop all modalities used in relation to a particular assertion and that they should accept and borrow this assertion as an established matter of fact, preferably by citing the paper."[27] This persuasive endeavor occurred in an "agonistic field," because other scientists were always putting forth rival claims for the attention of the research community. Not only is this "metalanguage" Latour and Woolgar developed *rhetorical* in form, it is also *martial*.[28] The *agonistic field* is "a political field of contention"; it is all about winning and losing (credibility).

Whereas in their final products, scientists sought to represent their conclusions as necessitated by experimental evidence, Latour and Woolgar insisted that in the microcontext of the laboratory, the phenomena were literally constructed: "It is not simply that phenomena *depend on* certain material instrumentation; rather, the phenomena *are thoroughly constituted* by the material setting of the laboratory."[29] They explain: "We do not wish to say that facts do not exist, nor that there is no such thing as reality. . . . Our point is that 'out-there-ness' is the *consequence* of scientific work rather than its *cause*."[30] This was the linchpin of the *constructivism* elaborated in *Laboratory Life*. Ian Hacking makes clear the radical form of the constructivism advocated here by distinguishing between the "antirealism" of what can be called the instrumental tradition in analytic philosophy, most recently exemplified by Bas van Fraassen, and a radical "irrealism" best represented by Nelson Goodman. Irrealism is the position that the "world" is entirely *made,* in other words, the restriction of all claims to a scheme or frame-work.[31] "Latour and Woolgar fit together with Goodman quite well."[32] In *Science in Action* (1987), perhaps Latour's most influential work, he offered a methodological formulation of this view of ontology and epistemology. As his "first rule of method" he proposed: "We will enter facts and machines while they are in the making; we will carry with us no preconceptions of what constitutes knowledge; we will watch the closure of the black boxes and be careful to distinguish between two contradictory explanations of this closure, one uttered when it is finished, the other while it is being attempted."[33]

If Latour and Woolgar scotched any conventional "realism," the attitude toward *rationality* was more subtle. To be sure, Latour dismissed any universal normative rationality, but he insisted that it was indeed "possible to reconstruct the research process" if one adopted a resolutely local or micro focus, if the endeavor at historical reconstruction was *contextual,* recognizing that "a statement draws its meaning from where, when, and by whom it is uttered."[34] Of course, "when you get closer to the research process, the multiplicity and the chaos increase."[35] In that context, the research process appears "opportunistic": "There are rules—borrowed from previous experience—but they are followed *or not* according to the circumstances. . . . On the other hand, it is wrong to think like Feyerabend that 'anything goes.' "[36] It "cannot be described as a chaos of random moves and lucky guesses," for "the process is not without reason, or more exactly, it is not without heterogeneous, short lived, circumstantial reasons."[37] But accordingly the research process "makes sense only if one looks at the local place of work."[38] While this might appear a dismal outcome to a philosopher of science seeking an *algorithm,* it is heartening indeed to a historian of science seeking an *account.* "If logic was taken out of the laudative meaning that it has since Aristotle and was understood as logos or path, then, we could say that the research process is to build paths or, to use another source of metaphor, to tell plausible stories."[39] That scientific papers reorder—usually quite radically—the path of discovery in their path of exposition—that they become, in Latour's phrase, "science fiction"—should have nothing scandalous about it.[40] Indeed, Latour insists that we "realize that there is in fact only one large literary genre: that of science fiction (the best part of which is *not* written by science fiction writers)."[41] "We constantly make sense of the world and build paths leading points to one another and convince people that a particular path is more straightforward than any other."[42]

As Knorr-Cetina observed, "an assertion which transforms into a 'fact' aligns other statements behind it, and appears to be used to reorganize scientific arguments."[43] That is, what is involved is a dynamic process of revisions and cumulations. "Previous scientific selections thus change the conditions of further selections."[44] Latour would come to call these "translations."[45] The point is made clearly by Knorr-Cetina: "a result in the literature is usually not just simply taken over by another scientist, but rather it will be taken over in order to be converted into a new result of the scientist making the transfer, and it will, in the process, itself undergo transformations and reconstructions."[46] The key idea is that in the pursuit of "credibility," scientists are engaged not only in a cognitive but in a social endeavor, what Latour and Michel Callon coined as "socio-logics."[47] Latour writes of "enrolling allies," but also of the important problem of keeping them tied to the original agenda, rather than simply transforming the ideas for their own projects.[48]

Knorr-Cetina was herself a decisive pacesetter in conceptualizing constructivism in science studies in the years 1977–1983.[49] She was among the earliest to find the Strong Program and especially the "interests model" inadequate.[50] The "macroscopic congruency claims" involved in the interests model, she averred, "do not specify the causally connected chain of events out of which an object of knowledge emerges congruent with antecedent social interests."[51] One key reason for this failure was that no transition from the social to the individual (from interest to belief) had been developed. To get the requisite concreteness, Knorr-Cetina argued, science studies needed to "adopt a *genetic* and *microlevel* approach" (original emphasis).[52] One must "specify social phenomena on a micro-level, and derive concepts of 'social structure' from the analysis of a multitude of micro-events."[53] In her own theoretical manifesto, Knorr-Cetina contrasted constructivism with descriptive notions of scientific investigation: "The constructivist interpretation considers the products of science as first and foremost the result of a process of (reflexive) fabrication."[54] She insisted, "one must move inside the epistemic space within which scientists work and identify the tools and devices which they use."[55] "In the laboratory, the 'texts' are provided by constantly accumulated combinations of measurement traces (graphs, figures, printouts, diagrams, tables, etc.). Objectification through traces in no way eliminates interpretation: such traces must first be recognized; moreover they can be ignored, selected, recombined or simply forgotten in the light of further results or the contingencies of another situation."[56] Her point was that "choices exist *throughout* the process of experimentation," resulting in openness and overdetermination of outcomes.[57] The laboratory site was "indexical and contextually contingent," there prevailed "an opportunistic logic which stresses the time-and-space-bound conditions of scientific work and the scientist's active role in reflexively organizing these conditions in terms of their potential resourcefulness."[58] That is, "what we have . . . are the practices of ongoing research within local research scenes in which certain selections get reproduced, and through reproduction are expanded, reinforced, and perhaps petrified to become the solid rock of what counts as true."[59]

She insisted that laboratories were about not contemplation but intervention, and that there was nothing *natural* in the laboratory, for everything got *constructed*.[60] In the laboratory "objects are subject to tens, and often hundreds, of separately attended to *interferences* with their 'natural' makeup, and so are the natural sequences of events in which objects take part. . . . Consequently the conclusions derived from such experiments are not justified in terms of the equivalence of the experiments to real-world processes."[61] Above all, experimental *data* should not be confused with natural phenomena. Thus Knorr-Cetina brought vividly home the old Baconian image of putting nature to the rack, extorting "truth" from her.[62] But Knorr-Cetina went further: laboratories were not about "finding the truth." They were about "making things work."[63] Opportunism was the key, "a process guided by success more than truth."[64] Laboratories were production sites, involved in the "process of fabrication." This "not only involves tools and materials which are highly preconstructed, [it] also involve[s] decisions and selections."[65] Thus, "the focus upon laboratories has allowed us to consider experimental activity within the wider context of equipment and symbolic practices within which the conduct of science is located."[66] Moreover, "it is perhaps the single most consistent result of laboratory studies to point to the indeterminacy inherent in scientific operations, and to demonstrate the *locally situated, occasioned* character of laboratory selections."[67] Theory, from the perspective of the laboratory, becomes "discursively crystallized experimental operations, [which] are in turn woven into a process of performing experimentation."[68] And the *products* of science "are contextually specific constructions which bear the mark of the situated contingency and interest structure of the process by which they are generated."[69] That is, the *path* matters. "Processes of fabrication involve chains of decisions and negotiations through which their outcomes are derived."[70] Moreover, "the products of science are not only decision-impregnated, they are also decision-impregnating, in the sense that they point to new problems and predispose their solutions."[71] For the interpreter, the *temporality* of innovation is decisive: "The occurrence of an 'innovative idea' is a hit-and-strike chance-event if one tries to predict it, and a step in a logical sequence of events if one reconstructs a research program after the fact."[72] This historicity is problematic for conventional philosophy of science, but not for a Hegelianism which remembers the great admonition about the "owl of Minerva."

There are crucial implications of the situated innovation for the question of acceptance, as well. "The standards against which the ideas of the laboratory are measured do not refer to the world of theoretical interpretation. Instead they refer us to a world of instrumentation and collaboration, of chances of publication and of the investments at stake."[73] That is, the way in which acceptance emerges in scientific practice is not through some grand ballot of all participants on what is "true," but rather the selective incorporation of results in "the ongoing process of research production."[74] That is, "selections realized in previous scientific work become both topic and resource for further scientific investigations."[75] In real scientific practice, "scientists appear to be actively engaged in building, solidifying and expanding resource-relationships."[76] The local site is part of a wider network, and the crucial question of dissemination of results is always nested in the question of sustaining resources for continuing work. It is in the "transepistemic arenas" of publication and funding that Knorr-Cetina finds the "locus in which the establishment, definition, renewal or expansion of resource-relationships is effectively negotiated."[77] Yet these "transepistemic negotiations" have a *rhetorical* component in the scientists' retrospective redescription of the very process that the microsociologist unearths: scientific writing is an effort to *efface the traces* of local construction in order to enhance the acceptance and dissemination of the product. As Bertolt Brecht would have put it, scientists follow the admonition, "cover your tracks."

Knorr-Cetina's initial formulations of the constructivist program for science studies were quite close to the instrumentalism of Bas van Fraassen's "constructive empiricism," but she swiftly moved away toward a more radical posture.[78] "Strong" constructivism, she urged, disdained accommodation with philosophy of science: "The constructivist thesis in the original laboratory studies . . . shifted the question from . . . realist, instrumentalist and such-like doctrines, to an enquiry into the constructive process of world making."[79] The resonances, here, of a departure from the antirealism of van Fraassen to the irrealism of Goodman's "world making," as Hacking noted in Latour and Woolgar, should not be missed. Knorr-Cetina offered a careful defense of the epistemic relativism implicit in her version of constructivism: "Epistemic relativism is not committed to the idea that there is no material world, or that all knowledge claims are equally good or bad. . . . It is only committed to the idea that what we make of physical resistances and of meter signals is itself grounded in human assumptions and selections . . . specific to a particular historical place and time. This neither precludes development over time, nor does it require one to subscribe to conceptions of incommensurable world views or coherent paradigms."[80] But there *were* strong ontological implications:

> Science changes its view about the character of natural objects, so the precise sense in which these objects are supposed to have pre-existed as scientifically delimited objects *independent* from us is not clear. . . .
> . . . From a sociological perspective, . . . it is precisely the reformulation of ontologies as a consequence of science that brings into focus the instru-

mental, symbolic and political work required to refurnish the world with new, scientifically derived objects. . . .

. . . While the existence of the world as a material, physical entity independent of us may be granted on principle, the existence of specific objects identified in terms of their character cannot so be granted . . . specific scientific entities like subatomic particles begin to "pre-exist" precisely when science has made up its mind about them and succeeds in bringing them forth in the laboratory.[81]

This ontological radicalism finds its counterpart in the ideas of Latour, as well.[82]

Harry Collins, like Knorr-Cetina, was intensely interested in the practice of scientists. He chose to focus on controversies, at least as a "first stage" of his approach, which he termed the "empirical programme of relativism."[83] He sought to demonstrate the insufficiency of experiment to settle by itself the question of acceptance. As Collins put it, "the essentially cultural nature of the local boundaries of scientific legitimacy." In other words, "the potential local interpretative flexibility of science," could be empirically shown to "prevent experimentation, by itself, from being decisive" in scientific controversies.[84] By examining the problem of "replication," Collins showed that all aspects of any particular experimental result could and did get questioned (he termed this "experimental regress"). As a matter of fact, "most experimental work is never replicated, or if it is replicated it will be done 'by the way' as part of some larger project."[85] It was only in the context of controversies that replication became a recourse, but it was hardly conclusive. Hence, controversy exposed the need for a decisive supplement which carried scientists beyond the impasse of contradictory findings. This, according to Collins, could only be a matter of social, not experimental, determination. "The physics and politics of experimentation are not separable." In settling controversies, "accumulation of experimental results" was not "decisive," instead "a variety of political and rhetorical strategies were mobilized." This was what authorized a sociological approach, "reference to rhetoric, authority, institutional positions, and everything that comes under the 'catch-all' terms *interests* and *power*."[86]

The "core-set" of scientists involved in the research specialty established what would be taken to be valid through an intense process of "negotiations."[87] Controversies created core-sets: "they are transient hot-spots in science," but they "do not tend to die out clearly."[88] Opposition got silenced, but not eliminated. "Members of the core-set . . . are aware of the possible redescription of their work, aware of the socially mediated nature of the closure of debate, and aware of the potential for reopening it."[89] It is essential to recognize that Collins believed "Scientific conclusions are as solid and reliable as anything can be, and the methods by which they are reached are perfectly proper."[90] It is simply a fact that "the *correct* scientific method for work within a controversial area cannot be known in many of its important respects prior to the outcome of the controversy."[91]

Most notoriously Collins asserted that "the natural world in no way constrains what is believed to be."[92] In another essay, Collins put it this way: "[T]he natural world has a small or non-existent role in the construction of scientific knowledge. Relativist or not, the new philosophy leaves room for historical and sociological analysis of the processes which lead to the acceptance, or otherwise, of new scientific knowledge."[93] There is a strategy behind this dogma. "What we need is radical uncertainty about how things about nature are known." That is why "the natural world must be treated as though it did not affect our perception of it." Thus, "though this point can be argued as an epistemological principle the important thing is to adopt it as a methodological imperative."[94] Collins was confident that "the models of rationality developed by philosophers do not seem generally useful in helping scientists to decide how to act scientifically."[95] More specifically, "Recent philosophies, coming to terms with Duhem, Quine and Lakatos, admit to the potential revisability of theoretical and even observational claims. The same message can be read from historical studies such as Kuhn's."[96] Crucial to his conception, as both Collins and his commentators have noted, was his construal of the Duhem-Quine thesis. "One of the more decisive results for earlier sociology of science . . . confirms the Duhem-Quine-Hesse philosophical view . . . that science does not have a set of methodological techniques that can quickly or decisively prove or disprove the existence of natural phenomena."[97] Yet even so sympathetic a commentator as Mary Hesse recognizes that Collins "has missed the point of the Duhemian conception of the holism of theory. The point is not that *all* individual replications can be reinterpreted or constructed at will and independently of each other, but that *some* can be while being constrained by others, and by the coherence of the whole in the theoretical network."[98]

Despite harsh criticism, Collins reasserted, "I do advise sociologists of science to act on the assumption that 'the natural world in no way constrains what is believed to be.' I believe it is good advice whatever one's epistemological preference."[99] Collins was using rhetorically forceful language to make an essentially *methodological*, rather than epistemological, point.[100] "The most important element in the relativist program is methodological rather than philosophical. Much is to be gained by taking up the relativist perspective as a heuristic, even if not as a constitutive rule. It is a question of the integrity of one's approach to the data of science and the vigour with which sociological analysis should be pursued."[101] That is, one should "push the relativistic heuristic as far as possible: where it can go no further, 'nature' intrudes."[102] As his commentators, Yves Gingras and Sam Schweber, astutely observe, "if one is seeking to investigate the *process* of 'construction' of scientific knowledge, one cannot seriously invoke the outcome of that process to *explain* the process itself without going into a circular argument."[103]

Offering an adequate conceptualization of scientific *change* constituted the explicit ambition of Collins and others in his vein of research. Collins recognized that "whether a change comes about is a consequence of the way that attempted innovations are treated by the larger scientific community," but the exact framework for that larger contextualization remained vague in his analysis.[104] How "closure" got achieved in scientific practice, as empirically there is no question that it does, Collins called the "second stage" for his "empirical relativist programme."[105] Latour and Callon questioned whether his program in fact ever got there: "Extremely good at showing the opening of controversies, the indefinite negotiability of facts, the skill necessary to transport any matter of facts, the infinite regress of underdetermination, Collins has nothing to say about the closing of controversies, the non-negotiability of facts, and the slow routinization that redistributes skills."[106]

In his review of Collins's book, Andrew Pickering applauded Collins for his distinction between an "algorithmic" and an "enculturational" model of scientific process and his insistence that only the latter made empirical sense. Pickering saw as Collins's key claim that " 'closure'—the production of consensual fact—is an achievement within a form of life, not something dictated from nature."[107] Collins offered a "detailed demonstration" that "uncertainty is endemic to experiment—a problem requiring continuous practical management."[108] Yet Pickering expressed reservations about the two principles of Collins's case studies: inertia and contingency. Because Collins presumed that scientists were entirely conservative and indisposed to change, he ascribed everything to contingency. "Contingency, for him, is ineffably particularistic and immune to general forms of explanation."[109] But to ascribe all change to such random eruptions was to "make history of science a singularly pointless exercise."[110] Pickering argued that "the inertial assumption is not just suspect, it is empirically false."[111] Instead, scientists demonstrated what Pickering called "contextual opportunism"; in other words, "response to innovation [is a function of] . . . opportunistically structured evaluation by situated actors."[112] Pickering referred to the work of Knorr-Cetina and his own monograph as examples of that alternative approach, which he dubbed "constructivist."

Like Collins, Pickering set out by rejecting the "scientist's account" of scientific practice, the account that professed to be compelled by the experimental data, the sheer "facts" of nature.[113] In one of his early papers, Pickering conducted an elegant demonstration of the closure of a controversy, which he also took to be constitutive of a paradigm shift from the "old" to the "new" particle physics. He gave a historical account of the defeat of the "color" theory by the "charm" theory of quarks. Pickering insisted that while significant changes in data played a part in that story, they were by themselves inconclusive.[114] Color theorists could always find a way to explain the data shifts, but what they could not do was find ways for their theory to open up rich and attractive theoretical and experimental avenues for further research. This was the winning element in the charm theory: it appealed to the "interest" of the theoretical and experimental physicists because it allowed them to deploy and advance the expertise they had already cultivated in new and exciting domains of work. In this early essay, Pickering stressed theoretical connections to Kuhn's notion of the exemplar and to Mary Hesse's notion of networks, in other words, to the intellectual horizon of the Strong Program, and he explicitly identified his approach with the "interest model," though he construed it idiosyncratically.[115]

In his own major work, Pickering followed the lines of microanalysis and constructivism that characterized the entire second generation. As he put it, the key was to reject the "scientist's account" of scientific process, with its propensity of *"putting the phenomena first."*[116] Pickering aligned himself with Latour, Woolgar, Knorr-Cetina, and Collins in insisting that the phenomena were produced, in other words, *constructed* in the scientific process itself: "interpretive practices and the natural phenomenon stood or fell together."[117] As Pickering elaborated: "In general it seems reasonable to assume that natural phenomena float on a sea of interpretative and instrumental practices, all of which are in principle vulnerable to counterargument."[118] Concretely, "the assessment of natural phenomena is itself conditioned by the dynamical aspect of scientific practice: that is, by the continuing process of choice by experimenters to perform one experiment rather than another and of theorists to elaborate one theory rather than another."[119] Pickering offered compelling historical evidence for this judgmental discretion, but he had a penchant at the same time to offer an abstract rationalization for his constructivism, namely the Duhem-Quine thesis, "the holistic proposition that scientific knowledge is an inter-connected network and that unexpected experimental observations can, in principle at least, be accommodated by appropriate adjustments anywhere within the system."[120] Thus, Pickering argued in the introduction to *Constructing Quarks*:

> In the scientist's account, experiment is seen as the supreme arbiter of theory. Experimental facts dictate which theories are to be accepted and which rejected. . . . There are, though, two well-known and forceful philosophical objections to this view, each of which implies that experiment cannot *oblige* scientists to make a particular choice of theories. First, . . . choice of a theory is underdetermined by any finite set of data. It is always possible to invent an

unlimited set of theories . . . capable of explaining a given set of facts. Of course, many of these theories may seem implausible, but to speak of plausibility is to point to a role for scientific *judgment*. . . .

The second objection . . . is that the idea that experiment produces unequivocal fact is deeply problematic . . . experimental reports are *fallible* . . . dependent upon theories of how the apparatus performs. . . . More far reaching . . . experimental reports necessarily rest upon incomplete foundations. . . . Again a *judgment* is required. . . . The determined critic can always concoct some possible, if improbable, source of error.[121]

The first philosophical argument is Quine's underdetermination thesis, and the second is the Duhem-Quine thesis. What is important about the explicit appeal to "philosophical" objections here, however, is that what Pickering tried to hold open is the *judgmental* participation of scientists in the process of experimentation. "My first objective has been to demonstrate the intervention of judgments throughout the developments at issue."[122] He refused to see scientists as merely passive: "It is useful to reformulate the objection to the scientist's account in terms of the location of *agency* in science."[123] This was what it meant to stress scientific practice, the process, not the finished product or theory. "Agency belongs to actors not phenomena: scientists make their own history, they are not the passive mouthpieces of nature."[124]

It is here that Pickering deployed his model of "opportunism in context," attributing to scientists individually and by specialty a particular expertise which "interests" them in its effective elaboration. "Resources may be well or ill matched to particular contexts. Research strategies, therefore, are structured in terms of the relative *opportunities* presented by different contexts for the constructive exploitation of the resources available to individual scientists."[125] "Judgments are best understood as situated within a continuing flow of practice. . . . Within that flow, judgments are seen to have implications for future practice."[126] The "perceptions of those opportunities were structured by the resources available for their exploitation."[127] Pickering acknowledged that "one can speak of a symbiosis between natural phenomena and the techniques entailed in their production, where each confers legitimacy upon the other."[128] As he noted, "I have no wish to deny reality—in the shape of experimental data—a role in the development of scientific knowledge. . . . However, . . . it seems . . . impossible to regard the data *alone* as forcing a particular outcome to the debate."[129]

In the critical reception, Pickering's *historical narrative* of the development of high-energy physics since World War II was often praised highly, while criticism centered on the *methodological and epistemological principles* that Pickering articulated in his work.[130] Gingras and Schweber concentrated their critique on "the philosophical assertions advanced by Pickering." There were, they noted, "less than forty pages of the book's 415" which are explicitly methodological.[131] They did recognize, however, that Pickering's theoretical commitments were—as is usual with historical narratives—implicitly "manifested by the manner in which the story is told and in the choice of the events on which the analysis focuses."[132]

Gingras and Schweber found Pickering too "eager . . . not to 'put the phenomena first,' . . . he does everything possible to show that the phenomena never limit the possible interpretations that the theory can offer. In so doing, Pickering either forces the argument or neglects some part of his own story."[133] They elaborated: "Every time data seem to impose themselves on theory, he insists that the physicists always had the *choice* to ignore or to interpret them in a different manner. But what does that mean? . . . The real questions are: What was the basis for the decision taken? Was the choice taken completely arbitrary (as Pickering implicitly suggests) or was it well grounded in the existing sets of data?"[134] For Gingras and Schweber, Pickering was making illicit use of the Duhem-Quine thesis to evade these questions. "Nowhere in this model is the reality of the phenomena studied by experimentalists invoked to *explain* the acceptance or rejection of a theory or of new 'facts.'"[135] They charge that this failure resulted from "confusion about the meaning of the Duhem-Quine thesis, and about the significance of the observation that all 'facts' are theory-laden."[136] They are correct that "in principle" invocation of the "Duhem-Quine thesis does not necessarily cut that much ice."[137] And they are correct, as well, to observe "that theories are implicated in the production of empirical facts does not necessarily entail that a theory cannot be absolutely constrained by experiments."[138] David Henderson makes the same point.[139]

But Gingras and Schweber muddle their own formulation. They insist that the proper question is whether judgment is "well-grounded in the existing sets of data." *That* is just where both post-positivist arguments *do* bind. And they compound the error by the demand that "phenomena" be invoked to explain adoption of a theory or a "fact." But the whole point is that phenomena are what theory and "fact" lead scientists in the course of their study to constitute as plausible entities to be ascribed to nature. They cannot be postulated in advance, as Gingras and Schweber themselves understand—but only as a sociological heuristic, not a natural scientific one. The strength of their criticism of Pickering's "philosophical" argument lies elsewhere. They observe: "meaning is not ascribed to data in isolation, but rather in the light of a whole network relating other explained facts and predicted phenomena. The overall consistency of this network also plays in the choice between alternative interpretations or theories."[140] Hence, "we believe that the rigidity of the network is much greater than Pickering allows."[141] It is not merely the "data"—despite their misguided protest—but the entire *process* (with data playing a significant but not exclusive role) that entails far greater determinacy than Pickering "in principle" allowed. *That* is the point.

Paul Roth and Robert Barrett returned to these matters in a vituperative debate in the pages of *Social Studies of Science* in 1990. They too set out by distinguishing the "detailed historical narrative"—which they found "interestingly and entertainingly recounted, richly detailed, and highly informative" but of "no central concern" to their critique—and "a methodological thesis" to the effect that "logical and epistemological considerations traditionally invoked by scientists to account for the acceptability of scientific theories are inadequate to that task, if not entirely irrelevant to it."[142] For them, the essential *philosophical* claim of Pickering was "that scientists do not discover what is real, but produce it," in other words, radical constructivism.[143] They disputed the *epoche* (methodological or epistemological) which Pickering, like Collins, conceived between the accounts that scientists offer for their theories and the accounts that sociologists offer about those scientists, precisely Collins's counsel to take the natural order to have no impact. "Both Collins and Pickering deny that considerations internal to scientific practice explain the outcomes of scientific controversies and scientific change."[144] That signified, for them, that "theoretical objects such as quarks, or any of the special entities indigenous to other scientific theories, are constructs and nothing more."[145] They elaborated: "As Pickering sees it . . . the entities whose study as causes or effects is in question are 'perceived realities,' and since the perceptions involved presuppose the interpretive changes, such entities have to be effects of those changes, and cannot be causes of them. . . . But, of course, it was never in question whether 'perceived realities' cause interpretive changes. What was in question was whether *actual* entities cause such changes, and that is quite another matter."[146] Roth and Barrett insisted on some kind of "constraint" on scientific knowledge in the nature of the world.

They noted that Collins's notion of "experimental regress" appealed to "tacit knowledge" among scientists to account for their behavior, but this was tacit *social* knowledge—"immune to epistemological but amenable to sociological characterization."[147] Since, *ex hypothesi,* experimental evidence could never settle the matter, *something* had to, and that something— "tacit knowledge"—was socially accessible, because socially constructed. Roth and Barrett were not convinced of the force of this inference. They hammered home that there was a poverty of explicitness in this idea of sociological explanation. "Pickering purports to *explain;* we take him at his word and look for an account of causes."[148]

Roth and Barrett articulated an "in principle" incredulity regarding empirical sociological accounts of process. In short, theirs was a global counterattack on empirical disciplines of the human sciences from the philosophers on the grounds of an implicit standard of justification drawn from formal epistemology. The philosophers pose the issue explicitly: "how are imputations of causal efficacy to be assessed in historical and sociological reconstructions?"[149] That is a very general question about "the whole business of explanation."[150] It catches up *all* historical (as well as sociological) practice.[151]

The Crisis of Sociological Explanation

By the mid-1980s, despite the enthusiasm of its early successes, the sociology of scientific knowledge (SSK) shared fully in the general crisis of orthodox sociology, the shattering of its faith in its own categories of causal explanation.[152] One crucial source of this crisis was the challenge of ethnomethodology.[153] In the words of Richard Hilbert, "[E]thnomethodologists have repeatedly announced their suspension of belief in social structural phenomena *per se* as objects of theoretical inquiry. . . . To adopt an ethnomethodological perspective is to leave the question of social structure entirely behind . . . in favor of another topic: social practice."[154] Rather than consider structure, an ethnomethodologist studies its "artful production" by "members."[155] As Michael Lynch puts it, "Rather than trying to explain a practice in terms of underlying dispositions, abstract norms, or interests, a task for sociology would be to describe the ensemble of actions that constitute the practice."[156] Thus, in the words of Steve Bruce and Roy Wallis, ethnomethodology "reformulates as topics of analysis what conventional sociologists employ as resources for analysing behaviour, e.g., talk. . . . The order that exists in the social world derives from members' procedures for making sense of their contexts and interactions."[157] Lynch explains this as the agenda of the grand master of ethnomethodology, Harold Garfinkel: "For Garfinkel, both the detailed methods for producing social order and the conceptual themes under which order becomes analyzable are members' local achievements. There is no room in such a universe for a master theorist to narrate the thematics of an overall social structure. Instead, the best that can be done is to closely study the particular *sites* of practical inquiry where participants' actions elucidate the grand themes."[158] When this message reached the inner circles of SSK, it resulted in a radical revisionism.

Thus Woolgar pronounced, "As the ethnomethodologists have reminded us, this presumption of structural reality is a travesty of a key sociological phenomenon: how are structurings of this kind managed and achieved, for what purposes, by whom, and so on? Treating them as a frame for analysis is a sorry violation of topic in favor of resource."[159] Woolgar used ethnomethodology to "jump frames."[160] He set out in his challenge by cleverly citing an important observation from the Strong Program's most empirical and historical ally, Shapin: "The mere assertion that scientific knowledge 'has to do' with the social order or that it is 'not autonomous' is no longer interesting. We must now specify how, precisely, to treat scientific culture as social product."[161] Woolgar argued that the Strong Program left it "unclear what precise form of explanatory account is envisaged" to fulfil Shapin's desideratum. The ostensible candidate was the "interests model." But there was simply too much slack in the interests model: "the same instance of action can be read as indicative of more than one desire, and conversely, . . . different actions can be interpreted as indicative of the same desire."[162] And there were indefinitely many plausible interests (desires) one could invoke.

As David Bloor recognizes, there is a totalizing foreclosure about this ethnomethodological critique: "For ethnomethodologists the expressive, internalist, nonrepresentational picture of discourse has universal application. . . . [T]alk is at once the subject and object of all discourse."[163] His colleague Barry Barnes makes the same observation: radical ethnomethodologists made "an ontological commitment to speech as the totality of what is."[164] Thus Garfinkel's ideas were "widely used critically to deconstruct theoretical discourse."[165] Another frequent source for the rationale of this critique was Wittgenstein's later philosophy. As Bloor notes, "the aspects of Wittgenstein's work that most appeal to ethnomethodologists are his determined contextualization of every feature of our speech and thought and his opposition to the construction of explanatory theories."[166]

Yet laboratory studies were demonstrating an unforeseen degree of indeterminacy in scientific practice. And that was just the beginning. Knorr-Cetina elaborates:

> In short, scientists' *ex situ* involvements are simply marked by the same indexical, occasional and socially accomplished logic which appears to characterize laboratory action. . . .
>
> Thus we are confronted with the somewhat annoying picture of an indeterminacy inherent in social action. The indexicality and idiosyncrasies of scientific work jeopardize the hopes of philosophers of science to find once and for all the set of criteria which rule scientific selections. The situational contingency and the social dynamic of scientific action resist the attempt of the sociologist to specify once and for all the (social and cognitive, internal and external) factors which determine scientific action.[167]

The summary verdict pronounced by Lynch has reverberated widely throughout the social study of science: "Sociology's general concepts and methodological strategies are simply overwhelmed by the heterogeneity and technical density of the languages, equipment, and skills through which mathematicians, scientists, and practitioners in many other fields of activity make their affairs accountable. It is not that their practices are asocial, but that they are more thoroughly and locally social than sociology is prepared to handle."[168] The whole structure of sociological explanation was being undermined. This grim conclusion was underscored by the impact of yet another, perhaps even more sweeping theoretical indictment, that of poststructuralism.

John Law penetratingly characterized the situation in 1986 as a "crisis in the sociology of knowledge." It had entered upon a new surge of energy in the postwar era through the ideological analysis of Louis Althusser, the grid-group theory of Mary Douglas, and above all the Strong Program of SSK, but now "the second phase is in crisis. A third phase is upon us in the form of work that has gone some way to eroding the basis of the sociology of knowledge as this has been traditionally conceived."[169] What had come under attack was "the idea that there is a backcloth of relatively stable social interests which directs knowledge or ideology." Such a stable backcloth had allowed sociological explanation of beliefs in terms of structure, but now, especially under the impact of Michel Foucault, there appeared "no room for a structure/ideology division."[170] Thus "structure has collapsed into knowledge in the form of discourse, and the sociology of knowledge (if this is still an acceptable title for an inquiry that has so extensively chopped away at its own foundations) has been refocused upon methods for the reduction of discretion and the constitution of power."[171] Since "the discovery of discourse has eroded the distinction between social structure and knowledge," interest has refocused on "the *technique* of power/knowledge—methods, if one might put it this way, for the reduction of discretion."[172]

As Bruno Latour put it in *The Pasteurization of France*, "the evolution of our field has made the notion of a 'social explanation' obsolete."[173] Michel Callon explains, "The theoretical difficulty is the following: from the moment one accepts that both social and natural science are equally uncertain, ambiguous, and disputable, it is no longer possible to have them playing different roles in the analysis. Since society is no more obvious or less controversial than nature, sociological explanation can find no solid foundation."[174] Latour is emphatic: "notions like 'context,' 'interest,' 'religious opinion,' 'class position,' are . . . part of the problem rather than of the solution."[175] Hence "sociology of science cannot always be borrowing from sociology or social history the categories or concepts to reconstruct the 'social context' inside which science should be understood. On the contrary, it is time for sociology of science to show sociologists and social historians how societies are displaced and reformed with and through the very content of science."[176]

Law sums it up: for Latour, "social scientists should stop trying to determine the *nature* of the social structure that they believe generates . . . conflicts, and instead treat the latter as data. In other words, society should not be seen as the referent of an ostensive definition, but rather seen as being *performed* through the various efforts to define it."[177] Knorr-Cetina put it very clearly: "Since what is to be counted as 'social' rather than 'scientific,' 'rational,' or most broadly speaking 'non-social,' is itself negotiated in scientific practice, the 'social' cannot be unproblematically presupposed as an analyst's resource in describing this practice. Instead of drawing upon the 'social' to account for scientific knowledge, one can see the 'social' itself grow out of the construction of this knowledge."[178]

A publishing event of 1986 nicely epitomizes all of this. In that year, Princeton University Press reissued Latour and Woolgar's classic laboratory study. The authors took that opportunity to make a statement. They changed the title in a very telling way by dropping the word *social*—thus, *Laboratory Life: The Construction of Scientific Facts*. They added a postscript which gave expression to all the equivocality of social constructivism that the crisis had engendered, situating their own work within the radically contingent stream of rhetorical claims which constituted "knowledge" as such. "Each text, laboratory, author and discipline strives to establish a world in which its own interpretation is made more likely by virtue of the increasing number of people from whom it extracts compliance."[179] Every knowledge-claim, including their own, they avowed, was "infinitely renegotiable."[180] Nonetheless, they sought to "extract compliance" from their readers by urging "a ten-year moratorium on cognitive explanations of science."[181] This performance was greeted, as one might expect, with indignation. Paul Thagard wrote of the "remarkably self-refuting postscript to *Laboratory Life*," and Peter Slezak protested that "Latour and Woolgar evade criticism by adopting deconstructivist double-talk and affecting a posture of nihilistic indifference to the cogency of their own thesis."[182]

Once "construction" is severed from "social," all one has left is discourse, language, text. The passage from Anglo-American post-positivist study of science to a strikingly different "philosophy of language" in French poststructuralism yawns open. Before venturing forth on this "continental" course, two monuments demand our attention—the grand exemplar of Edinburgh sociology of scientific knowledge, *Leviathan and the Air Pump,* and the grandfather of science studies, Thomas Kuhn himself, in his final, unhappy confrontation with his progeny.

The Grand Exemplar: *Leviathan and the Air Pump*

In the controversies that swirled around the Strong Program in the early 1980s, a key theme was the warrant that case studies provided for the overall stance of the sociology of scientific knowledge. In *Natural Order,* Barnes and Shapin gathered sources which they believed set the project of a social construction of scientific knowledge in the best light.[183] In his survey essay

"History and Social Reconstruction," Shapin presented a consideration of a wider range of literature that he thought made the same case.[184] In his contribution, "The Social Use of Science," to an important anthology on eighteenth-century science, he presented another historiographical essay, this time on "Newtonianism" in that epoch, which demonstrated the powerful interpretive resource that a social-political perspective could bring to the history of science.[185] That body of evidence, however, has nothing like the saliency that one text has assumed in the controversy over the fruitfulness of SSK. That one text, the *grand exemplar* of the Strong Program, is the monograph *Leviathan and the Air Pump: Hobbes, Boyle, and the Experimental Life*, which Shapin published in collaboration with Simon Schaffer in 1985.[186] As the subtitle indicates, what they produced was an analysis of the controversy between Hobbes and Boyle over the air pump and the status of experimentation in the generation of scientific knowledge. No other text in the field has had the canonical status—for friend and foe alike—that this one study has assumed. It deserves the most careful consideration. I propose to follow my standard procedure. I will offer an account of the claims of the text and juxtapose them with the critical reception. The one difference is that here I find myself in the pleasant circumstance of being professionally familiar with the substance under examination. Thus, I will take the liberty to comment substantively on both authorial claims and the claims of critics about historical evidence and historiographical interpretations.

No one—not even, I think, Cassandra Pinnick—questions that *Leviathan and the Air Pump* is a work of formidable erudition. Even hostile critics—and there have been many—tend to recognize the verve, candor, and consistency of the interpretive positions adopted by Shapin and Schaffer. While Pinnick has suggested that she was advised to read *Leviathan and the Air Pump* with the "sandwich strategy" that certainly emerged as a consensus of the reception of Pickering's contemporary *magnum opus, Constructing Quarks* (i.e., "omit the introduction and conclusion; the rest is straight history"), it really is a misguided bit of advice, which she disregarded anyway (for other reasons).[187] The work is a whole, and as a whole it aims to present "an exercise in the sociology of scientific knowledge."[188] In the context of this chapter, it is important to note that while Shapin has consistently upheld his longstanding affiliation with the Edinburgh school, identifying himself thus with the Strong Program, he and Schaffer acknowledge decisive methodological influence from Collins, Pinch, Latour, and Pickering—the "constructivist" second generation.[189] Shapin and Schaffer composed their text at a moment when the solidarity among these thinkers outweighed the fissional impulses that were about to explode them into warring factions. Significantly, virtually all the subsequent factions have claimed *Leviathan and the Air Pump* as support for their camp.

Thus, there is every reason to regard this as one of the most important achievements in science studies in the late twentieth century. But that is not to say it had no problems. Let me begin with one alleged problem: Cassandra Pinnick's claim that the work was simply "bad history" because the debate between Hobbes and Boyle was a trivial incident with no substantial impact upon the course of scientific development and because Hobbes and Boyle were so equivocal (and equivalent) in their thought on natural science that Shapin and Schaffer had no real warrant to discriminate them. In her words, "the supposed dichotomy between the rationalism of Hobbes and the experimentalism of Boyle is, to a large extent, an artifact of Shapin and Schaffer's highly selective filtration of the historical evidence."[190] She accuses the historical profession of practicing excess charity in suspending critical judgment about their historical failings. *Balderdash* on both counts. Historians have quarrels with Shapin and Schaffer, but not on that score. I find Pinnick's historical objections myopic and misguided. And I reject her inference—as I trust other professional historians who actually reviewed the work or use it in their coursework or research would, as well—that since "textual study shows that Hobbes's and Boyle's views of methodology were not always opposed," *therefore*, "it is by no means clear from the historical record that they were polarized on method."[191] Good history, Pinnick notwithstanding, requires efforts to situate and to integrate evidence, and above all, not to lose the forest for the trees! Pinnick should have stuck with her philosophical objections, which are far nearer the mark. The historical issue is not whether Hobbes and Boyle advocated rival philosophies of science; we—not just Shapin and Schaffer, but the professional community of historians of science—accept *that*. The point of controversy is what Pinnick calls the philosophical issue: their "equicredibility" at the moment. And an even greater philosophical issue: whether credibility is only a matter of momentary (culturally relative) contingency.

Another piece of historical housecleaning: Paul Gross and Norman Levitt offer their resounding judgment that historians like Shapin and Schaffer have missed the whole point: it is obvious to anyone today who knows mathematics that Hobbes was not taken seriously at the time, simply because he was a terrible mathematician. Nothing further need be said. That there might have been philosophical, that there might have been social-political elements in the exclusion of Hobbes from the Royal Society, Gross and Levitt find "hard to believe."[192] With great hermeneutic subtlety, they construe Shapin and Schaffer to be equating the Royal Society with "a kind of thought police."[193] That is all they can make of the delicate relationship between natural philosophy and social order in the Restoration. "How accurate and complete is Shapin and Schaffer's analysis of the dispute between Hobbes and his foes at the Royal Society?" they ask.[194] Clearly, since mathematics was not at the center of it, they missed the boat: they could/should have had "a concrete and substantive reason, *in contrast to an ideological one*, for Hobbes's notoriety in scientific circles."[195] Note that Gross and Levitt mean not only that a sociopolitical concern among the members of the Royal Society would be a *merely* ideological one, neither "concrete" nor "substantive" (take that, historians!), but that Shapin and Schaffer are *themselves* ideological in having taken such a lamentable course. They "unaccountably fail to link [Hobbes's defeat] to the question of Hobbes's doubtful mathematical competence," though this "can hardly have been irrelevant."[196] Failure to face up to these points compromised the history Shapin and Schaffer offered. Though Gross and Levitt avow elsewhere that neither is a professional historian, they "contradict not only Shapin and Schaffer, but some later work on the same subject from a similar social-deterministic perspective. . . . Again, out of eagerness to derive everything from 'politics,' a scholar has overlooked the contrast between Hobbes's high opinion of his own mathematical abilities and his manifest mathematical incompetence. Hobbes's contemporaries, however, were not deceived on this point."[197] Again, Gross and Levitt *know* what mattered to Hobbes's contemporaries better than professional historians of the period, because they *know* that mathematical competence, not "politics," is indisputably the decisive factor. It never even occurs to them that there might be other considerations besides mathematical competence, and they return to this adamance in their 1998 supplement, where they proclaim: "The reputation of Hobbes as *a mathematical scientist* was, by 1661, so deservedly low among important thinkers that his not having been taken seriously as a participant in scientific debate hardly needs explaining."[198] This smug a priori approach to history is then accompanied by the pronouncement: "the science studies community failed in an elementary duty to scrutinize sweeping claims."[199] It is little wonder that *Higher Superstition* provoked outrage.

What, then, did it mean to conduct an exercise in the sociology of scientific knowledge with materials from late-seventeenth-century English natural philosophy? There was a strong *presentist* agenda throughout, which found bald articulation at the close of the work: "As we come to recognize the conventional and artifactual status of our forms of knowing, we put ourselves in a position to realize that it is ourselves and not reality that is responsible for what we know. Knowledge, as much as the state, is the product of human actions. Hobbes was right."[200] Leaving aside for the moment the provocation of the last three words, what the rest of the passage affirms was anticipated in the last line of the introduction: "We argue that the problem of generating and protecting knowledge is a problem in politics, and, conversely, that the problem of political order always involves solutions to the problem of knowledge."[201] In short, this was an exercise in SSK because it derived the key moral of contemporary science studies from an assessment of a controversy in the seventeenth century. Less charitably, it projected the categories of late-twentieth-century social constructionism back upon the late seventeenth century. Charity aside, no historian of seventeenth-century science would mistake for a moment that what was being offered is a "stranger's account."[202] While "historians are in wide agreement in identifying Boyle as a founder of the experimental world in which scientists now live and operate," Shapin and Schaffer aim to problematize this feeling of "membership" in a common culture. "It is now time to move on from the methods, assumptions, and the historical programme" of conventional history of science. "We wish to adopt a calculated and informed suspension of our taken-for-granted perceptions of experimental practice and its products."[203]

One of the principles Shapin and Schaffer subscribe to, as followers of the Strong Program, is the pursuit of "symmetry" in accounting for knowledge-claims. True or false, received or discarded, the Strong Program asserted that all should be explained by the same causal approach. That did not, however, mean that Shapin and Schaffer approached Boyle and Hobbes impartially. Their "stranger's account" was largely an account that *restored* plausibility to Hobbes and problematized Boyle. As Margaret Jacob aptly phrased it, "reading late-twentieth-century 'doubts about our science' backward, they imagined that in the seventeenth century, contemporaries could see little political or intellectual difference between the rival 'games' offered by the experimentally based science of Boyle and the geometrically based science of Hobbes."[204] This presumption of "equicredibility" is an ahistorical imposition with drastic methodological, not simply historical, consequences.

But in addition to methodological revisionism, Shapin and Schaffer offer substantive revisionism as well. One of their key objectives is to retrieve Thomas Hobbes as an important theorist of natural philosophy in the late-seventeenth-century context. "Hobbes as a *natural philosopher* has disappeared from the literature," they note.[205] They seek to establish "Hobbes's true place in seventeenth-century natural philosophy," since "there is no lack of evidence of the seriousness with which Hobbes's natural philosophical views were treated in the seventeenth century, especially, but not exclusively, by those who considered them to be seriously flawed."[206] At the same time Shapin and Schaffer are clear that Hobbes's contemporaries "considered he decisively lost" *all* his scientific controversies.[207] Our authors believe that accounts for historians' neglect, since in "Whig" history, "losing sides have little interest."[208] They note that this has obliterated Hobbes even more thoroughly from the history of mathematics.[209] The question is, *why* resurrect Hobbes as a natural philosopher? What's in it for Shapin and Schaffer? The answer is that his contestation of Boyle resonates remarkably with their own contestation of late-twentieth-century philosophy of science, in particular via the Duhem-Quine thesis.[210] How much this is so and how much it is their wish that it be so proves the nub of the reception controversy over their work.

But there is another whole dimension to their work as substantive history of science, namely, their attention to the history of *experiment*. The opening of the book is, in a certain light, more noteworthy than its notorious close: "Our subject is experiment. We want to understand the nature and status of experimental practices and their intellectual products."[211] *Leviathan and the Air Pump* needs to be situated in a historiographical and theoretical context which was asserting with unprecedented fervor the centrality of experimentation for the history and sociology of science. Pickering's *Constructing Quarks* (1984) and its hostile sibling, Peter Galison's *How Experiments End* (1987), are two decisive exemplars. Ian Hacking's *Representing and Intervening* (1983) was perhaps the most vivid theoretical articulation of the approach, followed by Allan Franklin's *The Neglect of Experiment* (1987) and important collective endeavors like *The Uses of Experiment* (1989). Shapin and Schaffer offered one of the paramount treatments of historically crucial experiments, with detailed descriptions of apparatus, discussion of the material culture and social arrangements of their conduct, and so on. Furthermore, they made the question of the place of experiment in natural science a historically *open* one by going back to what historians have conventionally regarded as the crystallizing moment of its acceptance.

The late seventeenth century witnessed a crisis in the mechanical program for scientific knowledge, which has traditionally been couched as a crisis of Cartesianism or of rationalism, namely, the impropriety of absolute certainty as a standard and of deductive logic as a vehicle for scientific knowledge. *L'esprit géométrique*, for all its formidable strengths, had proven to be too prone to overhasty applications in accordance with *l'esprit de système*. Both on the continent and in England, natural philosophers were questing for a sustainable method for pursuing the mechanical philosophy. "The contest between Hobbes and Boyle was, among other things, a contest for the rights to mechanism," Shapin and Schaffer observe. "Evidently, nothing could be clearer than that both Boyle and Hobbes were *mechanical* philosophers. . . . Yet each was in a position plausibly to charge the other with serious violations of mechanism."[212] That is what it meant that the mechanist paradigm was in crisis, and since this was simply the most important issue in seventeenth-century science, Pinnick's claim that the Hobbes-Boyle debate was insignificant for the history of science is false.

The emergence of a fallibilist, probabilist approach to knowledge of the natural world is an epochal shift, and locating it in the context of the Hobbes-Boyle controversy is historically illuminating because it crystallizes the issues of a longer term problem into a self-conscious moment of controversy. Boyle and his associates self-consciously abandoned absolute deductive argumentation in natural science (the discourse of "real causes") and opted instead for an avowedly contingent and fallible approximation (probabilism). "Physical hypotheses were provisional and revisable; assent to them was not obligatory, as it was to mathematical demonstrations; and physical science was, to varying degrees, removed from the realm of the demonstrative. The probabilistic conception of physical knowledge was not regarded by its proponents as a regrettable retreat from more ambitious goals; it was celebrated as a wise rejection of a failed project."[213]

Hobbes, as Shapin and Schaffer aptly argue, had not given up on the geometric, deductive model for mechanical philosophy of nature, which did not mean, of course, that he was ignorant of the problem of nominal versus real causes and the limitations of human access. Conversely, Boyle was not insensitive to the power of a deductive proof, but his assessment of the problem of nominal versus real causes led him to take a dramatically different path, namely, to abandon deductive proof in matters of physical science. If one abandoned absolute certainty, where was one to begin? Boyle opted for the "matter of fact." But what was that, and how was it attested? The key terms for Boyle's revisionism for the mechanist philosophy were empiricism and probabilism. "Matters of fact were the outcome of the process of having an empirical experience, warranting it to oneself, and assuring others that grounds for their belief were adequate."[214] The loss of necessity had to be compensated by the recourse to intersubjective consent, with universalizability as its regulative ideal.[215] That made the "multiplication of witnessing" essential to the new experimental method. Shapin and Schaffer identify three "technologies" that Boyle and his allies developed to execute this program. There was the material technology of the apparatus, there was the literary technology of the disseminative reporting, and there was the social technology of establishing the "moral conventions" for presenting and accepting claims about "matters of fact." While conceptually distinguishable, these technologies "each embedded the others."[216] In effect, what Boyle and his colleagues created was a new physical and social space: the laboratory and the "experimental form of life," or community. The emergent form of the laboratory was "a disciplined space where experimental, discursive, and social practices were collectively controlled by competent members."[217] Both the physical form, the laboratory, and the form of life, the experimental community, were *emergents:* they constituted social novelties in an epoch, moreover, which was highly sensitive to disruptive threats. "Unless the experimental community could exhibit a broadly based harmony and consensus within its own ranks, it was unreasonable to expect it to secure the legitimacy within Restoration culture that its leaders desired."[218]

   This is the decisive historical link between the scientific and the political
which Shapin and Schaffer consider their prime theoretical concern. "There
are three senses in which we want to say that the history of science occupies
the same terrain as the history of politics. First, scientific practitioners have
created, selected, and maintained a polity within which they operate and
make their intellectual product; second, the intellectual product made
within that polity has become an element in political activity in the state;
third, there is a conditional relationship between the nature of the polity occupied by scientific intellectuals and the nature of the wider polity."[219] Note
that this formulation is theoretical, not historical. It is postulated as a general conception of the relation between science and society. The implied continuity from the endeavor of Boyle and his colleagues to current conceptions
of the relation of science to society (and politics) gets a bit of qualification
later, in the postmodern sentiment that ruptures must have intervened between this Enlightenment instauration and the fragments we have shored
against our ruin. But let us concentrate for the moment on the historical version. "At the Restoration it seemed clear that all free debate bred civil strife.
It seemed less plausible that *some* form of free debate might produce knowledge which could prevent that strife."[220] Therefore, "the restored régime
concentrated upon means of preventing a relapse into anarchy through the
discipline it attempted to exercise over the production and dissemination of
knowledge."[221] In order to survive, Boyle and his colleagues realized, they
needed to *insulate* their activities from the wider political fray and established *internal* order. But this insulation needed simultaneously to be negotiated with the wider polity in terms that the latter could accept. In fact, Boyle
and the Royal Society offered two recompenses. "Boyle and his allies made
two things available to Restoration society: the form of life practised within
the experimental space, and the matters of fact which experimenters helped
make."[222] As it happened, we have good reason to believe that these experimentalists provided precious little in the way of viable technological aug-
mentation of power or utility to Restoration society or the state. It was the
symbolic capital that proved decisive. "The experimental community could
be constituted as a model of the ideal polity."[223] That is, "the experimental
polity was said to be composed of free men, freely acting, faithfully delivering what they witnessed and sincerely believed to be the case."[224] More
graphically, "in the body politic of the experimental community, mastery
was *constitutionally restricted*. . . . The experimental polity was an organic
community in which each element crucially depended upon all others, a
community that rejected absolute hierarchical control by a master."[225] In
short, the experimental community became one of the models for emergent
civil society. But that model was hardly established in the wider polity.

There, by contrast, Hobbes's political theory seemed far more plausible: "He who has the most, and the most powerful, allies wins."[226] And that allows Shapin and Schaffer to offer us a dubious syllogism:

1. Hobbes's political theory is correct.
2. As a polity, the experimental community falls under the principles of political theory.
3. "Hobbes was right" about scientific communities.

Not one of these propositions, however, deserves our consent. Where did Shapin and Schaffer go wrong? Let us turn to the reception literature.

Margaret Jacob has argued that the Restoration settlement and the Royal Society are progenitors of more than the experimental method in natural science.[227] They are part of a secular turn to "civil society." Jacob's concern to reconstruct this tradition is at odds with Shapin and Schaffer's disenchanted view of the nascent Enlightenment in their postmodern aporia. The irony is that in both insisting upon the continuity between the social and the scientific and in privileging Hobbes's critique of Boyle's *scientific program,* Shapin and Schaffer never really reflect on what it means to attach plausibility to Hobbes's absolutism *as a political program.* To assert so nonchalantly at the close, "Hobbes was right," is a gesture of provocation, but not merely in *natural philosophy;* it is also one in *political theory.* Not to put too fine a point on it, the last thing we need now is affirmations of Hobbesian absolutism or the Hobbesian rationales for it, including his war of each against all. Shapin and Schaffer are so intent on their contestatory goals within science studies that they are stunningly blind to the politics of their own posture. It is here that Bruno Latour enters the story.

There is a curious interfiliation between the theoretical endeavors of Latour and of Shapin and Schaffer. The latter were expressly influenced by the turn that Latour made with Michel Callon in the essay "Unscrewing the Big Leviathan" (1981), and which Latour formalized methodologically (see my next chapter) in *Les microbes: guerre et paix* (1984).[228] Latour then took up Shapin and Shaffer's work and "translated" it in his provocative historiographical revisionism concerning modernity.[229] Here I wish to focus strictly on the relevance to *Leviathan and the Air Pump,* leaving detailed consideration of Latour's own views to later. I will bring to bear Margaret Jacob's commentary on the whole relationship to situate the historiographical issues.

In "Unscrewing the Big Leviathan," Callon and Latour suggested that science studies could no longer rely upon conventional sociological macrostructures for explanation. Instead, the field needed to confront the radical novelty of scientific knowledge as a constitution simultaneously of nature and of society. Categories for this radical emergence could not be derived

from nature or society; rather, terms for the construction of these larger orders had to be discriminated from the emergence itself. For Callon and Latour, the crucial moment at which this whole problem got enunciated came
with the writing of *Leviathan*. Latour discerned that the resolution Hobbes
proposed to the problem of the *Leviathan* was generalizable as a theory of
knowledge-generation as such. Hobbes proposed that the sovereign pur-
ported to "say nothing on his own authority" but was rather "authorized by
the multitude, whose spokesman, mask-bearer and amplifier he is."[230] This
theory of "representation" takes the general form Steve Fuller has captured
precisely:

> Thus, when one object, X, *represents* another, Y, three things happen:
> 1. X speaks in the name of Y.
> 2. Y no longer speaks for itself.
> 3. There is no longer any need to refer to Y.
> Thus representation is not only informative and economical, but it is also—
> for better or worse—repressive.[231]

Latour calls this *translation*: "By translation we mean all the negotiations,
intrigues, calculations, acts of persuasion and violence, thanks to which an
actor or force takes, or causes to be conferred upon itself, authority to speak
or act on behalf of another actor or force."[232] He works out this new theory
in detail in *Les microbes*. The point in the original essay is: "Everything is involved in these primordial struggles through which Leviathans are structured: the state of techniques, the nature of the social system, the evolution
of history, the dimensions of the actors and logics itself."[233]

While it would appear too simple to suggest that this essay directed the attention of Shapin and Schaffer to the specific historical moment of the
Hobbes-Boyle controversy, it is not too much to suggest that it aroused in
them the consideration that Hobbes was more important for science studies
than had appeared hitherto and that *Leviathan* deserved reading as a work
in *natural philosophy*. Ironically, Latour found that they had not quite got
his point. But he was clear that the reception of *Leviathan and the Air Pump*
was also off point. It "has been often mistaken for a book on the social history of 17th-century science," but "theirs is a book of social theory . . . lost
on historians of science and of the 17th century. It is a book about the theory
of the *co-production of science and its social context*."[234] "Both Boyle and
Hobbes struggled to invent a science *and* a context *and* a divide between the
two . . . since neither of them exist *before* Boyle and Hobbes achieve their respective goals and settle their disputes."[235] Latour is as interested in the natural philosophy of Hobbes as in the political philosophy of Boyle and above
all in reading both natural and political philosophy simultaneously, which is
the distinct achievement of Shapin and Schaffer. "Others have studied the

practice of science; others have studied the religious, political and cultural *context* of science, but none so far have been able to do the two at once."[236] Focus on experiment, on the material technology, "is where the book becomes so important. In what is no less than a *reverse* Copernican revolution, S & S make their analysis and that of their characters turn *around the object,* around *this* specific leaking and transparent air pump."[237] As Latour has it, "The triumph of Boyle is to transform a bricolage around a patched up air pump into a decisive way to win the partial assent of gentlemen about matters of fact; the triumph of S & S is to explain how and why discussions about the Body Politic, God and His miracles, Matter and its power, could be *made to go* through the air pump."[238] But Latour finds Shapin and Schaffer ultimately guilty of a lapse of symmetry. That is the irony implicit in their having "been trained after all in social studies of science and inside the Edinburgh school. . . . That there is no Nature 'out there' to account for the success of Boyle's programme is obvious to them; but they seem to believe that there is a Society 'out there' to account for the failure of Hobbes's programme."[239] Shapin and Schaffer show too much deference to Hobbes and his social theory. "No," Latour retorts, "Hobbes was wrong. How could Hobbes be right on that since he is the one who invents monist society in which Knowledge and Power are one and the same thing?"[240] Latour contends "no one has yet deconstructed his vocabulary of power, society, group, calculation of interests and sovereignty."[241] Thus Shapin and Schaffer have missed the essential issue: "Far from reacting against Kant's Copernican Revolution they have simply replaced his Transcendental Ego by the Transcendent Society."[242] What are we to make of Latour's remarkable "translation" of *Leviathan and the Air Pump?*

I think the key lies in Latour's phrase about Hobbes as envisioning a "monist society in which Knowledge and Power are one and the same thing." That is, what is going on in Shapin and Schaffer and in Latour is the conflation of Hobbes and Foucault. As Margaret Jacob puts it: "In the work of Shapin and Schaffer, Foucault decisively entered the field of science studies."[243] As she says, "At its core the Shapin-Schaffer linguistic reading of Hobbes and Boyle equates scientific discourses with strategies of power."[244] She goes on: "To make the story work, and to enhance the cunning attributed to Boyle, Shapin and Schaffer downplayed Hobbes's commitment to absolute monarchy and his refusal to countenance any social space not under its control. In their hands Boyle and his allies became the more active, aggressive promoters of science as an ideological prop for the state, while Hobbes and his philosophy came to be seen as less excluding of ordinary mortals with little access to laboratory or equipment."[245] But that is jaundiced history indeed. "Taking a look at the nature of monarchical absolutism as practiced throughout Europe and at its impact on science, as well as assessing the legal implications of Hobbes's philosophy, if enacted, exposes the ahistorical character of the Shapin-Schaffer account of Hobbes, showing it to be flawed precisely because of its lack of attention to the larger stakes as contemporaries would have understood them."[246] In assigning Hobbes this role, not only Shapin and Schaffer but Latour as well dramatically subvert the notion "that both the English Revolution and, less obviously, the science prescribed by first the Boylean and then the Newtonian syntheses helped in the evolution of civil society."[247] This historian's protest is not without a presentist political urgency.

Indeed, the projection of Foucault back upon Hobbes is a presentist cultural politics, as Dick Pels has noted.[248] Christopher Norris makes the same point. He argues that the Strong Program is "mistaken in principle" in "espousing the Hobbesian (also Foucauldian) belief that knowledge is *always and everywhere* a product of power-interests concealed by a rhetoric of 'disinterested' truth-seeking inquiry."[249] Norris puts it bluntly: "Hobbes stands to Boyle and the Royal Society as Foucault (or indeed Shapin and Schaffer) stands to the received 'Whiggish' tradition of thinking in history and philosophy of science."[250] That accounts, in his view, for Shapin and Schaffer systematically projecting late-twentieth-century ideas back upon the seventeenth century. "Our goal is to break down the aura of self-evidence surrounding the experimental way of producing knowledge," they write.[251] The tense is essential: this is a presentist contestation. By imagining that Hobbes might have prevailed in the seventeenth century, Shapin and Schaffer want to deny the idea of experimental knowledge *now*. With Hobbes and more pertinently with Foucault, they endeavor "an indiscriminate levelling of the various 'discourses' or 'regimes of truth' . . . on the standard power/knowledge scheme."[252] That is, Norris continues, they imply that "scientific truth can be reduced *without remainder* to the currency of in-place consensus beliefs or prevailing social values."[253] Norris faults the blurring of "ontological and epistemological issues." By invoking the Duhem-Quine thesis, Shapin and Schaffer feel entitled to the view that "there is nothing that could settle such issues except the appeal to a *force majeure*, a socialized nexus of power/knowledge interests." Ironically, like Latour, but for diametrically opposite reasons, Norris faults the *social* constructivism of this approach. If it is a question of credibility or "equicredibility" (today, even more than in the seventeenth century), "there is on the face of it, something highly suspect about the transfer of epistemic confidence [to] social facts. . . . To put it bluntly, we have progressed much further in our knowledge of the physical world through the methods of the natural sciences than we have in our knowledge of the human world . . . through sociology and allied disciplines."[254]

The issue that needs to be taken with *Leviathan and the Air Pump* is philosophical or theoretical, not historical. Pinnick zeroes in on the crucial point, Shapin and Schaffer's identification of Hobbes's strictures on experiment with the currently fashionable Duhem-Quine thesis. *That* is the Achilles heel of this work. Philip Kitcher notes that "Shapin and Schaffer take for granted the general doctrine of underdetermination, supposing that there is no way in which opponents could be forced to concede Boyle's experimental results and that there are always ways of avoiding Boyle's explanations of those results."[255] Kitcher maintains, with good reason, that the "apparent epistemic gains of the Boylean enterprise (the explanation of the barometers, the suspended weights, the mice, the feathers, in terms of the evacuation of the globe)" breaks any question of a standoff, and consequently "there is no underdetermination." Not only his contemporaries but we as historians (and philosophers) must find, "Hobbes was wrong."[256] Kitcher holds that Shapin and Schaffer "reconstruct the debate between Boyle and Hobbes without ever going through the details of Boyle's numerous experiments with the air pump or investigating the ways in which an opponent of vacua might have tried to account for Boyle's findings" because "they 'know' from the start that any experiment can always be interpreted in many different ways." They are guilty of "overextending the argument from underdetermination," or the Duhem-Quine thesis, as they call it.[257] Bad philosophy once again saps the force of science studies. Let us consider the decisive concluding sentence of the work: "As we come to recognize the conventional and artifactual status of our forms of knowing, we put ourselves in a position to realize that it is ourselves and not reality that is responsible for what we know."[258] Of course it is true that the forms of our knowing are artifactual, and in a quite strong sense of the term, they are also conventional. And to be sure, *responsibility* for what we know rests always and only with us. But all that having been said, the conclusion they seek still does not follow. It is not true, it is not even conceivable, that *only* we, and not as well the world with which we interact, participate in the generation of knowledge. And it is simply bad thinking to believe that "underdetermination" or the Duhem-Quine thesis allows us to *forget* the role of evidence, even if we are quite clear that evidence may not *suffice*.

Mutual Disownment: Kuhn and the Radical Social Constructivists

Thomas Kuhn was inadvertent godfather to the sociology of scientific knowledge. It is fitting to draw an interim balance of the proliferation of that field with a brief return to its namesake. It can hardly occasion surprise that the more radical descendants of SSK of the 1980s proved even more appalling to Thomas Kuhn than the original had in 1977. In one of the final publications of his life, Kuhn expressed his outrage at this last phase of science studies. In his 1991 "Robert and Maurine Rothschild Distinguished

Lecture" at Harvard, Kuhn devoted the bulk of his remarks to rebuking "people who often called themselves Kuhnians" but who were "damagingly mistaken."[259] In the baldest language, Kuhn asserted, "I am among those who have found the claims of the strong programme absurd: an example of deconstruction gone mad."[260] He complained that the "new kind of historical and more especially of sociological studies [which] dealt, in microscopic detail, with the processes within a scientific community or group from which an authoritative consensus finally emerges . . . deepen rather than . . . eliminate the very difficulty they were intended to resolve."[261]

He made it clear that "it's as a philosopher that I speak." This was the identity he always most coveted. This is one of the points where I disagree with Steve Fuller's assessment of Kuhn. In 1990, in an interview for the *Harvard Science Review,* Kuhn responded to the question whether he would at that point consider any fundamental revisions to his approach in *Structure* with the following statement: "I see no reason to suppose that the things I think I have learned about the nature of *knowledge* are going to be disturbed by the need to change the theory of *science*. . . . I would make this separation to explain why I'm less concerned about the question, 'Is science changing?' than I might be if studying the nature of science weren't in the first instance simply a way of looking at the picture of knowledge."[262] Fuller focuses on the fact that Kuhn "respecified his project at a level of abstraction that escaped having to decide" between whether contemporary science lived up to his model or his model lived up to contemporary science.[263] I think it is more accurately a reflection of Kuhn's fundamental effort to redescribe himself as a philosopher (concerned with *knowledge*) rather than as a student of *science*. In a final interview, Kuhn made an important observation: as contrasted with historians of science, "philosophers and scientists are much closer to one another, because they all come in being concerned about what's right and wrong—not about what happened."[264] Above all, "you are not talking about anything worth calling science if you leave out the role of [nature]."[265] That made Kuhn an ally of those disposed to see the whole turn in social studies of science as misguided. Kuhn in his final publication willingly enlisted in the "science wars" on the side of the scientists.[266] It is a small step indeed from his remark about the Strong Program to Gross and Levitt's *Higher Superstition.*

In return, radical social constructionism has quickly found Kuhn too "conservative" to remain anything more than a distant progenitor, part of the remote archaeology of the discipline. As one spokesperson, Trevor Pinch, puts it, "Perhaps the most striking development in the last decade and a half is that Kuhn's ideas have no longer provided the specific template upon which to build sociology of science."[267] Steve Fuller does not seem content to let Kuhn's work even become a fossil. He has been intent upon exhuming it for public tarring, feathering, and burning at the stake. At every occasion he denounces its pernicious influence, its Cold War shadiness, its lack of historical reflexivity.[268] In his full-length work, Fuller seeks to establish that "Kuhn's 'acritical' perspective has colonized the academy," in other words, that "the impact of *The Structure of Scientific Revolutions* has been largely, though not entirely, for the *worse*."[269] Fuller's goal is "to overcome *Structure*'s effects on its readers."[270] That is, "the monumental fecundity of misreadings attached to *Structure* begs for explanation."[271] For Fuller, "to understand the overarching significance of *Structure*, especially the sorts of projects it has helped and impeded, we need to start taking seriously that Kuhn's book constituted, pace Shapere, less a revolt against positivism than a continuation of positivism by other means."[272] For Fuller, the key to Kuhn is the elaboration of normal science and the "demonization" of revolutionary science. Kuhn was like the positivists in believing that rule-following was the essence of rationality, not the "critical" challenge to any position.[273] Kuhn, like Rudolf Carnap, made the distinction of internal from external questions a vehicle for escaping rational consideration of value conflicts. But Fuller is not content with an "internal" account. He proposes to write "philosophical history," to problematize Kuhn in the broadest conspectus of "our times."[274] In short, Fuller writes, "I urge that *Structure* be read as an exemplary document of the Cold War era."[275]

Given his "normatively charged language," Fuller regrets the implication that he might be passing judgment on Kuhn: "I have neither the interest nor the evidence to deliver a verdict on Kuhn's life, let alone indict the man of crimes of the intellect."[276] That comes a bit late, after almost four hundred pages which seek to demonstrate that *Structure* lies at the root of every intellectual evil of our times. Small wonder, then, that Kuhn felt like a hostage in the sociological camp. Small wonder, too, that that camp is ready to bury him. Kuhn is now *passé;* the new guru, Richard Rorty assures us, is Bruno Latour.[277]

CHAPTER 7

In our time, a mythic time, we are all chimeras, theorized and fabricated
hybrids of machine and organism.
— DONNA HARAWAY, "A CYBORG MANIFESTO"

"Real worlds out there" are the consequences of lines of stable force and
not the cause of their stabilization.
— BRUNO LATOUR, *THE PASTEURIZATION OF FRANCE*

Science is not politics. It is politics by other means.
— BRUNO LATOUR, *THE PASTEURIZATION OF FRANCE*

Women, ANTs, and (Other) Dangerous Things: "Hybrid" Discourses

Who's afraid of Bruno Latour? Well, Paul Gross, for one, now that he
doesn't have Thomas Kuhn to fear.[1] And Rachel Laudan, fearful for the future of history of science.[2] And the Institute of Advanced Study at Princeton,
I have heard, perhaps fearing that Latour's farce about Einsteinian relativity
blasphemed the *spiritus loci*.[3] And, saddest of all, David Bloor, who fears
that Latour is spreading "goodness knows what" misrepresentations of the
Strong Program.[4] Actually, I have to confess *I* was, too, at first, even when I
was reassured by colleagues that Latour was all just good fun.[5] Then, reading through the body of his work, something funny did happen. I had the
most uncanny sense of *déja vu*, reading *Science in Action*. By happenstance I
read this work before I read "Irreductions" (in *The Pasteurization of France*)
or my surprise might have been forestalled. The point is, reading *Science in
Action*, I found myself in the presence of something that at least since Heidegger the West had supposedly overcome: a *first philosopher*, someone who
proposed to rethink everything from the beginning. As I completed reading
Science in Action, the book that kept coming to mind was Hegel's *Phenomenology of Spirit*: "bootstrap" thinking indeed![6] With that, Latour wasn't so
scary after all. He was *really* funny, though perhaps not the way he meant to
be. I guess one could say he got inadvertently "enrolled" in another "network."

Actor network theory (ANT) is, of course, an industry, not an idiolect,
proudly boasting "production of six books, five edited volumes and about
sixty articles" already by 1992.[7] Still, there is no mistaking the signature of

Bruno Latour. In order to situate ANT within science studies, I propose to take seriously—to play up, acting on my sense of *déja vu*, my confusion of Latour with the tradition of first philosophy—the representations Latour offered of himself and his project in *The Pasteurization of France*. To be sure, I do not seek to "reduce" ANT to Latour, or Latour to his antics, but, as he might well understand, one has to start somewhere.[8]

"Pseudoautobiographically," Latour writes of a moment "at the end of the winter of 1972, on the road from Dijon to Gray" when an epiphany befell, one that was as transfiguring for "him" as Paul's on the road to Damascus or Descartes's in that "stove-heated" room. "Suddenly I felt that everything was still left out." Contemplating a very blue wintry sky, pseudo-Latour "no longer needed to prop it up with a cosmology. . . . It 'stood at arm's length,' . . . it alone defined its place and its aims. . . . And for the first time in my life I saw things unreduced and free."[9] So was forged the gospel of "irreduction." We are invited to receive this vision (*ironically*, of course, humorously, "pseudoautobiographically") as the inscription of Latour into the lineage of provisioners of *grands récits*. Latour offers us nothing less than a "Tractatus Scientifico-Politicus" in the spirit of Spinoza's *Tractatus* and in the form of his *Ethics*.[10] It literally proposes to begin entirely anew, at 1.1.1. (Though of course neither he nor we any longer even for a nanosecond believe that can be done.) What could possibly motivate such a gesture—even as a jest?

"To escape, we have to eliminate almost everything."[11] That is the key. Latour, in fact, is not joking. He is cutting his losses. Let us move from "pseudoautobiography" to existential psychology and ask after the *motive* behind his *beau geste*. "We would like to be able to escape from politics," Latour tells us.[12] "We would like science to be free of war and politics. At least, we would like to make decisions other than through compromise, drift, and uncertainty. We would like to feel that somewhere, in addition to the chaotic confusion of power relations, there are rational relations."[13] Western civilization, at least in its self-professed "modern" moment, sought science as the refuge from politics, reason as the refuge from force, right as the salvation from might. Science was perhaps our last hope for a refuge, a "transcendental world."[14] But that was the folly of Enlightenment, and Latour has grown "a-modern" enough to be disabused of that hope.[15] Enlightenment has simply supplied might with more "radiant" equipment.[16] The point of departure for Latour is "our disappointment in the redeeming virtue of science."[17] The "link between science and democracy has become tenuous."[18] We can no longer ignore "how science and war have come to be so intermingled."[19] Latour's mentor, Michel Serres, called this sinister imbroglio "thanatocracy."[20] The lineage of disillusionment is longer, of course; World War I quashed the soteriological pretense of science. With great eloquence and pathos the verdict was pronounced in the classic essay "*Crise de conscience*" (1918) by Paul Valéry: "Science is mortally wounded in its moral ambitions and, as it were, put to shame by the cruelty of its applications."[21] More recently, Paul Fussell summed it up quite nicely: after the "Great War," we could not think the word "machine" without adding the word "gun."[22] Science and/as war: Latour ends his book with an apocalyptic nightmare, cruise missiles descending upon France: "In the few seconds that divide illumination from irradiation I want to be as agnostic as it is possible to be for a man who is present at the passing of the first Enlightenment."[23]

Agnosticism motivated by *Angst:* that is the path into ANT. "Agnosticism in matters of science is the only way to start without being trapped on one side of the many wars being fought by the guardians of science's borders."[24] Latour proclaims, "all social studies of science are thought to be *reductionist*" given "the assumption that force is different in kind from reason[, that] right can never be reduced to might." So he has "decided to see how knowledge and power would look if no distinction were made between force and reason."[25] "My 'Tractatus Scientifico-Politicus' instead of clearly dividing science from the rest of society, reason from force, makes no a priori distinction."[26] Agnosticism mandates that "we cannot explain a science by paraphrasing its results . . . we must describe it without resorting to any of the terms of the tribe."[27] Nor is Latour prepared to resort to already available terminology from the sociology of science, whose project has been "to construct a strict chain of command going from macrostructures to the fine grain of science."[28] "*The congenital weakness of the sociology of science is its propensity to look for obvious stated political motives and interests in one of the only places, the laboratories, where sources of fresh politics as yet unrecognized as such are emerging*" (original emphasis).[29]

If we are to start all over, the issue boils down to this: "where can we find the concepts, the words, the tools that will make our explanation independent of the science under study?"[30] Epistemological reassurance is in order: "The fact that we do not know in advance what the world is made up of is not a reason for refusing to make a start, because *other* storytellers seem to know and are constantly defining the actors that surround them."[31] In "Irreductions," this gets its own number (3.1.2): "I know neither who I am nor what I want, but *others* say they know on my behalf, others who define me, link me up, make me speak, interpret what I say, and enroll me . . . they impose an interpretation of what I am and what I could be."[32] In the "first-philosophical" vein, we could read this as a version of Aristotle's claim that philosophy emerges out of ongoing discourse, as the ambition to achieve consensus through the clarification of its topics. But this would be altogether the wrong trope, for the object in Latour's world is not *agreement* but *self-*

defense—or rather, *conquest*. We are in an *agonistic* world. And that undercuts the aboriginality of Latour's metaphysic: he is *always already* in the discourse of force, war, battle. "To understand simultaneously science and society, we have to describe war and peace in a different way."[33] The obsession with the imbrication of science and war cannot be suspended, even at the ostensible origin.[34] "We never bow to reason, but rather to force" (4.7.5). Such is his vision, and Latour finds us an *ethical* reason to accept it: "As soon as 'right' is divided from 'might,' or 'reason' from 'force,' right and reason are weakened because we no longer understand their weaknesses, and we steal the only way of becoming just and reasonable that is available to those who are scorned. These two losses leave the field free for the wicked" (4.7.7). Here is balm in Gilead indeed for all those who hunger and thirst after justice.[35]

Latour feels free to "start from the assumption that everything is involved in a relation of forces" even if he has "no idea at all of precisely what a force is."[36] Force = x. It designates the initial unknown. "In place of 'force' we may talk of 'weaknesses,' 'entelechies,' 'monads,' or more simply 'actants'" (1.1.7). "We should not decide a-priori what the state of forces will be beforehand or what will count as a force."[37] Still, it is time for *ontology*: "Whatever resists trials is real" (1.1.5); in other words, a force evinces "gradients" of resistance (1.1.5.1). Reality is experienced—it presents itself primordially—as resistance to intervention or "enlistment" (1.1.8). Metaphoricity charges this metaphysics from the outset: *trial* or *enlistment* aims at *control*. The governing trope is *strength*. "All of these forces together seek hegemony by increasing, reducing, or assimilating one another."[38] "No actant is so weak that it cannot enlist another" (1.1.8). "An actant can gain strength only by associating with others" (1.1.9). "There are winners and losers" (1.1.10). Hence, *irreversibility* and *asymmetry* result from these trials. "To create an asymmetry, an actant need only lean on a force slightly more durable than itself" (1.1.12). The measure of strength or control is *durability*. "Forces that ally themselves in the course of a trial are said to be durable" (1.2.5). There is *only* duration, never permanence, "because permanence costs too much and requires too many allies" (1.2.5.2).

Latour is now in a position to mobilize his metaphysic: "Every entelechy makes a whole world for itself. It locates itself and all the others; it decides which forces it is composed of. . . . It translates all the other forces on its own behalf" (1.2.8). Latour finds first-philosophical precedents in Nietzsche's idea of "evaluation" and Leibniz's notion of "expression."[39] In this "state of nature," if I may invoke another of his founding fathers—omnipresent but largely unmentioned in "Irreductions"—there is *a war of each against all* yielding "two states, dominating or dominated, acting on or acted upon" (1.3.4).[40] "The one who defines the nature of the association without being contradicted takes control" (1.3.5). "A force establishes a pathway by making other forces passive" (1.4.4). It "creates *lines of force*. They keep others in line. They make them more predictable" (1.4.5). That is what a *network* is. To *connect* is to *command* by *translation* (1.4.6), "speaking for" the passive elements coordinated. "Certainly we can *extend* this network by recruiting other actors, and we can also *strengthen* it by enrolling more durable materials" (2.5.4). "For an entelechy there are only *stronger* and *weaker* interactions with which to make a world" (2.4.7). "Anything can be reduced to silence, and everything can be made to speak. Thus any force may appeal to an inexhaustible *supply* of actors *who may be spoken for*" (3.1.6). And "some actants become powerful enough to define, briefly and locally, what it is all about" (3.2.1). In the contest that constitutes "reality," "a force becomes potent only if it *speaks for* others, if it can make those it silenced *speak* when called upon to demonstrate its strength, and if it can force those who challenged it to *confess* that indeed it was saying what its allies would have said" (3.1.12).

What happens, I think it time to ask, when one so promiscuously *metaphorizes* practice into discourse, action into speech? Granting for the moment that the harvest of all this metaphysics is (as we shall explore at length when we get back from metaphysics to science studies) to suspend our normal discrimination of agents from things, of human from nonhuman, what exactly happens when we ascribe to a virus, for example, *not* the capacity to impose its code on victim cells and to imprint its entelechy upon these suborned others, but rather the power to "speak" for these others? *It's just metaphor,* I can imagine the retort. But that is just the problem. With personification run amok, Latour can play the devil with the line between ontology and epistemology. "It is not possible to distinguish for long between those actants that are going to play the role of 'words' and those that will play the role of 'things'" (2.4.4). "A word can thus enter into partnership with a meaning, a sequence of words, a statement, a neuron, a gesture, a wall, a machine, a face . . . anything, so long as differences in resistance allow one force to become more durable than another" (2.4.2). "Everything is negotiable" (1.2.2). "Before negotiation we have no idea what kind of trials there will be—whether they can be thought of as a conflict, game, love, history, economy, or life. Neither do we know whether they are primordial or secondary before we enter the arena" (1.2.10).

Is there no difference? Ontology and epistemology seem to be perilously blending. "The interpretation of the real cannot be distinguished from the real itself" (1.2.7.2). "Nothing is, by itself, either knowable or unknowable, sayable or unsayable, near or far" (1.2.12). "There are *acts* of differentiation and identification, not differences and identities" (1.3.7). "A sentence does not hold together because it is true, but *because it holds together* we say that it is 'true'" (2.4.8). "The word 'true' is a supplement added to certain trials of strength to dazzle those who might still question them" (4.5.8). Yet Latour admonishes us, "do not accuse me of nominalism" (4.4.4). "Semiotics remains inadequate because it persists in considering only texts or symbols instead of also dealing with 'things in themselves'" (2.4.1). "Recently there has been a tendency to privilege language. . . . [T]he attempt was made to reduce all other forces to the signifier. The text was turned into 'the object.' This was the 'swinging sixties,' from Lévi-Strauss to Lacan by way of Barthes and Foucault. What a fuss! Everything that is said of the signifier is right, but it must also be said of every other kind of entelechy" (2.4.5). Agnosticism, again: neither language nor referent, only inscrutable actants. "Demonstrations are always of force, and the lines of force are always a measure of reality, its only measure" (4.7.5). "Words are forces like others with their own times and spaces, their 'habitus' and their friendships" (2.4.4). Indeed, "languages . . . are entelechies like all others" (2.4.4). "We neither think nor reason. Rather, we *work* on fragile materials—texts, inscriptions, traces, or paints—with other people" (2.5.4). "Though there is no proper or figurative meaning, it is possible to appropriate a word, reduce its meanings and alliances, and link it firmly to the service of another" (2.3.2).

All well and good, but let me rephrase my question. Is this grand metaphysic not a *façon de parler*, a metaphor? Given: "If a message is transported, then it is transformed" (2.2.5). And: "Commentary is never faithful. Either there is repetition, which is not commentary, or there is commentary, which is said *differently*. In other words there is translation and betrayal" (2.1.5). Then: does reformulating every event or act as *speech* not "translate/ betray" its ontological character as *event* or *act*?[41] No, it is simply *symmetry*, first proposed in a restrictive sense by Bloor and the Strong Program but now dramatically generalized, rendered *orthogonal* to the line of prior discourse.[42] "Is it a force of which we speak? Is it a force that speaks? Is it an actor made to speak by another? Is it an interpretation or the object itself? Is it a text or a world? We cannot tell, because this is what we struggle about, the building of a whole world" (1.2.9). "We must not believe in advance that we know whether we are talking about subjects or objects, men or gods, animals, atoms, or texts" (1.2.11). "Entelechies cannot be partitioned into 'animate' and 'inanimate,' 'human' and 'nonhuman,' 'object' and 'subject,' for this division is one of the very ways in which one force may seduce others" (3.1.4). "If something resists, it creates the optical illusion among those who test it that there is an object that can be seen and described causing this resistance. But the object is an effect, not a cause. . . . 'Real worlds out there' are the consequences of lines of stable force and not the cause of their stabilization" (4.5.10). Now what sounded like epistemological *epoche* has turned into ontological dogma.

Like the Buddhist philosopher Nagarjuna, Latour pronounces a litany of negative ontology: "It is not a matter of *economics*" (3.4.2). "It is not a matter of *law*" (3.4.3). "It is not a matter of *machines* or *mechanisms*" (3.4.4). "It is not a question of *language*" (3.4.5). "It is not a matter of *science*" (3.4.6). "It is not a matter of *society*" (3.4.7). "It is not a matter of *intersubjective relationships*"—Latour is contemptuous of "people so impoverished as to try to explain nuclear reactors, nation-states, or stock exchanges on the basis of 'interactions'" (3.4.8).[43] "It is not a question of *nature*" (3.4.9). And most emphatically, "it is not a question of *systems*" (3.4.10). "The notion of system is of no use to us, for a system is the end product of tinkering and not its point of departure" (3.2.3). We must—this is the point, after all—start from scratch.

"'Science'—in quotation marks—does not exist" (4.2.1). It "has no standing of its own" (4.2.2). "Belief in the existence of 'science' has its reformers, but it does not have its skeptics, even less its agnostics" (4.1.8). Yet "'science' exists no more than 'language' or 'the modern world'" (4.1.6). "Belief in the existence of 'science' is the effect of exaggeration, injustice, asymmetry, ignorance, credulity, and denial. If 'science' is distinct from the rest, then it is the end of a long line of coups de force" (4.2.6). That assimilates "science" pretty closely to politics. But Latour rejects that as well. "Science is not politics. It is politics by other means" (4.6.2.1). "If it were possible to explain 'science' in terms of 'politics,' there would be no sciences, since they are developed precisely in order to find other allies, new resources, and fresh troops" (4.6.2.1). This, Latour believes, was the great insight of *Leviathan and the Air Pump* by Shapin and Schaffer: they recounted the moment when science presented itself as a new potency: "What could be better than a fresh form of power that no one knows how to use?" (4.6.2.1).[44] I think here we get closest to what Latour finds most interesting in technoscience, namely, the generation of new sources of power in society.

Of course, "science" is not autonomous. Latour spares little breath for the conventional pieties of the "context of justification." "'Science' is a sanctuary only so long as we treat the winners and the losers asymmetrically. Nobody can separate the 'internal' history of science from the 'external' history of its allies. The former does not count as history at all. . . . The latter is not the history of 'science,' it is history" (4.2.5). Taking him up on these last claims, I am interested in what Latour has to say about stabilization, irreversibility, duration: when do "forces" take on facticity, historicity, even "hegemony"?

"A well-defined state of affairs is the work of *many forces*." Individual "networks reinforce one another and resist destruction . . . isolated yet interwoven" (3.2.5). "Each network is sparse, empty, fragile, and heterogeneous. It becomes strong only if it spreads out and arrays weak allies" (3.5.1). Latour offers this to solve the mystery of the European conquest of the rest of the world despite the palpable frailty of each of its agents considered singly.[45] "Each group thus lent its strength to the others without admitting it, and therefore claimed its purity."[46]

Tacit collaboration all the while under strict professions of "purity"— this turns out to be the one "mystery" (read: mystique) of Western "modernity." Latour finds the whole pretense suspect: "There is no such thing as a modern world," he writes.[47] "For years we have *voluntarily* granted to the 'modern world' a potency that it does not have."[48] "Our most intelligent critics have done nothing for the last 150 years but complain of the damage caused by progress, the misdeeds of objectivity, the extension of market forces, the march of concrete in our towns, and dehumanization." "They say this 'modern world' is *different* from all the others, absolutely and radically different."[49] But Latour will have none of that. It is a mystification which demands dispelling. That is the calling of ANT: "We have to be the anthropologists of our own world."[50]

"Is it our fault if the networks are *simultaneously real, like nature, narrative, like discourse, and collective, like society?*"[51] Anthropologists are used to weaving such syncretic narratives about Others. "As far as foreign collectivities are concerned, anthropology has been pretty good at tackling everything at once."[52] "Are they not able to do for our societies what they do so well for savage ones?"[53] Yet they hesitate before our own, because "our fabric is no longer seamless," they believe.[54] That is the modern mystique, a "Constitution" that guarantees the "proliferation of hybrids," all the while conserving the "purity" of discursive domains.[55] "It is with Kantianism that our [modern] Constitution receives its truly canonical formulation. . . . Things-in-themselves become inaccessible while, symmetrically, the transcendental subject becomes infinitely remote from the world."[56] Kant's "phenomena" are really only *epi*phenomena, arising only "through the application of the two pure forms, the thing-in-itself and the subject."[57] "Moderns do differ from premoderns. . . . [T]hey refuse to conceptualize quasi-objects as such. In their eyes, hybrids present the horror that must be avoided at all costs by a ceaseless, even maniacal purification, . . . [yet] this very refusal leads to the uncontrollable proliferation [of hybrids]."[58] That is "the paradox of the moderns," that "the more we forbid ourselves to conceive of hybrids, the more possible their interbreeding becomes."[59] That may be the "secret of the modern world," but "is purification necessary to allow for proliferation?"[60] "Either it is impossible to do an anthropological analysis of the modern world . . . or it is possible . . . but then the very definition of the modern world has to be altered."[61] Latour and ANT pursue this anthropology, and therefore they must maintain "We Have Never Been Modern."

History can serve this anthropological mission in disclosing the *scalar* difference between the Western modern and its Other: "In the past only small collectivities were fortified. . . . With the building of bigger Leviathans it became necessary to pursue more things for longer, to be more exact, more meticulous, and to reach out into the middle of more forces with more laboratories" (4.6.7). Laboratories were the key: "The supplement of force gained in the laboratory comes from the fact that lots of small objects are manipulated many times, that these microevents can be recorded, that they can be reread at will, and that the whole process can be written for people to read" (4.5.2). A scientist will have tried "arguments out dozens of times on small-scale models and made all possible mistakes" (4.5.3). That is why he "confidently emerges from hiding *at the end of the day*" (4.5.3). "Prediction is the repetition of something that has already taken place, scaled up or down" (4.5.4). There is never any "science" the first time. There is only trial (and error). Science works only by accretion, the termite-like tunneling into the pith of things: " 'Science' has no outside, but only narrow galleries which allow laboratories to extend and insinuate themselves into places that may be far away" (4.5.7). ANT might well be termed the termite theory of science.[62]

Science in Action (1987) is Latour's manual for the conduct of science studies.[63] It incorporates and extends the conceptualization of *Laboratory Life*: the theory of *inscription devices*, the transformation of *modalities* in statements, the *agonistic field*—all these reappear at the outset of *Science in Action*. What Latour offers is a comprehensive program: a *general theory of knowledge-production*. The key to the whole work is: "what is called 'knowledge' cannot be defined without understanding what *gaining* knowledge means."[64] I will strongly misread Latour on the basis of my overweening confusion of Latour with a Hegelian processual-learning approach to knowledge in rebellion against a Kantian a priori one.[65] Emergence is the key: "an original event . . . creates what it translates as well as the entities between which it plays the mediating role."[66] Latour proposes that "explanations no longer proceed from pure forms toward phenomena, but from the centre toward the extremes."[67]

The argument of *Science in Action* is a movement from local to global claims and structures of knowledge. One could—playfully—suggest that the exposition traces a "dialectic."[68] We are led from "subjective" to "objective" to . . . *something* we might call "definitive" (of course, "absolute" won't do); that is, from a *private* (and therefore weak) assertion through its *collective* mediation to its *theoretical* entrenchment as knowledge. Latour, a good son of the French 1960s, loathes *Aufhebung*. Yet his Janus figure embodies temporality, contradiction, advance—elements which strongly approximate *Aufhebung*! Above all, the Janus figure epitomizes historicity: prospect and retrospect simultaneously on the cusp of an open present.

Latour, with an exquisite attunement to what historicity entails, constantly seeks to rescue *openness* from the *always already* of the a priori. In Hegel's *Phenomenology,* the "philosopher" knowingly accompanies the groping "subject" on the path to (self- and world-) discovery that the "philosopher" has already traced. That fits: for Latour science or knowledge is always only *recognition.* The risk in recapitulation is: we misremember how we got here. Latour urges we resist the overweening temptation to tell the story as if we knew all along what we only came to know. Philosophy (not just Hegel's)— and science, too—always goes about covering its own tracks, tidying all-too-human process into "pure" products.[69] Latour (and Hegel in what is most memorable in his *Phenomenology*) wants to *un*cover the making.

"When we follow facts in the making . . . we do not need to know what Society is made of and what Nature is; more exactly, we need *not* to know them."[70] "We need to get rid of all categories like those of power, knowledge, profit or capital, because they divide up a cloth that we want seamless in order to study it as we choose."[71] In addition to the charming frankness of its cognitive imperialism, this passage offers an essential methodological premise: we have to *unlearn* if we are to retrieve the historical "gaining" of what is known. We always already "know" too much. Of course, we can't *really* unlearn, not even Latour, and so we only *suspend* or "bracket" to approximate the original openness. But we can always find situations where we really don't know yet (a crucial warrant for the endeavor) to help us remember. Philosophers and scientists don't much like that. Their business is to codify the known: or, at least, that's their business when people like Latour come round asking questions about what they are up to. Catch scientists in the act, at the frontier of their ignorance, and they will *show* you a different order from what they *tell.* That is what Latour keeps trying to get at, what all science studies is trying to get at: not to debunk science but to *reconstruct* it "in the making." "The first time we encounter some event, we do not know it; we start knowing something when it is at least the *second* time we encounter it, that is, when it is familiar to us."[72] Knowledge really is *recognition.* "When the out-thereness is really encountered, when things out there are seen for the *first* time, this is the end of science, since the essential cause of scientific superiority has vanished."[73]

Latour is a post-positivist in his basic premises, like everyone else: "Facts and machines in the making are always *under-determined.*"[74] And the "construction of facts and machines is a *collective* process."[75] That is, these get "collectively stabilized" into "black boxes." "We need others to help us transform a claim into a matter of fact."[76] If something "becomes a fact, it will be included in so many other papers that soon it would not be necessary to write it at all."[77] It would be *tacit knowledge.* It is the same as the machine: we just use it (unless it breaks). "Confronted with a black box, we take a series of decisions. Do we take it up? Do we reject it? Do we reopen it? Do we let it drop for lack of interest? Do we make it solid by grasping it without any further discussion? Do we transform it beyond recognition?"[78] What others *do* with a fact or a machine establishes its durability, its facticity. "A statement was fact or fiction not by itself but only by what the other sentences made of it later on."[79]

"Scientists and engineers invariably argue that there is something behind the technical texts which is much more important than anything they write."[80] But "Nature is not directly beneath the scientific article; it is there *indirectly* at best." So, "what is behind a scientific text? Inscriptions. How are these inscriptions obtained? By setting up instruments."[81] Laboratories *manufacture* facts and machines: "Laboratories generate so many new objects because they are able to create extreme conditions and because each of these actions is obsessively inscribed."[82] Intervention combined with documentation: "Laboratories are now powerful enough to define *reality*."[83] "Nature . . . always arrives late, too late to explain the rhetoric of scientific texts and the building of laboratories."[84] Latour has his Janus make the indicative utterances: the old face proclaims, "Nature is the cause that allowed controversies to be settled," while the young face rejoins, "Nature will be the consequence of the settlement."[85] In trying to understand science in the making "we do not try to undermine the solidity of the accepted parts of science."[86] It is pointless to gainsay this enormous "process of reification." Simply, "the new object emerges from a complex set-up of sedimented elements each of which has been a new object at some point in time and space."[87]

Latour does not want to deny that we know a lot by now. He does not even mind if we call that Nature and if we say that we know it because that's how Nature is. He only wants to remind us that before we knew any of these things that are-what-they-are-because-that's-how-Nature-is, it would not have helped us just to say Nature is that way. Nobody would believe us. And they would be right. Scientists get angry, philosophers shake their sage heads: either it's true or it's not. Either Nature is that way or it isn't. Of course it is—*after* we have agreed, and even then, we *could* be wrong. That is, "as long as controversies are rife, Nature is never used as the final arbiter since no one knows what she is and says. But *once the controversy is settled*, Nature is the ultimate referee."[88] There is an important difference between what Latour is after with this argument and what Collins had in mind with his claim that nature plays no role. Latour has been criticized by his science studies cronies for falling back into *realism!* His whole argument about the nonhuman as actant is an argument for the *real* participation of natural "forces" in the constitution of scientific knowledge and of the "reality" (Nature) which it conceives. Latour does, it bears insisting, recognize that there are experiments which generate the "inscriptions," and that what gets written comes not from the scientist reading the data off but from some "force" on the other side of that inscription, a "force" that can confute. He does play insufferably fast and loose with the distinction of the physical and the discursive, the ontological and the epistemological. He is always already situating the question in terms of what we can *say* to one another of the world, and never conceding a world apart from our discourse on it. He systematically "translates" practice into discourse. There is method mixed in with the "madness" here, but it is certainly off-putting, and not just for natural scientists.[89] As Jan Golinski has noted, Latour does not want to make things easy for historians, either.[90] He is all too comfortable in a cocoon of ironies that appear more irenic than they are. One can read him intolerantly, as Paul Gross and Norman Levitt, John Huth, Alan Sokal, and Oscar Kenshur have. Then, indeed, he is guilty of "gross and blatant error," to use Sokal's disparagement.[91] One can simply find him clever, or worse, *warranted* in "translating" all into discourse. (I believe his colleagues Callon and Law follow that course too far, to say nothing of his postmodern fan club, starting with Richard Rorty.) I suggest that we handle him with a combination of irony and sympathy and take from him what we can use.

Having trained his guns on Nature for so long, Latour surprises us by turning them upon a new target. "After three chapters," Latour suddenly interpolates, "there has been not a word yet on social classes, on capitalism, on economic infrastructure, on big business, on gender, not a single discussion of culture, not even an allusion to the social impact of technology. That is not my fault."[92] "Society" is just as misleading as "Nature" if you want to retrace the development of science, or rather of "technoscience" or "sociologics." It's the "seamless web" again; technology cannot be disaggregated from science, nor can science be from society. To understand anew, we have to erase how we have already understood. The point is, "Society" is as much a result of how we collectively construct what we know as "Nature" is. That is why Latour erased "social" from the phrase "Social Construction of Scientific Facts" in the title of the second edition of *Laboratory Life*. "We have never been interested in giving a social explanation of anything, but we want to explain society, of which the things, facts and artifacts, are major components."[93]

Everything has been constructed together. It has been going on for some time: by now, of course, there are entrenched social classes, economic infrastructures, genders, and all the rest, and Latour is hardly denying any of that. Again, he simply wants to remind us that we *made* (and continually *un*make and *re*make) these things. There is nothing *always already* about them. In its precise sense, in fact, Latour maintains *technoscience* is pretty recent: "as far as numbers are concerned, technoscience is only a few decades old."[94] And it is not evenly distributed in space, either: "Half of technoscience is an American business."[95] Latour can be a hard-boiled sociologist of the present, and when he is, he cuts to the quick: "Technoscience is a military affair. . . . By and large, technoscience is part of a war machine and should be studied as such."[96] (That takes us back to my beginning.) "Scientists have succeeded only insofar as they have coupled their fate with industry, and/or that industry has coupled its fate to the state's."[97] That is because "resources are concentrated in very few hands."[98] The delusion of scientific autonomy gets very short shrift from Latour. "By and large, scientists and engineers have been able to gather support only when they do *not* do basic research. . . . Technoscience is on the whole a matter of development."[99] More cruelly, Latour points out that most scientists don't matter: "the vast majority of the claims, of the papers, of the scientists, are simply *invisible*."[100] One might suspect at this point that scientists would be happier if Latour just went on doing "first philosophy." The funny thing is, sociologists haven't liked him much either. When Latour is done, "the resources the sociologist can use to say intelligible things about science and technology are drastically curtailed."[101] ANT is an insurrection within science studies: an insurrection *against* the sociology of scientific knowledge. To grasp that, we need to extend our consideration from Latour personally to his whole school.

The "Hybrid Collectif" of Latour et Cie.

The emergence of the "actor-network theory" of Bruno Latour, Michel Callon, and John Law proved the key consequence for science studies of the *crise de conscience* of sociological explanation in the early 1980s. The founding manifesto was "Unscrewing the Big Leviathan," which Latour composed with Callon in 1981.[102] In the aftermath of the controversy between Steve Woolgar and Barry Barnes over the interest model in SSK, Callon and Law declared that ANT took a position on neither side, though "closer to Barnes than . . . Woolgar."[103] While they proposed two key departures from the Edinburgh School in focusing on more concrete processes and maintaining agnosticism about imputed interests, they sought not to develop a reflexive critique of all explanation, as in Woolgar, but rather "to discover how it is that actors enrol one another, and why it is that some succeed whereas others do not."[104]

Callon's contribution to the state-of-the-art anthology, *The Social Process of Scientific Investigation* (1981), spelled out more concretely what ANT was after. The title of Callon's essay is itself indicative: "Struggles and Negotiations to Define What Is Problematic and What Is Not: The Sociologic of Translation." Callon set out from the observation that the sociology of science had "radically shifted" focus to the "content of science itself."[105] And "the deeper we delve into content, the more the legitimacy of black box-ism seems questionable."[106] Attacking "the chopped-up, compartmental-ized world the scientists are so patiently building up," this social approach to science made "the most solidly-based concepts dissolve, revealing their ambiguity."[107] Nowhere was that clearer, for Callon, than in the domain of scientific problem formation. "The study of problematisation is vital for un-derstanding the rules governing the mysterious chemistry, the constantly re-newed fusions, which permanently produce the social and the cognitive."[108] Like naturalized epistemologists, Callon directed his attention at the context of discovery, but with a widely different agenda. The language of Callon is loaded with implications of illegitimacy. "All we need to say here about the structure of the un-analysed is this: its structure resembles that of the uncon-scious. It represents what is kept silent so that the rest may be stated."[109] Black boxes, in short, are sinister. A close consideration of problem forma-tion revealed that: "first of all an initial frontier is traced between what is analysed and what is not, between what is relevant and what is suppressed, kept silent. The problematisation carves out a territory which it then cuts off from the outside, forming a closed domain with its own coherence and logic."[110] In short, "consensuses are reached, lasting for shorter or longer periods of time, concealing balances of power."[111] "Setting up a black box" was creating "private 'hunting grounds.'"[112]

I would like to juxtapose a passage from Kuhn's *Structure* for contrast. Kuhn wrote: "When the individual scientist can take a paradigm for granted, he need no longer, in his major work, attempt to build his field anew, starting from first principles and justifying the use of each concept introduced. . . . [T]he creative scientist can begin his research where [normal science] leaves off and thus concentrate exclusively upon the subtlest and most esoteric as-pects of the natural phenomena that concern his group."[113] Bruno Latour has suggested that sociologists of science let the *scientists* decide which black boxes to open and which simply to use.[114] But Callon set out from the "hermeneutic of suspicion." All black boxes were sinister to him *on prin-ciple.* "Problematisation must of necessity rest on elements of reality (con-cepts, proposals, matchings up, results . . .) which are considered irrefutable and firmly established." They are "given the force of certainties and thus to-tally escape suspicion."[115] For those who operate with the presumption of a web of belief, it appears hardly surprising, much less inherently sinister, that some things must be held constant in order to place others under scrutiny. But Callon seeks an *aqua regia* whereby all black boxes must dissolve at once. In terms of Latour's metaphysics, he seeks to catapult us ever again to the (impossible) origin of inquiry.

Callon offered a paradigmatic exemplar of the "socio-logic" through which a network of "actants" simultaneously "enrolls" and "translates" knowledge and social actors in his study of the scallops of St. Brieuc Bay.[116] To understand the process of interaction between the human and the nonhuman in that context, Callon insisted, the scallops had to be recognized fully as "actants," as coproducers of that event of technoscience. Next to nothing was known about them: "though the scallops acted, they acted in mysterious ways."[117] Notoriously, Callon insisted that what was needed was a *spokesperson* for the scallops, "to give scallops the power of speech."[118] A more conventional reading might have been: it was necessary to gain more reliable data on scallop behavior, but ANT systematically styles actants as speech acts, the better then to render nugatory any distinction between a state of affairs and a report of that state.

In the vein of ethnomethodology, Callon and Law insisted that the distinction between content and context was always "negotiated and renegotiated by the actors themselves."[119] The point of ANT is the simultaneous constitution of the social and the cognitive ("socio-logic"), but its metaphysic lies deeper, in the rejection of the Nature-Society polarity for conceptualizing scientific process. Neither the official rhetoric of science, in which Nature dictates and scientists merely take dictation, nor the dogma of SSK, in which Nature has nothing to say but Society constitutes what science is *really* about, offers purchase on the intricacies of actual scientific practice. As an alternative, "the sociology of translation builds on two very simple principles," Callon and Law write. The first is "generalized agnosticism," and the second is "generalized symmetry."[120] In another essay, Callon and Law find that "the trajectories of technological projects are contingent and iterative."[121] The dynamics of these networks can give rise to *convergence* and to *irreversibility.*[122] As Latour and Callon put it, "instead of swarms of possibilities, we find lines of force, obligatory passing points, directions and deductions."[123] They "rely on the notion of translation, or network. More supple than the notion of system, more historical than the notion of structure, more empirical than the notion of complexity, the idea of network is the Ariadne's thread of these interwoven stories."[124] Callon conceives of the elements in network interactions in terms of four ideal types: texts, artifacts, humans, and money, any of which can merge with any other to form "hybrids."[125] All the various possibilities get mobilized in dynamic interaction, in which any can become an "actor," in other words, inaugurate a new impulse ("action") in the network, which then gets circulated, mutating all the other elements and feeding back upon itself via their assimilations. "Quite minimal changes may transform intermediaries into actors, or actors into intermediaries."[126] "In a universe of innovations solely defined by the associations and substitutions of actants, and of actants solely defined by the multiplicity of inventions in which they conspire, the translation operation becomes the essential principle of composition, of linkage, of recruitment, or of enrolment."[127]

The dissolution of the distinction between the human and the nonhuman inaugurates the regime of what Callon called the "*hybrid collectif.*"[128] The premise is that "non-human actors seem to be multiplying."[129] That calls for a reassessment of the very idea of agency, taking up "the *possibility* that agency is an emergent property."[130] Agency should be regarded not merely as *performative* but as *collective*: "A *collectif* is an emergent effect created by the interaction of the heterogeneous parts that make it up."[131] *Material heterogeneity* is crucial, here, because it insures the abandonment of old-fashioned humanist ideas of agency caught up in the obsolete idea of an "actor." Callon and Law embrace the "semiotic turn" for which "there is no useful distinction to be made between content and context. . . . Which is how we get to the slogan, the rallying cry, 'Everything is text; there is only discourse.'"[132] The project is a continuation, in short, of poststructuralism's "decentering of the subject."[133] What must go, apparently, is *intention*: "Once we divorce the notion of agency from the panopticism of strategic intention—once we no longer imagine that the former necessarily implies the latter—then we open up a whole field of empirical and theoretical possibilities."[134]

Somehow, Callon's version, which affirms the semiotic, and Latour's version, which faults it, too, (ironically?) for excluding referents, both collapse into a seamless monism. Steven Shapin got this precisely: "It is the world of the seamless web, a world in which everything is connected to everything else, in which even the discrete existence of things and the categorization of processes cannot be used to interpret or to explain the actions of those who are said to produce them. There is much to be said in favour of monistic impulses and the close inspection of seams, but there is little to be said from within a seamless web. Ultimately, those that truly inhabit the seamless web can say nothing intelligible about its nature, even, if they are consistent, that it is seamless and that it is a web."[135] If all conventional "explanatory enterprises are fundamentally misconceived," Shapin notes, "it is never made clear what sort of enterprise we are being invited to put in the place of explanation."[136]

Latour's reply is: "please simply consider that the history of non-human actors is on a par with that of humans."[137] One must begin with the *event* and find one's way—gropingly, in a new tongue—toward building both nature and society as results. "Instead of providing the explanation, Nature and Society are now accounted for as the historical consequences of the movement of collective things."[138] One might compare this to the "interaction theory" of metaphor: a metaphor is a novelty in which both elements in the analogy get transformed, take on attributes of the other. So it is with the human and the nonhuman. The result is a new actor, actually, an *actant*—a hybrid of both human and inhuman elements. This "hybrid actor" arises through "translation," that "displacement, drift, invention, mediation, the creation of a link that did not exist before and that to some degree modifies two elements or agents."[139] Latour's three homely homilies of the door opener, the speed bump, and the seatbelt all aim to jar us into recognizing the way in which our meanings—indeed, our morals—get "delegated" to the nonhuman for enforcement.[140] "Knowledge, morality, craft, force, sociability are not properties of humans but of humans *accompanied by* their retinue of delegated characters."[141] "Action is simply not a property of humans but of an association of actants."[142] "Purposeful action and intentionality may not be properties of objects, but they are not properties of humans either. They are properties of institutions, *dispositifs*. Only corporate bodies are able to absorb the proliferation of mediators, to control their expression, to redistribute skills, to require boxes to blacken and close."[143] "To *substitute* for the unreliable humans a *delegated nonhuman character*" results in a new "*distribution of competences* between humans and nonhumans."[144] In another homely homily, Latour described the impact of cumbersome weights attached to hotel keys in Europe to make the point that our artifacts alter our acts, that these acts are not only *shared* but *changed*.[145] That is why ANT insists on the hybrid *collectif*: they propose to "substitute *collective*—defined as an exchange of human and nonhuman properties inside a corporate body—for the tainted word *society*."[146] "The prime mover of an action becomes a new, distributed, and nested series of practices."[147] "The distinctions between humans and nonhumans, embodied or disembodied skills, impersonations or 'machinations,' are less interesting [than] the complete chain along which competencies and actions are distributed."[148]

This paradigm shift, Latour suggests, incites reconsideration of our relation with *primates* and with *primitives*. Latour has long been working with primatologist Shirley C. Strum on "Redefining the Social Link: From Baboons to Humans."[149] The project derived from the reconsideration of baboon sociality, discovering that it was not a "stable structure," but rather a process of group members constantly "negotiating what that structure will be."[150] This, of course, was the argument of ethnomethodology about human society, as well. "Social order, the ethnomethodologists argue, is not a given, but the result of an ongoing practice."[151] Yet what Latour and Strum want to underscore is the *difference* of the human communities from the baboon ones in the powers of *entrenchment* which are available to the humans. "The primatologists omit to say that, to stabilize their world, the baboons do not have at their disposal any of the human instruments manipulated by the observer." Conversely, "the ethnomethodologists forget to include in their analyses the fact that ambiguity of context in human societies is partially removed by a whole gamut of tools, regulations, walls and objects of which they analyse only a part."[152]

"What do human collectives have that those socially complex baboons do not possess? Technical mediation."[153] "The more elements one can place in black boxes—modes of thought, habits, forces and objects—the broader the construction one can raise."[154] And this is what constitutes the contrast of "modern" to "premodern" or primitive humanity. If there is an asymmetry about modernity, it is only in the accretion of this retinue of artifacts. That is, the modern "translates, crosses over, enrolls, and mobilizes *more elements,* more intimately connected, with a more finely woven fabric."[155] Indeed, this is Latour's key to all of human history: "why we so constantly recruit and socialize nonhumans. It is not to mirror, inscribe, or hide social relations, but to make them through fresh and unexpected sources of power."[156] History—in more fashionable terms, "genealogy"; in more Latourian neologistics, "pragmatogony"—seeks "to identify, inside the seamless web, properties borrowed from the social world in order to socialize nonhumans, and, vice versa, borrowed from nonhumans in order to naturalize and expand the social realm."[157]

There are irreversibilities, asymmetries. One of the most important is that "at some point in history, human interactions come to be mediated through a large, stratified, externalized body politic."[158] Simultaneously, Latour believes, emerged the modernist project of "creating that peculiar hybrid: a fabricated nonhuman that has nothing of the character of society and politics and yet builds the body politic all the more effectively because it seems completely estranged from humanity."[159] He means *technoscience.* "Nonhumans stabilize social negotiations. Nonhumans are at once pliable and durable; they can be shaped very quickly but, once shaped, last far longer than the interactions that fabricated them."[160] This, however, is too genial a construction. It fails to regard the "primordial structures through which Leviathans are structured."[161] "The Leviathan is such a monster that its essential being cannot be stabilized in any of the great metaphors we usually employ. It is at the same time machine, market, code, body, and war."[162] "Monstrous is the Leviathan in yet another way. . . . [T]here is not just *one* Leviathan but many, interlocked one into another like chimera, each one claiming to represent the reality of all, the programme of the whole."[163] "Each totality is *added* to the others without retrenching itself, thereby producing a hybrid monster."[164] And "the sociologist's language has no privileged relationship with the Leviathan."[165] "Yes, society is constructed, but not *socially* constructed."[166] That is the gauntlet ANT throws down.

It has been taken up. Nick Lee and Steve Brown call "ANT's most controversial feature" the "enfranchisement of nonhumans."[167] It has drawn spirited protest from Harry Collins and Steven Yearley in their key essay, "Epistemological Chicken." They protest that "symmetry between all kinds of actants once more removes humans from the pivotal role" that social studies of science had striven to retrieve in scientific practice.[168] Mocking Callon's essay on the scallops, Collins and Yearley scoff: "Would not complete symmetry require an account from the point of view of the scallops?"[169] As Simon Schaffer puts it, the effort to dissolve the distinction between humans and nonhumans, between language and world, between agency and entity, is a species of "hylozoism" which "directs our attention towards the items whose action is in dispute. It directs our attention away from the forces which help close that dispute. It therefore disables understanding."[170] Schaffer argues that Latour's "hylozoism stifles an account of laboratory life."[171] That is, "sociologists of scientific knowledge have worked hard to understand the complex transitions between the privacy of the laboratory and the publicity of the agonistic field." This "translation of private lab. work to a public domain is just what happens in controversy, and these controversies are deliberately omitted from Latour's story."[172] In trying to make the mystery of concrete scientific practice more accessible, Schaffer suggests, Latour has mystified it with an impossible ontology.

Worse, according to Lee and Brown, by virtue of its "foreclosure on all alternative descriptions and complete ontological monism . . . ANT offers no critique and countenances neither alternative nor supplement."[173] As "the only game in town," ANT offers "a monstrous genesis without offering an exodus."[174] Oscar Kenshur charges that Latour "seems to be postulating an alternative essentialism, one that renounces metaphysical dualism in favor of what appears to all the world to be a metaphysical hybridism."[175] As Yves Gingras puts it, "it is impossible to write or even think without making distinctions," and Latour and ANT take away all the means to do so.[176] What Gingras protests against, I think aptly, is that ANT, vis-à-vis others, always proposes that all the black boxes be dissolved at once. Thus "one would be obliged to reconstruct society (or, more exactly, 'the great whole') from scratch every time one writes a paper on any subject."[177] "This insistence in finding new words . . . suggests that those who use 'traditional' disciplinary categories cannot but succumb to their reification."[178] Yet at every moment in actual historical development there is already in place a configuration—and a lexicon of identification—of actors and intermediaries from which the next event sets off: "It is clear that a *previous* state of the distribution of actors and institutions can be used in the explanation."[179] Indeed, in their own case studies, ANT authors glibly employ all such actants and intermediaries without going into ontological *Angst*.

Have the critics made a valid point? Well, yes: even Latour admits that: "I now realize that no one is ready to abandon an arbitrary but useful dichotomy, such as that between society and technology, if it is not replaced by analytical categories which have at least the same discriminating power as those just jettisoned."[180] "I need to convince you," he writes, "that I can differentiate *much finer* details with the new paradigm than with the former one."[181] "I could take two different paths, one reasonable, the other unreasonable. The first path would be to describe with as many details as I could some modern sociotechnical imbroglios and to show you in what sense machines and machinations participate. . . . Unfortunately, I am going to take a totally unreasonable and speculative path."[182]

What else should one expect from Bruno Latour? Yet he can be more prosaic: "The new paradigm is not without its problems."[183] There is a "lack of narrative resource." Though we "try to alternate between context and content . . . we are not yet expert at weaving together the two resources into an integrated whole."[184] What Latour wants to capture is "the historicity of innovations ever dependent on the socio-logics of actors."[185] Though Latour despises the term, *dialectic* seems to express a similar ambition. Instead of being simply enrolled in his network, I will seek to enroll him in mine: here is stuff for the naturalizing, moderate historicism which I see as the prospect for methodical renewal in empirical inquiry in the (*pace* Latour) "*human* sciences." We can adopt Latour's decisive intervention: to begin with an *event*. "The activity of nature/society making becomes the *source* from which societies and natures originate. . . . History . . . is back in the center. . . . Historicity is back, and it flows from the experiments, from the trials of force."[186] "An experiment is an event and not a discovery," he writes.[187] For an experiment to matter, something has to happen, something that is *incremental* vis-à-vis everything that has gone before. That is why it is an *event*. The world *after* an event is different from the world before. It cannot be *derived* from that prior world, for the meaning of the event is to have changed it. The problem with the "principle of sufficient reason," Latour accepts from Isabelle Stengers, is that it is not a predictive but only a *retrodictive* principle. It is incapable precisely of predicting *emergence*.[188] Hegel, amazingly enough, was more modest for philosophy than Leibniz: he found it possible to comprehend only in arrears.

Latour goes hyperbolic in his lust for origins, for the primordially seamless beginning of knowledge, of history, of language. He forgets what he surely knows: that we have only *situated knowledge*. We cannot suspend present language, we cannot forsake what few narrative resources we have, for the sake of this homage to emergence. To interpret is always to situate in the context of the resources we already possess. Latour missteps in the hypermodernity of his impatience with available language, with ongoing disciplinary frames, with the extraordinary accretion of indispensable black boxes, without which we just do not travel well, even in networks.

Denouement: Another Hedgehog and Fox Story

In 1992, there is good reason to believe, science studies hit a wall. The volume *Science as Practice and Culture* documents the outbreak of a nasty civil war. One destructive engagement documented there was the debate of Collins and Yearley with Latour and Callon.[189] We visited that battleground earlier. In that same year, Latour published an essay that spawned an even more fractious debate between himself and David Bloor.[190] A striking feature of this controversy proved the contrast in styles between the two men. These fit all too readily the old frame of the hedgehog and the fox.[191] Latour saw it clearly: "I have continuously changed my topics, my field sites, my style, my concepts and my vocabulary. . . . David has not moved an inch."[192] Latour is nothing if not a fox; and Bloor is verily a hedgehog. Already in the essay that launched the controversy, Latour set these terms for the debate: "To this day, Bloor has not realized that his principle cannot be implemented."[193] In *We Have Never Been Modern*, Latour praised and damned the Edinburgh School in a single breath:

> It is the glory of the Edinburgh school of social studies of science to have attempted a forbidden crossover. . . . They used the critical repertoire that was reserved for the "soft" parts of nature to debunk the "harder" parts, the sciences themselves!
>
> What had started as a "social" study of science could not succeed, of course, and this is why it lasted only a split second—just long enough to reveal the terrible flaws of dualism.[194]

Bloor did not understand that his limited notion of symmetry "disguised the complete *a*symmetry of Bloor's argument. Society was supposed to explain Nature."[195] Latour insisted on "one more turn after the social turn."[196] The "only way to go on with our work is to abandon this frame of reference and to set up another standard."[197] ANT was designed to displace SSK. And it needed to do so, because, so Latour claimed, "after years of swift progress, social studies of science are at a standstill."[198] "The field is cornered in a dead end from which we want to escape."[199] "We have to make a ninety-degree turn from the SSK yardstick and define a second dimension."[200] "Science in action" and "society in the making" had to be "studied simultaneously."[201]

Bloor was not one to take such criticism without responding. He did in "Anti-Latour."[202] He began with the assertion: "Bruno Latour is a vehement critic of the sociology of knowledge in general, and the Strong Program in particular."[203] As far as Bloor was concerned, Latour's criticisms were "based on a systematic misrepresentation."[204] "Latour's criticism . . . starts by ignoring the fact that the Strong Program is part of a naturalistic and

causal enterprise. From the standpoint of the Strong Program, society itself
is part of nature. . . . Knowledge itself is just one more natural phenomenon."[205] This consistent naturalism, Bloor maintained, sufficed to blunt the
essential charge of Latour's criticism. But then he counterattacked: Latour's
proposed alternative was really a "step backwards."[206]

Because, according to Latour, "the resources of sociology are too crude,"
he offered a "second symmetry principle [which] restores agency to
things."[207] But "Latour never succeeds in giving a clear account of the
process he calls the co-production of science and society."[208] In his effort to
understand the "very basic process of emergence," he made "recourse to the
obscure terminology of monads, entelechies and the like."[209] Latour's new
vocabulary, according to Bloor, was "obscurantism raised to the level of a
general methodological principle."[210] "Nobody can turn every resource
into a topic without finishing up with topics which they have no resources
for tackling."[211] "Latour's attempt to get to a metaphysical bedrock doesn't
work; he can't get away from a pre-existing nature and a pre-existing society. . . . It seems that, after all, we have to begin our investigations into the
nature of knowledge from where we are standing. Our feet are on the
ground of nature, and our position is in the midst of an existing culture, our
own culture. . . . Given that our culture is complicated and pluralistic, and
equipped with a sense of its own history, and divided into opposing traditions, these resources for achieving the necessary role-distance are, I suggest,
as rich as we are ever likely to need."[212] Bloor still confidently affirms
"Durkheim's powerful naturalistic and sociological re-working of Kantian
themes."[213] He remains the unequivocal, scientistic sociologist of 1976.

Latour historicizes Bloor instantly in his rebuttal: "The Strong Program
was useful and still is against the few remaining epistemologists. It has become an obstacle for the continuation of science studies."[214] He tries to "explain why David cannot see those limitations which are so glaring to all of
us."[215] "In spite of twenty-five years of science studies, Bloor has not yet understood that scientists don't observe, nor see the world 'out there.' They are
much more involved than that."[216] "The Edinburgh School has not even begun to understand the first thing about the philosophical originality of science studies, their metaphysics is that of Voltairian materialism. What they
mean by a naturalistic enquiry takes its inspiration from the same type of nature as scientism."[217] Latour avers that he has "learned over the years that
all methodological questions are based on metaphysics, and that every metaphysics is at heart a moral and political issue."[218] He argues that without a
philosophical critique of the "modern Constitution," as epitomized by
Kant, and without elaboration of an alternative metaphysic, science studies—and the study of nature and society more generally—will be at an im-

passe, and a politically dangerous one.[219] He accepts the general methodological point Bloor made, so long as he can add his own twist in the end: "Bloor is of course right in saying that we cannot topicalize every concept and that some should be used as resources and some others as topics. The strategy in any research program is to distribute topics and resources in the most intelligent and fecund way—and, I would add, to move fast and to change tack often."[220] In short, a fox must dance his capers.

In his brief reply, Bloor again challenges Latour's philosophical acumen. He argues that Latour misunderstands the underdetermination thesis. But most of all he derides Latour's "extraordinary and ill conceived metaphysical exercise" involving " 'entelechies,' 'monads' and 'actants' which are neither in nature or society, nor in language."[221] Generally, however, Bloor has difficulty responding to Latour because he cannot distinguish irony from argument. What he reads as an "unstable mixture of concession and defiant dismissal" is really a combination of pity and disdain.[222] Bloor ends up calling Latour a false friend: "My conclusion, and it has saddened me to reach it, is that for many years he has been unwittingly spreading misinformation about the sociology of knowledge. Given his considerable influence in the field, then goodness knows what damage has been done."[223] There is, indeed, a great deal to be saddened by, much of which Bloor can't even see.

Feminism and Science

Nor Quine nor Kuhn, nor Bloor nor Latour seem to have been much concerned about the question of feminism in science studies. As Sandra Harding put it curtly, "gender is no more an analytical tool for the post-Kuhnian thinkers than it was for the more traditional observers of science."[224] Feminists themselves appear to have taken up the "science question" only late in their movement, whose "second wave" began in the 1960s.[225] "Only in the late 1970s did feminists begin to bring to bear on the theories and practices of science and technology the distinctive approaches that had been developing in the social sciences, the humanities, and, more generally, the women's movement."[226] In a very important review/intervention in 1981, Donna Haraway announced: "feminists have now entered the debates on the nature and power of scientific knowledge with authority."[227] Earlier, of course, had come recognition, consistent with wider currents in the movement, that women were systematically discriminated against in the professions of natural science and engineering.[228] But a particularly feminist stake in the question of the meaning and method of natural science only began to appear urgent for feminism around 1980. Haraway's review stated the issues clearly:

Do feminists have anything distinctive to say about the natural sciences? Should feminists concentrate on criticizing sexist science and the conditions of its production? Or should feminists be laying the foundation for an epistemological revolution illuminating all facets of scientific knowledge? Is there a specifically feminist theory of knowledge growing today . . . ? Would a feminist epistemology informing scientific inquiry be a family member to existing theories of representation and philosophical realism? Or should feminists adopt a radical form of epistemology that denies the possibility of access to a real world and an objective viewpoint? Would feminist standards of knowledge genuinely end the dilemma of the cleavage between subject and object or between non-invasive knowing and prediction and control?[229]

Haraway discerned dilemmas in the feminist stance as it took shape at the end of the 1970s. "Feminists want some theory of representation to avoid the problem of epistemological anarchism. An epistemology that justifies not taking a stand on the nature of things is of little use to women trying to build a shared politics."[230] The emergent feminist "critique of bad science" all too swiftly "glides into a radical doctrine that all scientific statements are historical fictions made facts through the exercise of power." But "showing the fictive character of all science, and then proposing the real facts results in repeated unexamined contradictions."[231]

Sandra Harding's *The Science Question in Feminism* (1986) and her 1991 reprise, *Whose Science? Whose Knowledge? Thinking from Women's Lives* stand as the most widely disseminated formulations of feminist critique of science. If the first book sought to situate itself at the culmination of a movement "from the woman question in science" to the "science question in feminism," the second book explicitly situated itself "after the science question in feminism," looking forward to a global integration of feminist critiques with those of other marginalized groups, especially non-Western ones. Harding set out in 1986 to articulate a shift "from a reformist to a revolutionary position" and to take up that position.[232] "Where the Woman Question critiques still conceptualize the scientific enterprise we have as redeemable, as reformable, the Science Question critiques appear skeptical that we can locate anything morally and politically worth redeeming or reforming in the scientific world view, its underlying epistemology, or the practices these legitimate."[233] Harding adopts the latter stance, since "science today serves primarily regressive social tendencies. . . . [It is] not only sexist, but racist, classist, and culturally coercive."[234] Insisting that science cannot insulate itself from responsibility for its uses in wider social processes, she asserts that science "has become the direct generator of economic, political, and social accumulation and control."[235] She holds it in principle impossible to isolate a "pure" science, innocent of all consequence, from "applications" that are mired in the politics of the world. Thus feminism sets out, in Latour's terms, from *technoscience* as the reality of scientific practices. If there can be no purity about *outcomes*, neither can there be about *internal* processes of science. The distinction of the context of discovery from the context of justification, collocating all scientific righteousness in the latter and using it as the bastion of imperviousness to social interests, becomes a central target of feminist critique. In this, Harding and her feminist colleagues take up and elaborate critiques which had already been powerfully deployed in earlier post-positivist thought, though without consideration of gender. Accordingly, it is the question of gender that must properly stand at the outset of an understanding of the distinctively feminist critique of scientific knowledge.

Evelyn Fox Keller summarizes the course of the feminist intervention in science studies as the sequential pursuit of three questions. First, the question arose, why were women excluded from participation in scientific inquiry? Next came the question of the scientific ratification of gender stereotypes through a physiological reductionism in biology. Finally, feminists turned to the influence of gender on the construction of science.[236] Feminist theory in general set out from the effort to distinguish firmly between *gender*, as a social construction, and *sex*, as a biological fixture.[237] That is, feminists mounted theoretical resistance to the practices whereby women were explained/controlled as a function of their biology without regard to their conscious election. This drew feminist attention directly to the science of biology in which this reduction got articulated—to theories of reproduction, to neuroendocrinology, evolution, and cytology.[238] Systematically, they critiqued the presumption that sex determined women's fate, insisting, rather, upon the wide sway of gender, of social construction in the actual constitution of women's lives.[239] But they began to discern a converse pattern. Sex itself was not so fixed, so "natural," as the model of reduction had implied; indeed, the science of biology itself, they claimed, was thoroughly permeated by gendered projections.[240] Thus, feminism advanced from the clarification of a theory of gender in juxtaposition to sex (the repudiation of physiological reductionism) to a clarification of sex (and other elements in biology) as a gendered projection upon the natural order. With that, the wider question of gender-bias in the epistemological and methodological presuppositions of natural science assumed centrality.

Even what Harding termed the "woman question" in science posed grave issues for a complacent view of science-as-usual. The discrimination against women in science recruitment is historically blatant. The rationalizations for it historically have been shamefully lame. Thus, the earliest and apparently least problematic "feminist" point was to insist upon the "liberal" idea of formal equity, of equality of opportunity for women in science. "Focuses on equity," Harding observed, "have been regarded by many as the least radical . . . because they appear not to challenge either the logic of the inquiry process or the logic of scientific explanation."[241] But, paradoxically, the *failure* of the equity effort revealed the *radical nature of the problem:* "it is only because of the fierce struggle waged in the nineteenth and early twentieth centuries to gain formal equality for women in the world of science that we can come to understand that formal equality is not enough."[242] What had been involved went beyond *exclusion;* it had to do with the very *conception* of science itself. "The concepts of women and of knowledge—socially legitimated knowledge—had been constituted in opposition to each other in modern Western societies."[243] That is, "the particular methods and norms of the special sciences are themselves sexist and androcentric . . . constructed primarily to produce answers to the kinds of questions an androcentric society has about nature and social life."[244] "The conceptual scheme of male scientists matches far too comfortably the dominant concepts of ruling."[245] In short, " 'scientific' and 'masculine' are mutually reinforcing cultural constructs."[246] "Masculine bias is evident in both the definition of what counts as a scientific problem and in the concepts, theories, methods and interpretations of research."[247]

It was the convergence of gender *symbolism* (or "totemism"), a gendered division of labor, and a gendered construction of individual identity that made science hostile to women.[248] A way of conceptualizing *order* (gendered symbolism) compelled and preyed upon a *division of labor* ("women's work" vs. prediction and control) and authorized male psychological *identity* ("abstract masculinity"—the "man of reason").[249] Furthermore, it not only excluded women from the position of the knower; it defined them as/ with the object of inquiry/control, "nature." Harding credited Evelyn Fox Keller for the core argument "that it is in the association of competence with mastery and power, of mastery and power with masculinity, and of this constellation with science that the intellectual structures, ethics, and politics of science take on their distinctive androcentrism."[250] Such "abstract masculinity" evinced itself in "excessive preference for quantitative measures," and a "preference for dealing with variables rather than persons."[251] Such "focus on quantitative measures . . . is both a distinctively masculine tendency and one that serves to hide its own gendered character."[252] This "suspicious fit" resulted in the "seamlessness of science's participation in projects supporting masculine domination."[253] Generally, "discourse of value-neutrality, objectivity, social impartiality appears to serve projects of social control."[254]

The question that feminists faced was whether they could work to correct "bad science" while affirming *some* elements of the scientific enterprise— even, for some, accepting without qualification its *ideal*—or they had to re-

WOMEN, ANTS, AND (OTHER) DANGEROUS THINGS 209

pudiate science-as-usual in its entirety. Given the extent of the contamination, the latter seemed more consistent, but it was not without grave costs, not least to the *political* agenda which most essentially constituted feminism.[255] Harding distinguished "feminist empiricists" as those who sought to work within and improve science by discriminating and denouncing "bad science," and "feminist standpoint" theorists as those who sought to challenge the whole edifice of science-as-usual, producing a "successor science" on feminist principles.[256] (She conceived, too, a third category, "postmodern feminists," to whom we will turn soon enough.) Her sympathies lay with the standpoint feminists, and her own theoretical endeavor was to shore up these arguments in the face of challenges from rivals. Her point of departure was the claim that simply *tacking on* feminist concerns to an ongoing and unchanged body of knowledge/practice in science (science-as-usual) could not get the job done. "In the humanities, biology, and the social sciences, it turned out to be impossible to 'add women' without challenging the foundations of those disciplines."[257]

But, as Haraway formulated the dilemma in 1981, how could a radical critique of *all* science leave any possibility for a *feminist* science? Here feminist epistemology had to advance from a critical to a constructive program. The key for Harding was "standpoint" theory. "Knowledge emerges for the oppressed through the struggles they wage against their oppressors."[258] The argument derives from Hegel's master-slave dialectic and its Marxist, especially neo-Marxist elaboration.[259] In general, "more complete and less distorting categories [become] available from the standpoint of historically locatable subjugated experiences."[260] "In a socially stratified society the objectivity of the results of research is increased by political activism by and on behalf of oppressed, exploited, and dominated groups."[261] This is the key claim of feminist "standpoint" theory: "only through political struggle could women get the chance to observe the depth and extent of male privilege."[262] The struggle of women in their situation of subjugation opened up critical (objective and incremental) knowledge which could be transformative for science (and society). "Women's subjugated position provides the possibility of more complete and less perverse understandings . . . a morally and scientifically preferable grounding for our interpretations and explanations of nature and social life."[263]

It must be observed, however, that, in addition to the obvious danger of romanticizing gendered oppression as the path to insight, the same problems regarding the theory of "imputed consciousness" that plagued Lukács in his ideas concerning class consciousness seem to arise here as well. *Who* speaks for women? Feminism is not simply *women*'s unmediated experience, even presuming the counterfactual homogeneity of that social congeries. Feminism cannot be collapsed to *female*, much less *femininity. Critical* intervention—theoretical clarification—is inescapable.[264] But who is *authorized*?[265] The horrors of a "vanguard party" seizing that role have played out fully in the sorry history of "applied" Marxism. Feminism could have no ambition to replicate such a recourse.[266]

That is the political problem for feminism. What concerns me here is the *science question* in feminism, or, more precisely, what feminism can illuminate in the science question. What does it contribute to post-positivist science studies? Perhaps most emphatically, feminism makes "externalist" issues salient *within* science. " 'Science versus society' is a false and distorting image," Harding argues.[267] There is, especially in the world of the late twentieth century, no justification for the presumption of insularity in science. This has been the decisive insistence of science studies. "The sciences are part and parcel, warp and woof, of the social order from which they emerge and which supports them."[268] Thus, "all scientific knowledge is always, in every respect, socially situated."[269] The emphatic repetition of universalizing grammatical forms in this sentence gives indication of the importance of the proposition. Yet Harding, facing up to Haraway's dilemma, insisted that while science was, in Latour's phrase, "politics by other means," it could still "produce reliable information."[270] The crucial question is how Harding could uphold both sides of this assertion. Her strategy entails not so much a construction of the feminist standpoint as a deconstruction of the idea of scientific autonomy.

Harding launches two familiar post-positivist arguments. First, she challenges the conception of the "context of justification" that "is supposed to be powerful enough to eliminate any social biases" via methodological rules.[271] Second, she insists upon the pervasive significance for science of the "context of discovery." With reference to the first, she argues that "there is no impartial, disinterested, value-neutral, Archimedean perspective" from which to adjudicate.[272] Nor will she countenance the recourse to the nonhuman as a "pure" domain, "the false belief that because of their nonhuman subject matter the natural sciences can produce impartial, disinterested, value-neutral accounts of a nature completely separate from human history."[273] *All* scientific knowledge, she insists, is socially situated.[274]

Focusing on the ostensible "context of justification" only mystifies the way in which social interest—in her view, androcentric bias—permeates the "context of discovery," especially in the *definition of problematics,* which she terms "the chief culprit in creating the racism, classism, and androcentrism of science."[275] "Whoever gets to define what counts as a scientific problem also gets a powerful role in shaping the picture of the world that results from scientific research," she writes.[276] "Scientific problematics are often (some would say always) responses to social needs that have been defined as technological ones."[277] Even this does not capture the situation adequately, for this formulation implies *intention* and *reasons*. But "there are some *causes* of scientific beliefs and practices that are to be found outside the consciousness of individual scientists; that is they are not *reasons* for the acceptance or rejection of these beliefs and practices."[278] This leads Harding to criticize laboratory ethnography: "important causes of scientists' everyday activities and experiences are to be found far distant from the laboratory or field site."[279] *Background assumptions*—especially those shared by an entire research community—will have entered their endeavor *unexamined*. "Our cultures have agendas and make assumptions that we as individuals cannot easily detect."[280] That is what makes scientific inquiry into science necessary.[281] Scientists are not and cannot be aware of all the elements that constitute their practices. It is for the science of science to take up the "task of critically identifying all those broad, historical social desires, interests, and values that have shaped the agendas, contents, and results of the sciences."[282]

Indeed, "extending the notion of scientific research to include systematic examination of such powerful background beliefs" is what Harding advocates as "strong objectivity," her personal answer to Haraway's dilemma and its relativist implications.[283] "A strong notion of objectivity requires a commitment to acknowledge the historical character of every belief or set of beliefs—a commitment to cultural, sociological, historical relativism. But it also requires that judgmental or epistemological relativism be rejected."[284] While recognizing the inevitability of historical and cultural contingency, there are, for Harding, "rational or scientific grounds for making judgments between various patterns of belief."[285] We can "apply rational standards to sorting less from more partial and distorted belief."[286] She rejects the disjunctive formulation of judgmental relativism that "if one gives up the goal of telling one true story about reality, one must also give up trying to tell less false stories."[287] Harding believes that conducting the "science of science"—exploring and retrieving the background beliefs and subjecting them to a critique (on *political* grounds of democratic inclusiveness)—will provide criterial perspective: "a scientific account of the relationships between historically located belief and maximally objective belief."[288] Not only are scientific beliefs "socially situated, but they also require a critical evaluation to determine which social situations tend to generate the most objective knowledge claims."[289] While such avowals evoke great sympathy, they remain at the very least in need of a good deal of epistemological elaboration.

In the 1980s, two highly divisive controversies broke out within feminism. The first disputed the "woman's way of knowing" advocated by "standpoint" theory as a dangerous and insupportable "essentialism."[290] The second invoked a radical relativism to sever feminism from any residue of the Enlightenment. Both challenges invoked "postmodernism." The result was what Linda Alcoff called an identity crisis for feminist theory.[291] The category *woman* shattered into that of many different *women*.[292] This fracturing occurred not merely at the hands of the "theoretical" critique of poststructuralism but also and indeed primarily at the hands of a sustained critique of the ethnocentrism of white Western feminism by "women of color."[293] This essentially *political* critique, while it fragmented feminism into a set of potentially conflicting ("hyphenated") feminism*s*, remained committed to a fundamentally liberatory agenda in which claims to knowledge, grounded in *experience* (*inter alia*, of subjugation), remained possible and necessary.[294]

The theoretical critique, launched from within Western feminism by assimilation of high European poststructuralist theory, challenged the very idea that there was any nature, however diffracted, to be found in the category *women*. Indeed, *categories* altogether suffered from inveterate "phallogocentrism" best combated, poststructuralism insisted, by the dissolution of any subject position, any claim to authority, resituating all such notions in the infinitely disseminative field of textuality.[295] Poststructuralist or postmodernist feminism found the position of cultural relativism inevitable and politically appropriate, insisting that the proper role of feminism was *critique:* disruption of authoritative discourses rather than constitution of a "successor science" with its own authoritarian posture.[296] Kenneth Gergen, a male postmodernist, suggested this was the inevitable outcome for feminist critique of science.[297] Obviously, here Haraway's dilemma got raised to a dogma, and it disturbed not only standpoint feminists like Harding but even Haraway herself.[298]

In the early 1980s Haraway discerned in the field of primatology a "feminist scientific revolution."[299] Primatology acted at once "as natural science, political theory, and science fiction."[300] "Monkeys and apes have been enlisted in Western scientific story telling to determine what is meant by human," to talk "about origins, about the nature of things."[301] Primatology recounted "the simultaneous and repetitive constitution and breakdown of the boundary between human and animal."[302] Since "women and animals are closer epistemically to each other within the tortuous logic of nature and culture than are man and animal," it followed that "the possible meanings of being female in the primate order are at the center of primatological discourse."[303] With the decisive intervention of feminism in primatology, "female sex has become very active, social, and interesting—not to mention orgasmic across the primate order—in the last 15 years."[304] That was the scientific revolution.[305]

Haraway insisted upon situating primatology within the larger transformations of knowledge/power in the late twentieth century. Fieldwork in primatology really began only in the late 1950s and took off dramatically after 1975. And this fieldwork took place far from the metropolitan West. Thus, for Haraway, primatology should be construed as "simian orientalism," part of the history of colonial discourse, "the history of race, sex, and class in a world capitalist system."[306] But the world capitalist system itself was mutating, and with it scientific culture in the metropole. That is, modes of analysis in biology got "subordinated to explanatory strategies from the market." More specifically, by the late twentieth century, "both sex and mind have been recast from organismic molds into technological-cybernetic ones."[307] Accordingly, it has become plausible to "image such bodies and communities in metaphors, as hybrids of organism and machine, of animal and human." Hence Haraway's signature metaphor of the *cyborg*: "the sciences and science fictions of the late 20th century are full of these cyborg metaphors."[308]

In 1985, Haraway published an imaginative and influential "Cyborg Manifesto" to elaborate this notion.[309] It set out from the claim that in "our time, a mythic time, we are all chimeras, theorized and fabricated hybrids of machine and organism."[310] Indeed, Haraway contended, "late 20th-century versions of nature are more about simulacra than about originals."[311] Three boundaries had decisively blurred to occasion this situation. First, the boundary between human and animal had been "thoroughly breached." Second, "late twentieth-century machines have made thoroughly ambiguous the differences between natural and artificial, mind and body, self-developing and externally designed." Third, the "boundary between physical and non-physical is very imprecise for us."[312] All three dissolutions were part of a larger technological-geopolitical transformation: "*the translation of the world into a problem of coding,*" which Haraway termed the "informatics of domination," a "movement from an organic, industrial society to a polymorphous, information system" directing "rearrangements in world-wide social relations tied to science and technology."[313] In this brave new world, "it is not clear who makes and who is made in the relation between human and machine. It is not clear what is mind and what body in machines that resolve into coding systems."[314] In the leading sector, "communications sciences and biology are constructions of natural-technical objects of knowledge in which the difference between machine and organism is thoroughly blurred."[315]

In such a context, Haraway concludes, "we need fresh sources of analysis and political action," and she finds them in Bruno Latour. She cites Latour's *Pasteurization of France* as her source.[316] Earlier she used Latour's phrase, "politics by other means," to title an important essay on primatology.[317] Indeed, there is a strong and sustained impact of Latour on Haraway's thinking. "Technologies and scientific discourses can be partially understood as formalizations, i.e., as frozen moments, of the fluid social interactions constituting them, but they should also be viewed as instruments for enforcing meaning," she writes, encapsulating the thesis of *Science in Action*.[318] In particular, Haraway admired Latour's discussion in "Irreductions," the second, theoretical, part of *The Pasteurization of France*: Latour "does not 'reduce' science to politics, to arbitrary power rather than rational knowledge . . . his is an argument against reduction of any kind and for attention to just what the 'other means' are."[319] His "brilliant and maddeningly aphoristic polemic against all forms of reductionism makes the essential point for feminists: *'Méfiez vous de la pureté; c'est la vitriol de l'âme.'* . . . Latour is not otherwise a notable feminist theorist, but he might be made into one."[320] Haraway's cyborg metaphor is built upon Latour's hybrid vision, "a matter of fiction and lived experience."

What Haraway proposes to do is to use all this to construct an "ironic political myth" of socialist-feminist resistance. "My cyborg myth is about transgressed boundaries, potent fusions, and dangerous possibilities," "a cyborg world [where] people are not afraid of their joint kinship with animals and machines, not afraid of permanently partial identities and contradictory standpoints."[321] This "postmodernist, non-naturalist stance" she proposes to be "in the utopian tradition." "My hope is that cyborgs relate difference by partial connection rather than antagonistic opposition, functional regulation, or mystic fusion."[322] "What kind of politics could embrace partial, contradictory, permanently unclosed constructions of personal and collective selves . . . ?"[323] Certainly, she avers, it would be one that would have "no truck with bisexuality, pre-oedipal symbiosis, unalienated labor, or other seductions to organic wholeness." Rather it would take *"pleasure* in the confusion of boundaries" and *"responsibility* in their construction."[324] For cyborgs, writing was the key technology: "the struggle for language and the struggle against perfect communication, against the one code that translates all meaning perfectly, the central dogma of phallogocentrism."[325] For Haraway, "production of universal, totalizing theory is a major mistake," but so too is "anti-scientific metaphysics, a demonology of technology."[326] Instead, "destabilizing an origin story is perhaps more powerful in the deconstruction of the history of man than replacing it with more progressive successors."[327] In short, "feminism must be opposed to holistic organicisms if it is to avoid logics and practices of organic domination."[328] Instead, one should opt for "partial realities that value serious difference."

The "Cyborg Manifesto" was one of the most brilliant briefs for postmodernism in the feminist movement—indeed, *tout court*. All the more telling, then, is the revisionism which marked Haraway's next major essay, "Situated Knowledges: The Science Question in Feminism and the Privilege of the Partial Perspective" (1988).[329] This essay, composed as a commentary on Harding's *The Science Question in Feminism,* offered a meditation on the "inescapable term 'objectivity.' "[330] Haraway described her own intellectual development in terms of the absorption of two powerful theoretical-critical perspectives: social constructivism in science studies and poststructuralism in the human sciences. The first taught that "all knowledge is a condensed node in an agonistic power field," that "all drawings of inside-outside boundaries in knowledge are theorized as power moves, not moves towards truth." The second presented "always already absent referents, deferred signifieds, split subjects, and the endless play of signifiers," engendering a "hyper-real space of simulations." In sum, "the strong programme in the sociology of scientific knowledge joins with the lovely and nasty tools of semiology and deconstruction to insist on the rhetorical nature of truth, including scientific truth." But, she now acknowledged, "the further I get with the description of the radical social constructivist programme and a particular version of postmodernism, coupled to the acid tools of critical discourse in the human sciences, the more nervous I get." Pragmatically, politically—and therefore cognitively—"we would like to think our appeals to real worlds are more than a desperate lurch away from cynicism." Accordingly, Haraway proposed a new sobriety: "We cannot afford these particular plays on words— the projects of crafting reliable knowledge cannot be given over to the genre of paranoid or cynical science fiction." Taking up the dilemma she had earlier enunciated herself, she wrote: "no matter how much space we generously give to all the rich and always historically specific mediations through which we and everyone else must know the world," it now appeared that "feminists have to insist on a better account of the world; it is not enough to show radical historical contingency and modes of construction for everything." She called for "some enforceable, reliable accounts of things not reducible to power moves and agonistic, high status games of rhetoric or to scientistic, positivist arrogance." More concretely, the "problem is how to have *simultaneously* an account of radical historical contingency for all knowledge claims and knowing subjects, a critical practice for recognizing our own 'semiotic technologies' for making meaning, *and* a no-nonsense commitment to faithful accounts of a 'real' world."[331]

Haraway then makes a decisive contribution to solving this problem: "feminist objectivity means quite simply *situated knowledges.*"[332] Renouncing the all-or-nothing "god-trick" of a view from nowhere, Haraway nonetheless refuses to sink into relativism. For Haraway, "relativism is the perfect mirror twin of totalization in ideologies of objectivity." "The alternative to relativism is partial, locatable, critical knowledges, sustaining the possibility of webs of connections called solidarity in politics and shared conversations in epistemology."[333] "Only partial perspective promises objective vision. . . . Partial perspective can be held accountable for both its promising and its destructive monsters."[334] Haraway does not see this as an abandonment of postmodernism: "We seek not the knowledges ruled by phallogocentrism (nostalgia for the presence of the one true Word) and disembodied vision, but those ruled by partial sight and limited voice."[335] Still, there is the prospect for actual knowledge. "The knowing self is partial in all its guises . . . it is always constructed and situated together imperfectly, and *therefore* able to join with another, to see together without claiming to be another."[336] Moreover, "better accounts of the world, that is, 'science,'" require that "the world encountered in knowledge projects is an active entity," that "the object of knowledge be pictured as an actor and agent, not as a screen or a ground or a resource, never finally as a slave." The world is not passive: "feminist objectivity makes room for surprises and ironies at the heart of all knowledge production; we are not in charge of the world."[337]

Such situated knowledge must resist "unequal translations and exchanges—material or semiotic—within the webs of knowledge and power."[338] Ideal speech conditions hardly prevail. Instead, "politics and ethics ground struggle for . . . what may count as rational knowledge."[339] Apropos the argument of feminist standpoint epistemology, the invocation of the vantage of the subjugated, Haraway observes, "there is no unmediated vision from the standpoints of the subjugated," and she cautions against "romanticizing and/or appropriating" that vantage.[340] Not identity—even that of the subjugated—but "critical positioning" occasions objectivity.

With this revisionism, I submit, Haraway makes possible a more fruitful dialogue with "feminist empiricism" in defining this "critical positioning" of situated knowledge. In particular, Helen Longino's "contextual empiricism" offers important resources for the feminist approach to science studies.[341] The history of Longino's relation to Harding and to Haraway is not entirely congenial. Harding had targeted one of Longino's early essays as her exemplar of "feminist empiricism" and its inadequacies.[342] That piqued Longino. In her own work, accordingly, a trail of rejoinders followed, in which Harding came in for criticism both for misunderstanding what a genuine feminist empiricism stood for and for articulating a standpoint theory with deep problems of its own.[343] Similarly, Longino found herself at some distance from the posture Haraway had taken by 1980 as a radical postmodernist feminist. Yet what is striking is the convergence in their positions that appeared possible—at least to Longino—by the early 1990s. In an important distillation of her own viewpoint in 1992, Longino endorsed "an account of knowledge as partial, fragmentary, and ultimately constituted from the interaction of opposed styles and/or points of view," which she claimed was "postmodernist in spirit."[344] The challenge for a feminist approach to science, she went on, was "to reconcile the claim that scientific inquiry is value- or ideology-laden *and* that it is productive of knowledge," in other words, to resolve the dilemma Haraway had identified in feminist theory of science from the beginning.[345] That seems a clear gesture of accommodation toward Haraway. In her book of 1990, Longino had recognized a shift in Haraway's position from the early "disdain, if not outright contempt for the rhetoric of truth and objectivity that is the mark of scientific texts," to the position in "Situated Knowledges" that reasserted objectivity as "recognition of the local, mediated, situated, and partial character of one's knowledge."[346] Still, Longino found it necessary to object that "it is not individual recognition of partiality or, as used to be said, of one's subjectivity but the subjection of hypotheses and theories to multivocal criticism that makes objectivity possible."[347] Indeed, specification of a concrete social epistemology along these lines is the distinguishing feature of her own feminist approach, which Longino calls "contextual empiricism."

Science as Social Knowledge is a work equally of post-positivist philosophy of science and of feminist critique of science. Longino sees the two projects as mutually implicative: "by focussing on science as practice rather than content, as process rather than product, we can reach the idea of a feminist science through that of doing science as a feminist."[348] Longino has no interest in making a case for a uniquely feminist *method*. She fears that standpoint feminism ineluctably confused feminism with femininity, a political strategy with an existential "way of knowing" ascribed to women—and indeed to women constructed in a sexist society.[349] The correct way to carry forward the political strategy, she suggests, is to situate feminist critique of science within a broader science studies frame. That is, feminism should not presume that the theory of gender suffices for the critique of natural science; instead, it should take up what post-positivism had already achieved and *use* it for the feminist agenda.

That was what Longino proposed to do. Longino's monograph systematically recapitulated the trajectory of post-positivism in science studies in terms of a juxtaposition of two traditions: "positivist empiricism" (what this book has termed the "Received View") and "Kuhnian wholism"—specifically, "the version of Kuhn . . . that has found its way into the social studies of science, that is, a Kuhn that licenses, or even mandates, accounts of knowledge construction that appeal to causes other than what the philosopher would recognize as good reasons."[350] While the first tradition approached science predominantly from a normative vantage, the second did so from a primarily descriptive one. The Received View took its normative prescription as at the same time an accurate description of actual science, but that had been exploded by the work in history and sociology of actual science. Post-positivism in science studies had established that "while empiricism with respect to knowledge may provide constraints on justification in empirical science, it is *not* a description of how inquiry proceeds or theories are developed."[351] But the issue, as Longino viewed it, lay not in accurate description but in adequate *adjudication,* in other words, the revindication of the normative in actual science. The essential tension was over the place of *value* in science. She minced no words about her view: it was "nonsense to assert the value-freedom of natural science."[352] What that claim actually meant was a contrast between intrinsic (in her terms, "constitutive") or cognitive values and external (in her terms, "contextual") or social values. The Received View insisted that "good" science admitted only constitutive values. It offered two rationalizations, which Longino identified under the rubrics of the *autonomy* and the *integrity* of science.[353] The claim to autonomy was that science proceeded without concern for its social embeddedness. In an age of technological capitalism, that was laughably false in terms of both inputs and outcomes. In effect the defense of value-free science fell back on the argument for integrity, in other words, "internal practices" in the "context of justification," where "inquiry [was] protected by methodology from values and ideology."[354] Here the key notions were objectivity and rationality, and the question was: "Are there criteria or standards of truth and rationality that can be articulated independently of social and political interests?"[355] As such, Longino argued, "the question of the integrity of science is misconceived."[356] Not only had post-positivism demonstrated that value-neutrality was *empty,* but feminism had demonstrated that it was *pernicious.*[357] Her monograph aimed not only to substantiate each of these two claims, but to propose a basis to get beyond the whole idea.

Taking up post-positivist philosophy of science as a feminist meant for Longino registering that the theory-ladenness of data and the underdetermination of theories by data left a "logical gap" that could not be covered by purely internalist reasoning ("constitutive" values). What careful analysis of the two key tenets of post-positivism revealed was the decisive role of "background assumptions" in both the constitution of data and the adjudication of argument. In short, "evidential reasoning is context dependent."[358] More explicitly, "there is always much more going on about us than we are aware of, not just because some of it is beyond our sensory thresholds or behind our backs but because in giving coherence to our experience we by necessity select out some facts and ignore others."[359] Not only is experience *active* in this selectivity, but it is *social* in its principles of organization: "experience itself must be rethought as an interactive rather than a passive process."[360] This has to be taken into account to formulate an adequately post-positivist idea of objectivity. "What we are looking for in the account of objectivity is a way to block the influence of subjective preference (read: ideology) at the level of background assumptions involved in observation and inference, and of individual variation in perception at the level of observation."[361] The only way forward, Longino insisted, was to "understand the cognitive process of scientific inquiry not as opposed to the social, but as thoroughly social."[362] The crucial point is that objectivity is a collective/communal, not an individual achievement.[363] "It is not the individual's observations and reasoning that matter in scientific inquiry, but the community's."[364] That is, objectivity is "a process of critical emendation and modification of . . . individual products by the rest of the scientific community."[365] In science what objectivity there is must be "secured by the social character of inquiry."[366] That makes objectivity a very complicated matter: "contextually located background assumptions play a role in confirmation as well as in discovery," and "it is the social character of scientific knowledge that both protects it from and renders it vulnerable to social and political interests and values."[367]

Why should such a social epistemology be accepted by philosophers of science or scientists still convinced of the integrity of science? Longino offers a series of nested claims: "Empirical arguments support the claim that science just *is* a social practice; conceptual arguments support the claim that the cognitive practices of scientific inquiry are best *understood* as social practices; and logical and philosophical arguments support the claim that if science is to be nonarbitrary and minimize subjectivity, it *must* be a social practice."[368] The evidence in the empirical arguments derives from all the work done in history and sociology of science, and it is compelling.[369] The notion of conceptual argument is even more crucial. What Longino proposes is that strictly evidential issues do not suffice in actual scientific practice, in light of the two key tenets of theory-ladenness and underdetermination, and that consequently, there must be supplementary arguments about how data get constituted and about how interpretations get evaluated.[370] These arguments are ubiquitous in actual science and they necessarily entail the introduction of "background assumptions."[371] "Scientific knowledge . . . rests on a bed of presuppositions about what questions are important, what sorts of connections are meaningful, about the general direction of causal relations (or more precisely, about which causal relations are worth investigating or establishing)."[372]

The strictly logical concomitant is that scientific inference is more than logical derivation. The logical gap is both real and constitutive of actual controversy and uncertainty in science; in other words, it makes contingency and fallibility inevitable. "A consequence of embracing the social character of knowledge is the abandonment of the ideals of certainty and of the permanence of knowledge."[373] But that makes all the more salient the italicized word "*must*" in her formulation: the very constitution of a community *of inquiry*—in other words, one in which the demand is for something beyond subjective expressionism—entails a specific normative element to which all members of a scientific community acknowledge allegiance.

To achieve the maximal degree of objectivity. Longino identifies a set of four normative conditions. First, a scientific community must have recognized avenues for criticism. Second, there must be at least some shared standards. Third, the community must be responsive to criticism. Finally, intellectual authority must be equal.[374] These are *normative* conditions, that is, regulative ideals: their fulfillment is a matter of *degree*.[375] But their presence is indispensable. Particularly important are the terms of the second and fourth conditions. "In order for criticism to be relevant to a position it must appeal to something accepted by those who hold the position criticized . . . public standards or criteria to which members of the scientific community are or feel themselves bound."[376] Concretely, this is what constitutes *membership* in the community. "It is the existence of standards that makes the individual members of a scientific community responsible to something besides themselves."[377] But these standards are diverse, open-ended and nonhierarchical enough to ensure considerable slack and controversy.[378] Condition four is clearly "Habermasian."[379] The important point is not the full attainment of the ideal speech situation but that "objectivity is dependent upon the depth and scope of the transformative interrogation that occurs in any given scientific community."[380] That depends on the "interactive dialogic community," such that "effective criticism of background assumptions requires the presence and expression of alternative points of view."[381] In short, "the greater the number of different points of view included in a given community, the more likely it is that its scientific practice will be objective." That is a "necessary but not a sufficient condition," of course.[382]

If objectivity is a contextual (socially constituted) normative commitment encoded in background assumptions, and it is also through these same background assumptions that ideology and bias intrude upon science, Longino argues that "we cannot restrict ourselves to the elimination of bias but must expand our scope to include the detection of limiting interpretive frameworks and the finding or construction of more appropriate frameworks."[383] That is, "a method of inquiry is objective to the degree that it permits *transformative* criticism."[384] A scientific community must always take cognizance of its "object of inquiry," Longino argues, taking up a term from Michel Foucault. "The kind of knowledge sought and represented in a specification of the object of inquiry functions as a goal determining constitutive values. It stabilizes inquiry by providing assumptions that highlight certain kinds of observations and experiments and in light of which those data are taken to be evidence for given hypotheses. It also provides constraints on permissible hypotheses."[385] As Longino notes, "contextual analysis shows that such objects are constituted in part by social needs and interests that become encoded in the assumptions of research programs."[386] In acknowledging the contextual constitution of the object of inquiry, a scientific community reckons with its "accountability and choice."[387] Conversely, the perniciousness of the idea of value-free inquiry is that "by concealing the reliance of inquiry on a background of assumptions of very mixed character, it discourages the investigation of alternative frameworks."[388] Always, "social interactions determine what values remain encoded in inquiry and which are eliminated."[389]

Again, what Longino develops is a tenuous balance between agreement and dissent in the scientific community. "To say that a theory or hypothesis was accepted on the basis of objective methods does not entitle us to say that it is true but rather that it reflects the critically achieved consensus of the scientific community."[390] That is, "we can read consensus in a community as signalling belief that certain fundamental assumptions have endured critical scrutiny."[391] This is essential to pragmatic scientific endeavor: "the utility of scientific knowledge depends on the possibility of finding frameworks of inquiry that remain stable enough to permit systematic interactions with the natural world."[392] But "when . . . background assumptions are shared by all members of a community, they acquire an invisibility that renders them unavailable for criticism."[393] Nevertheless, while these assumptions "are largely invisible to practitioners within the community," they are "articulable and hence in principle public," in other words, "available to critical examination, as a consequence of which they may be abandoned, modified, or reinforced."[394]

The danger of consensus motivates "embracing multiple and, in some cases, incompatible theories that satisfy local standards."[395] In that context, Longino suggests that some constitutive values proposed by feminists for science might be better alternatives than those which have hitherto been articulated within mainstream thought, for instance in Kuhn.[396] Kuhn mentioned five constitutive values: accuracy, simplicity, internal and external consistency, breadth of scope, and fertility.[397] Longino observes that accuracy really had to do with empirical adequacy, and that only the others could serve in the context of full underdetermination. Her claim is that these values do not offer the same prospect for an objective science as a set of constitutive values developed by feminist critics: "novelty, ontological heterogeneity, mutuality of interaction, applicability to human needs, and diffusion or decentralization of power."[398] Longino juxtaposes ontological heterogeneity to the conventional value of simplicity, arguing that the former is a far superior guard against reductionism. Similarly, she finds novelty a superior standard to external consistency. "I don't want to say the traditional virtues are always politically regressive," she writes, "but that the fact that they sometimes are means that we cannot treat them as value-neutral grounds of judgment."[399] Longino is explicit that the constitutive values proposed by feminists are "neither uniquely nor intrinsically feminist." But they are "more likely to satisfy feminist cognitive aims," that is, to expose any "asymmetrical power relation that both conceals and suppresses the independent activity of those gendered female."[400] That is what she has all along meant by "doing science as a feminist." "We should worry more about the concealing of political agendas behind the mantle of scientific neutrality than about the consequences of abandoning the illusion of neutral arbiters of our cognitive practices."[401] While that is certainly plausible, this reader remains troubled by the question, "what is left to adjudicate scientific disputes?"[402] That, after all, is the dilemma for feminist theory of science which Longino's "contextual empiricism" was designed to address.

The Rejection of Feminist Epistemology

Feminist epistemology consists of theories of knowledge created *by* women, *about* women's modes of knowing, *for* the purpose of liberating women. By any reasonable standard, it should have expired in 1994. Working independently, Gross and Levitt in *Higher Superstition*, Sommers in *Who Stole Feminism?* as well as Patai and Koertge in *Professing Feminism* all identified the fatal flaws in the feminist epistemological program. More detailed analyses appeared in *Feminist Epistemology: For and Against*, a special issue of the *Monist*, edited by Haack. The simple bottom line of all these critiques is succinctly expressed by Pinnick in a 1994 issue of *Philosophy of Science*: "No *feminist* epistemology is worthy of the name, because such an epistemology fails to escape well-known vicissitudes of epistemic relativism. The central thesis of this article is that *feminist* epistemology should not be taken seriously."[403]

With this summation, Noretta Koertge announced the total fruitlessness of the feminist approach to epistemology and science studies. What is the force and the outcome of this counterattack?

A particularly virulent instance of this attack is to be found in Paul Gross and Norman Levitt's *Higher Superstition*, in the chapter "Auspicating Gender." This chapter begins with sweeping concessions on two of the three issues raised in the feminist critique of science, only to repulse the third in a manner which goes far toward withdrawing the very concessions of the outset. That is, Gross and Levitt do not contest the drastic discrimination against women in the recruitment into science (though they make statements about current equity which are quite problematic). They add: "at times, baseless paradigms in medicine and the behavioral sciences have been pretexts for subordinating women. Pseudoscientific doctrines of innate inferiority and moral frailty have been used to discount female capacity for achievement and to confine women to subservient roles. All this is beyond dispute."[404] But what Gross and Levitt dispute is "claims to go to the heart of the methodological, conceptual, and epistemological foundations of science."[405] They assert, in italics: "*We would have to be shown that there are palpable defects, due to the inadequacies of a male perspective, in heretofore solid-looking science and that the flawed theories can be repaired or replaced by feminist insights.*"[406] I would respond with the passage I cited earlier from them: did medicine and biology entertain—and for sustained periods extending all-too-close to the present—those *now* disparaged "pseudoscientific" ideas, or not? And what was it that led scientists like Gross and Levitt *now* to dismiss these *as* pseudoscientific? I submit that a substantial factor was the feminist critique. To maintain that feminist criticism of science amounted to nothing more than "metaphor mongering," and to conclude that "feminist linguistic criticism has not yet produced a single revision of the body of serious science" is to show such a callous disregard for evidence that it has led to some stark inferences about their intentions, notwithstanding their avowal that antifeminism is now gone from the academic and specifically the scientific scene.[407] To be sure, they are able to pick up on some unfortunate excesses from the feminist literature to dismiss the whole enterprise. Thus they take up Harding's rash and unnecessary stipulation that "feminism will not succeed in 'proving' that science is as gendered as any other human activity, unless it can show that the specific problematics, concepts, theories, language and methods of modern physics are gender-laden."[408] That is hardly called for, but Gross and Levitt trumpet it as the proper standard for the debate. More to the point, they can discriminate no difference between Harding's "standpoint feminism" and the positions of Keller and Longino, whom they credit with being "anything but inept" and therefore possibly acceptable.[409] *Helas!*—to use their own rhetoric—no feminist could ever satisfy them, because the very idea that social and political concerns could have normative bearing on a scientific claim, for good or ill, is something that they refuse to consider. Anyone who will consider it, no matter in how carefully constructed a framework, must be written off as just another ideologist, "neither stronger nor more convincing than . . . other feminist epistemologists."[410]

Similarly Noretta Koertge, speaking with assurance on behalf of "friends of science and reason," conducted a defense of science from the intrusion of any concern for democracy, for fear it would "damage the ethos of free inquiry."[411] As far as Koertge could discern, Helen Longino was an "inveterate ideologist" since she maintained that "feminist science practice admits political considerations as relevant constraints on reasoning."[412] All the careful qualifications in which this particular point was situated in Longino's argument get even less consideration from Koertge than they did from Gross and Levitt! It sufficed: scientific purity had no place for democratic concerns.[413] Longino's four carefully articulated normative constraints for the objectivity of a scientific community's internal practices never even get mentioned. Frankly, it is hard to tell "ideologues" apart here.

What had feminist critique of science achieved by the mid-1990s? The question that seemed apposite was: "has feminism changed science?" Evelyn Fox Keller and more extensively Londa Schiebinger addressed themselves to this question.[414] Schiebinger found little to be pleased about in either "liberal feminism" applied to science ("feminist empiricism") or "difference feminism" in science ("standpoint" theory). She treated both of them in a section revealingly entitled "Blind Alleys."[415] At least on this point, she aligned herself with the postmodernists who insisted on diffraction. She concluded: "women's historically wrought differences from men, then, cannot serve as an epistemological base for new theories and practices in the sciences. There is no 'feminist' or 'female' style ready to be plugged in."[416] What about "doing science as a feminist"? Schiebinger observed: "Women should not be expected to succeed happily in an enterprise that at its origins was structured to exclude them."[417] For the moment, "the hypothesis that women will do science differently remains just that—a hypothesis in need of testing."[418] More generally, "feminism has suffered from attempts to claim too much—all that is good and true—in its name."[419] Schiebinger seeks "to effect a shift away from abstract critique toward the more positive task of asking what useful changes feminism has brought to science."[420] She concludes that women have made significant inroads into many areas of science.

Feminism has decidedly shifted the norms of academic culture away from blatantly sexist styles (though not from more insidiously circumspect ones). On the other hand, a notable withdrawal of women from science and engineering has also emerged, and women still complain—and not just in natural sciences and engineering—of the persistence of a sexist climate in scholarship. The cultural-political problem cannot, of course, be separated from the theoretical one; it is naive to believe that acting upon the latter would automatically remedy the former any more than the converse. This having been said, the *theoretical* dilemma of feminism is central to this study. The political one, for all its democratic urgency, must remain at the periphery. There are no uniquely feminist methods of explanation. There is no distinctly feminist science. But there is trenchant feminist critique of practices and conceptions in science. And doing science as a feminist has begun to impact a variety of fields, even in the natural sciences. Susan Haack asks, rhetorically, "What has 'science as social' to do with feminism?" She answers: "Nothing. It is either a genuine insight but not a feminist one, or it is no insight at all." It is a genuine insight. Nonfeminists have had it. But

Longino has made one of the most compelling cases for it, and her motivation for doing so, and her ability to characterize descriptively and normatively what is important about it, derive substantially from the fact that she is "doing science as a feminist."

The "Mangle of Practice"

We have been verging on the realm of science fiction for the bulk of this chapter, encountering the monstrous Leviathans of the *hybrid collectif,* or the "inappropriate(d) others" called cyborgs. But the alienation effect can be just as aptly attained with a more homely metaphor. A "mangle" in English English (note for Americans) is just a laundry press. But it can hurt if you get stuck in it. And a mangle, according to Andrew Pickering, is just what science in practice turns out to be. Pickering is, in my view, one of the most important methodologists working through the problems besetting contemporary science studies. His work, both empirical and methodological, has drawn intense scrutiny precisely for this reason. Pickering's major achievement is *Constructing Quarks: A Sociological History of Particle Physics* (1984). But in defending that work and elaborating its implications, he has also written a series of important methodological essays that have culminated in a second book, *The Mangle of Practice* (1995).[421]

By the end of the 1980s, Pickering came to reject the "science as knowledge" agenda of SSK as "thin, idealized, and reductive."[422] More concretely, "SSK simply does not offer us the conceptual apparatus needed to catch up the richness of the doing of science, the dense work of building instruments, planning, running, and interpreting experiments, elaborating theory, negotiating with laboratory managements, journals, grant-giving agencies, and so on."[423] He questioned "whether analytic repertoires developed in the service of a problematic of knowledge can serve as the basis for understandings of practice."[424] He especially found the "interest" theory of the Strong Program to present the problem of scientific development too simplistically. For Pickering, the problem even with Kuhn and Feyerabend as historians of science or Bloor and Barnes as sociologists of science arose largely from their orientation to the macrolevel and to theory as opposed to practice.

Pickering's "practical realism" or interpretation of "science as practice" offers a robust appreciation for the *complexity* of science, its "rich plurality of elements of knowledge and practice," which he has come to call "the mangle of practice."[425] Indeed, as Ian Hacking has noted, it is the "richness, complexity and variety of scientific life" which has occasioned the widespread new emphasis on science as practice.[426] As against the "statics of knowledge," the frame of existing theoretical ideas, Pickering situates the essence of scientific life in the "dynamics of practice," that is, "a complex process of reciprocal and interdependent tunings and refigurings of material procedures, interpretations and theories."[427]

To access the dynamics, Pickering prefers a detailed specificity of focus, for example, on some locally situated scientist such as Luis Alvarez or Giacomo Morpurgo.[428] Their practices get contextualized in an extraordinarily complex "topology" of relations: the disunity, patchiness, and fragmentation of actual science, a far cry from its mythic "unity."[429] One must "see scientific culture as endemically *patchy,* composed of all sorts of *heterogeneous elements.*"[430] It is against the backdrop of this "openness" that the scientist enters upon negotiations. Pickering emphasizes, in the context of high-energy physics, a vast and tenuous "symbiosis" among theorists and experimenters which leads each of these internally heterogeneous groups to be sensitive to the "opportunities" that the other provides to enable ongoing research.[431] To take advantage of such opportunities entails a pragmatic mustering of resources, and "in a patchy, localized culture *not all resources are equally mobile.*"[432] Pickering conceives this mobilization in two stages, planning and implementation.[433]

The decisive feature of the planning phase is the articulation of goals. This is for Pickering a highly contingent and creative element, because, while it generally involves modeling after exemplars given in the research tradition, such modeling is always by analogy, and it remains an "open-ended concept."[434] Analogy could carry scientific imagination off in myriad directions. What occasions the specific one that emerges can never be fully explained. The element of chance, of radical historical contingency, cannot be evaded, and a historian must admit, at last, that "it just happened."[435] Yet the modeling, the goal articulation, is not entirely random. It has its own immanent structure, namely, the effort to achieve *coherence,* in other words, to overcome that patchiness of the scientific environment.[436] The inspiration for scientific practice is "the possibility of making *connections* between disparate cultural elements."[437] It does so by envisioning "associations" or "intersections," points where there would be a "closure" of the theoretical and the experimental elements floating in some measure of disaggregation in the field.[438] For Pickering, this element of coherence is crucial for an understanding of science as practice. Yet the element of contingency cannot be ignored.

This contingency multiplies as the enterprise moves from plan to implementation, from model to execution. Here Pickering's conceptualization of experimental practice is at its most dense and complex. He discerns three elements: a "material procedure," which involves setting up, running, and monitoring an apparatus; an "instrumental model," which conceives how the apparatus should function; and a "phenomenal model," which "endows experimental findings with meaning and significance . . . a conceptual understanding of whatever aspect of the phenomenal world is under investigation."[439] The "hard work" of science comes in trying to make all these work together. What emerges is a "gap between plans and their implementation in the material world. . . . [T]he implementation of such plans is continually upset by unforeseen blockages, obstacles, resistances encountered in material practice."[440] As Pickering sees it, "resistances have to be understood as *situated* with respect to goals; they are defined and exist only reciprocally with respect to the associations that they block."[441] That is, "unlike constraint—which seems somehow 'already there'—resistance *emerges* in the real time of practice."[442] This property of radical *emergence* is crucial for Pickering. "Resistance 'just happens.'" It is "truly emergent in time."[443] Hence his insistence upon attending to *temporality,* to "real time struggles to make things work."[444] This leads Pickering to argue for a distinctive *historicity* of scientific practice.[445] Emergence in real time, this utter unpredictability, this radical contingency, this "relativity-to-chance," marks what Pickering calls "a culturally situated historicism."[446]

Faced with these resistances, scientific practice must make "accommodations." These accommodations are triply contingent—upon the goals, upon the resistance, and upon the resources which the local situation entails. And the experimenter can choose to alter any of the elements involved to cope with the problem (an elaboration of the Duhem thesis).[447] For all the "forced moves," there remain a plethora of "free moves." This confutes the argument that scientists are determinately "constrained" in their choice, though this hardly suggests that they can be completely cavalier either.[448] There are so many variables pulling in so many directions that the wonder is that "closure" arises at all. It does happen, however, and it happens not by the supervenience of any one element but by the mutual constitution of all, which Pickering finds best conceived as a "stabilization."[449] "Recognition of the temporality of practice in terms of a goal-oriented dialectic of resistance and accommodation thus makes it possible to inscribe a statics of reason and knowledge within a dynamics of practice."[450]

I consider Pickering's efforts to conceptualize this model of scientific change as a powerful variant of empirical "thick description." The entire conception of practice which Pickering deploys, his "mangle of practice" or "pragmatic realism," represents an extraordinary instance of concrete interpretation. This links him up with a number of interpretive or contextualist hermeneuticists. Thus, Pickering's methodological reflections are resources for robust hermeneutic practice in the light of post-positivism. Indeed, it would be a worthwhile endeavor to redescribe Pickering's model in Hegelian language (of the pragmatist-revisionist, Deweyan variety) as a form of dialectical objectification, and to enlist it as a substantial contribution to a general theory of historical change.

The controversy that developed between Pickering and Peter Galison over *constraint* is, in my view, crucial for the elaboration of a moderate historicist methodology in the human sciences, though of course it is situated more concretely in the study of science. Constraint was the decisive term, as we have seen, in a series of harsh critiques of Pickering's major work, *Constructing Quarks*. As he noted, the reception of that work "induced in me a profound allergic reaction to the word 'constraint.'"[451] In particular, he resented the fact that "Peter Galison devoted an entire chapter of his *How Experiments End* to 'constraint' and made me the villain."[452] The rivalry between these two authors is real. "Galison and Pickering genuinely disagree," Ian Hacking observed in the introduction to a volume in which they debate their respective positions.[453] Brian Baigre, commenting on the dispute, observed, "we should be wary of Jan Golinski's suggestion that constraint- and resistance-talk are interchangeable ways of speaking. . . . [T]hese discourses are suited to different kinds of cultural perspectives."[454] In short, beyond the personal, there appears to be a conceptual difference here that makes a difference. I wish to understand that difference, then ask what difference it really makes, from the vantage of the elaboration of a moderate historicist methodology.

Pickering, in expressing his "allergic reaction" against *constraint*, helpfully distinguished two distinct species of "constraint-talk." He called them *traditional* and *postmodern*.[455] The traditional version of constraint-talk, exemplified in the critique of *Constructing Quarks* by Yves Gingras and Sam Schweber, really had to do with the effort, in Pickering's terms, "to reassert the objectivity of scientific knowledge, its distance from us," and unsurprisingly he believed it "fails miserably" in that endeavor.[456] Galison, in *How Experiments End*, criticized Pickering for his allegiance to SSK and the argument for "interest-theory" in *Constructing Quarks*. That view, Galison wrote, "denigrates the role of nature and supposes that scientists' presuppositions—bolstered by their interests—condition the admissible phenomena in such a way as to render a particular theory and its associated experiments closed and self-referential."[457] He faulted this view for three sorts of inadequacies. First, it failed to recognize the force of experimental *results*. Second, it "exaggerates the flexibility of theory." And, finally, it failed to "attend to the constraints on experimentalists' conclusions that are imposed by the skills and techniques of their work."[458] Pickering, as noted, resented this criticism. First, he had made great efforts to distance himself, as he saw it, from the traditional interest-theory of SSK, yet Galison lumped him with it. Second, he found Galison to be imputing to him the view that theory overrode experiment, whereas he had been at great pains to demonstrate their symbiotic, or mutually constitutive, relationship. Finally, he insisted, and here we get to the "constraint controversy" proper, that the model of practice that Galison and others proposed under the rubric of "constraint" was inadequate to the problem both sides sought to address, namely, how to characterize properly the complexity and difficulty of real-time scientific practice.

That was the dimension of "constraint-talk" that Pickering called "postmodern," in other words, recognition that the "topology" of scientific culture was "patchy, interrupted, and heterogeneous" rather than unified.[459] This patchiness and heterogeneity, this discontinuity created "a space in which all sorts of moves can be made."[460] Indeed, this is a common starting point for both Galison and Pickering, but like the proverbial glass of water, their perception of the situation is inverse. Galison sees the *structure* which constitutes the patchy topology; Pickering sees the *contingency* which leaves it incomplete and heterogeneous. Pickering insists: "to identify cultural elements as constraints is precisely to lose sight of the openness of their future extension."[461] Therefore, he argues, "whatever obstacles do arise in practice are not 'already there' to begin with; instead they genuinely emerge in time."[462] That is why he prefers the concept *resistance:* "resistances emerge in the real time of practice," therefore contain "an irreducible element of chance and *contingency,*" and leave the interpreter in the end with nothing more to say than "it just happened."[463]

But this is not quite sufficient. Pickering was very critical of Harry Collins for leaving everything to contingency.[464] He wants to reckon with historicity as emergence, but that means he has to acknowledge structure as well: "I do indeed want to defend a historicist understanding of scientific knowledge . . . contingency is not all that there is. . . . There *is* structure . . . and it is by virtue of this structure that one can make sense of the obvious fact that different cultures do sustain different practices and produce different knowledges."[465] I propose that if we elaborate the manner in which Pickering allows for structure and the manner in which Galison allows for spontaneity, we can, Baigre notwithstanding, find some crucial common ground.

Galison wrote in *How Experiments End*, "I want to use the notion of constraint the way historians often do—to designate obstacles that while restrictive are not absolutely rigid."[466] He drew explicitly on Fernand Braudel and the *Annales* school discrimination of orders in time: the geographical or *longue-durée*, the social, and the individual.[467] Each of these structures configured the space of events, but they did not, even collectively, *determine* them. Galison invoked Carlo Ginzburg's microhistorical approach to highlight that crucial point. At the level of the concrete, the historian must "grapple with the multiple subcultures within a single society."[468] It is in that complex interface that individual (and larger) elements of creativity enter the historical process, that real emergence and novelty infuse history. The challenge for historical methodology is to find a *language* to articulate this extraordinarily rich dialectic in its concreteness without losing purchase on its situatedness in larger nests of causality and meaning.

Galison has offered two powerful initiatives in that vein. First, he has challenged the "central metaphor" of the history of science.[469] And second, he has developed a model for the manner in which concrete practice orchestrates itself: the idea of the "trading zone" and pidgin language. The "central metaphor" of the history of science, Galison alleges, continued across the entire post-positivist rupture of theory from Rudolf Carnap through Kuhn to the social constructivists.[470] It took the form of "framework relativism" in theory and of "island empires" in historiography, in other words, the idea that scientific specialties formed closed communities which were internally homogeneous while externally incommensurable. Kuhn's general theory of paradigm change postulated that theory and experiment (and instrumentation, as well) changed in lockstep. Galison offered the analogy of a brick wall in which the bricks, instead of being intercalated, were piled level upon level one above the other, such that they would sever along the same seam. Instead, he insisted on seeing "science as an intercalated enterprise, in which several, often quite disparate subcultures follow their own pace and practices."[471] As a result, the historical picture would become far more complex, with varying strata of continuities overlaying moments of rupture. Thus, theories have often changed without substantial alteration in the experimental milieu. The same instruments have become resources for widely disparate experimental regimes. The *emergence* of a new instrument can have direct impact on theory, as of course upon experiment. The upshot is a view of the history of science which is both more complex and more realistic. The non-synchronicity of the periodization of theory, experiment, and instrumentation allows far more complex accounts of historical process. Above all, it concentrates attention at two theoretical margins: first, it betokens the way in which no subculture exists autonomously. Instead, it is always thrust upon the wider contexts for resources, but these are, by that very virtue, also constraints. This is how Galison believes his contextualism subverts framework relativism. "In a world in which the solution to scientific questions is often simultaneously the solution to problems in the wider culture, it becomes more, not less, difficult to see scientific knowledge as carved neatly into isolated frameworks. Contextualism works against framework relativism."[472] His view explodes the metaphor of "island empires" of science: "my real quarrel, then, is with the central metaphor of science as an amalgam of unrelated, internally homogenized frameworks."[473]

But if attention focuses on the permeability of any given scientific culture to a variety of contexts, it must also focus on the concrete negotiations of "local coordination" or "settlement."[474] And here Galison offers his model of the "trading zone" and "pidgin" languages. The essential point he offers is that in the complex, discontinuous world of scientific practice, actual concrete regions "(conceptual and spatial) of partial coordination" are continuously emerging and disintegrating.[475] No total integration of the disparate subcultures arises; instead, "a trading language" emerges, "characterized by a *local* agreement about the use of terms."[476] If the relation persists fruitfully, these short-term "pidgins" become expanded and entrenched into "creoles." This occurs, Galison observes, not only between large-order disciplines like biology and chemistry, but within disciplines, as in the symbiosis of theory and experiment in high-energy physics, and even in specific laboratory sites, where constantly the heterogeneity and patchiness of scientific practice need to be negotiated. When we get to these "trading zones" with their "pidgins," we get to the domain of emergence in real time that Pickering identifies as his domain of "resistances." In short, what putting Pickering and Galison together offers is yet another pidgin useful for theorists of moderate historicism. That is, the *hybrid* (even better than either in isolation) "allows us to speak about the domain of allowable actions without referring to an external scientific method valid always and everywhere *and* to avoid a representation of laboratory practice as relying purely on extrinsic forces to silence skepticism."[477] Complexity and highly local, concrete mediations call for "a historical vocabulary that leaves scientific practice neither utterly divorced from its cultural context nor relegated to a mere puppet of other forces."[478]

I think we need a theory which registers the *entrenchment* of practices, apparatus, and concepts as structures—heterogeneous and patchy, but nonetheless real and binding—hence a theory of *constraints*. And I think we need a theory which registers *emergence:* the radical novelty that erupts at the concrete level of event and agency in history. We need to develop a historical language with the dialectical richness to articulate what Pickering has called the topology, the temporality, and the materiality of practice.[479] As we do so, the old dualisms and the old "sticking points" of philosophy of science *and* of social constructivism will become increasingly *pointless*. It is naive to think they will go away, but it is plausible to believe they will become background noise that a trained experimental ear can readily ignore.

CHAPTER 8

Typically, . . . the sociologist knows less than the natural scientist, while the sociologist of science knows still less. Those engaged from day to day with the problems of reflexivity would, if they could achieve their aims, know nothing at all. We might say that SSK has opened up new ways of knowing nothing.

— HARRY COLLINS AND STEVEN YEARLEY,
"EPISTEMOLOGICAL CHICKEN"

As the new reflexivity, initially a welcome aid to the disenchantment of the sociological world, spiralled through the discourses, it consumed not only "ideology" but "science" itself.

— HILARY ROSE, "HYPER-REFLEXIVITY — A NEW DANGER FOR THE COUNTER-MOVEMENTS"

A Nice Derangement of Epistemes: Radical Reflexivity and the Science Wars

There is a narrative that practitioners within science studies have a tendency to tell about themselves. It is a narrative professed with varying degrees of irony, but on the whole that irony is disingenuous.[1] We learn that this fairly young research specialty has gone through a series of revisionisms, each more radical than the last. The question of exactly who superseded whom is a shade different in different accounts, though there is a remarkable degree of convergence. What limns their difference is not primarily what transpired, but how to evaluate it. In any case, the upshot is both drastic and monitory. In the words of Harry Collins and Steven Yearley, "the escalation from relativism through discourse analysis and 'new literary forms' to reflexivity [is] in the end . . . [a] regress [which] leads us to have nothing to say."[2] That is, this (self-told) story of science studies can be glossed as a nice derangement of epistemes. It is an instructive *reductio ad absurdum.*[3]

The narrative begins with the Strong Program heroically and successfully overthrowing both the old philosophy-of-science account of scientific knowledge (rational reconstruction) and the old sociology of science (Mertonian institutionalism). In the early 1980s, the Strong Program went on a major offensive, engaging in all-out warfare with philosophers of science. There were three major engagements. In 1981, David Bloor exchanged fire with Larry Laudan in *Philosophy of the Social Sciences* and spawned a sus-

232

tained discussion at a conference that same year, recaptured in James R. Brown's text *Scientific Rationality: The Sociological Turn* (1984).[4] Roughly simultaneously, Bloor presented a vindication of Emile Durkheim in an essay for *Studies in History and Philosophy of Science* which roused a lively debate.[5] Finally, at the very same moment, Barry Barnes and Bloor collaborated on a provocative essay revisiting the Rationality Dispute and defending relativism, eliciting yet again a lively discussion. This one was gathered within the pages of *Rationality and Relativism*, edited by their opponents, Martin Hollis and Steven Lukes.[6] These debates represented the high tide of the Strong Program, when it took itself to represent the vanguard of science studies.[7] It cannot be said that the ideas propounded by Barnes and Bloor came off well in these debates. Somehow, that did not matter. Science studies went on as if the Strong Program had won a decisive victory. Furthermore, it considered that victory old news and with it the Strong Program itself as something over the hill.

The displacement of the Strong Program in science studies came at the hands of its own progeny, but all these swiftly proliferating revisionist successors took for granted that the ancient rival, the rationalist philosophy of science, had been put definitively aside. These more radical formulations showed a total indifference to the claims of philosophy of science, which seemed quite literally to have been refuted by the Strong Program. Instead, the new social constructivists pursued the agenda of discrediting with the same radicality the claims of the *natural scientists themselves* that their knowledge somehow found corroboration from nature.[8] For this new and drastic school, science was *only* construction—initially *social*, later *discursive*.

Diversity marked science studies over the course of the 1980s, with the displacement of the Edinburgh School and the emergence of a variety of competing successor programs, the powerful entry of feminist critique, and a wider "cultural studies" challenge to the authority of science in Western society and politics.[9] But with proliferation also came striking divisiveness. "The philosophical, methodological and metaphysical commitments built into the practice of sociological observers of science fragmented and diversified," and "this fragmentation of the field . . . shifted the focus of attention. . . . The 'enemy' is no longer positivist and empiricist philosophy without, but heretical social theories within, the field."[10] From the vantage of ethnomethodology, from the vantage of actor network theory, or from the vantage of radical reflexivity, the claim of the sociologist to "an epistemic authority and 'ontological reality' (for social phenomena) which s/he denies the natural scientist" for the physical world appeared an insupportable "conceit."[11]

With that, a new epoch of post-positivism in the theory of science came

on the scene. All of sociology, including the sociology of scientific knowledge, came to crisis by the late 1980s. The fundamental plausibility of social
explanations came to be called into question. Consequently, within science
studies a drastic challenge to SSK came to be presented by "discourse analysis" as developed by Michael Mulkay and Nigel Gilbert. These analysts denied the possibility of reconstructing scientific practices and settled for an
immanent analysis of scientific *texts*. This challenge displaced the attention
to praxis in science studies with an exclusive concern with discourse. Yet the
narrative has it that discourse analysis only "paved the way" for the most
drastic "turn" of all: away from any hope of constructing an empirical explanation or description of science toward reappraising the very discourse
attempting to do so, the move of radical reflexivity.[12] This hyperbolic gesture marked a climax in the narrative. The denouement was pure farce: the
Sokal affair.

Discourse Analysis

Within science studies, one main current had moved toward the interpretation of scientific practice, especially via a new attentiveness to *experiment* as
the complement of theory.[13] Laboratory studies had played a significant role
in this, but they had also, especially in the paradigmatic work of Bruno Latour and Steven Woolgar, *Laboratory Life,* pointed in other directions. Latour went off down the path that led to ANT. Woolgar, as we shall see, set off
down the path to radical reflexivity. What both had done, in *Laboratory
Life,* was to "translate" practice and experiment into discursive techniques:
"scientists' *praxis* is reconstructed as a process for adding or removing
modal qualifiers attached to scientists' discourse."[14] No sooner had experiment become salient for students of science than it seemed to be displaced by
textuality, a neglect that some commentators took to be "symptomatic of a
prejudice against practical activity and in favor of speech acts."[15] More
fairly to Latour, Jan Golinski situated the "analysis of scientific discourse as
a rhetorical construction" in the larger question of the "transition of knowledge from private to public spaces," in which "a variety of experimental,
representational, and discursive strategies" required attention.[16] In *Science
in Action,* Golinski correctly observes, Latour made it clear that "we have to
see science as different from other forms of persuasion in its reliance upon
manipulations of the material world."[17]

Indeed, the relation between discourse and practice is far more intricate
than would allow either to be reduced to the other. "*Praxis* is not mere action. It is action which is informed by theoretical understanding," and "it is
not from *praxis per se* that 'scientific objects' and 'real world events' are constructed, but rather from *reasoning about* such events."[18] Thus, "the rela-

tion between discourse and *praxis* in an experimental context is . . . dialectical . . . *discourse both defines and is defined by praxis and the ongoing contingencies of the conduct of inquiry*" (original emphasis).[19] Michael Lynch elaborates this dialectical relationship in a semiotic lexicon: "We first encounter the sign in use or against the backdrop of the practice in which it has a use. It is already a meaningful part of the practice, even if each individual needs to learn the rule together with the other aspects of the practice. . . . [T]he sign is already embedded *in* a practice, and meaning arises through the very placement of the sign in accordance with the grammar of the practice."[20] Nonetheless, the impulse to reduce practice to discourse went forward.

Michael Mulkay had initially been in sympathy with the agenda of the Strong Program.[21] But in 1981 he proclaimed it was "simply impossible to produce definitive versions of scientists' actions and beliefs," because "there is no way you can get from this collection of incompatible statements [participants' accounts] to a conclusion about action."[22] He went on to suggest a wholesale defection from the sort of account the Strong Program proposed. In its place he offered the "analysis of scientific discourse . . . as an *alternative* to the more traditional concern with describing and explaining action and belief."[23] The advantage, Mulkay proclaimed, was that with discourse analysis "one is no longer trying to use observable evidence to explain unobservables such as past actions or ideas in people's heads. Instead one is concerned only with interpreting given documents or recorded utterances."[24]

Mulkay and Nigel Gilbert noted within SSK an "increasing intellectual diversity, even disarray" arising out of the dawning realization that "it is impossible for sociologists to 'tell it like it is in science.'"[25] The source of the "impossibility" lay in the variation among the participants' documentary accounts which the interpreter had to draw upon in constructing an interpretation. "Each analyst presents his version of events as definitive in that, if he has interpreted his evidence correctly, *this is the way things actually happen(ed)*. It is our contention that there is no satisfactory way of establishing such definitive analysts' accounts of action and belief."[26] And so they recommended a change of subject to what they called "discourse analysis." A year later (1983), reinforced by Steven Yearley, Gilbert and Mulkay reasserted their claim: "The potential diversity of participants' accounts, their lack of temporal precision and the variable meaning of central terms" made interpretative synthesis "fundamentally inadequate."[27] Hence, "virtually the whole of the sociology of science, insofar as it is empirically based, is undermined by basic methodological faults associated with its failure to consider the nature of scientific discourse."[28] In the abstract to another piece, Gilbert and Mulkay succinctly epitomized their view: "it is concluded that it is impossible to obtain definitive evidence of how theories are actually chosen and that a new form of sociological analysis is required."[29] Finally, in 1984, Gilbert and Mulkay made their most drastic statement: "we suggest that it is impossible to search for *any* kind of data that can be used to provide a firm bedrock for historical description and analysis. . . . [I]t is impossible for the analyst to devise any satisfactory techniques for reconstructing 'what actually happened.' "[30] They went on: "[M]uch of the interpretative work whereby historians or sociologists extract one coherent story from scientists' discourse is unsystematic and infrequently subjected to close examination, and it usually remains almost entirely hidden from the analyst and his audience. . . . [I]t may be necessary for historians to attend more explicitly to the way in which their craft is exercised in compiling histories from the interpretative work of participants."[31] They hastened to add, "We are not suggesting, of course, that historians should cease constructing their own historical accounts."[32] It sufficed that such historical accounts would remain if not "impossible" then "unsystematic"—in other words, *naive*.

Despite their glib denial, Steven Shapin recognized the radical nature of this imperious intervention. "Gilbert and Mulkay's strictures . . . apply to all descriptive and explanatory enterprises whatsoever."[33] They were claiming that "the goal of describing and explaining action and belief is . . . chimerical."[34] This was indeed drastic; Shapin would have none of it. "So far as Gilbert and Mulkay's view of *historians'* work is concerned, it is either trivially true or fundamentally false. We do not label our accounts as imaginative fictions because we hope they are not. Our goal is indeed telling it *wie es eigentlich gewesen*. . . . However, having this goal does not commit us to being simple-minded about how it is realized or about the nature of the accounts that are its end products. Historians routinely accept that their accounts are theoretical and interpretative in character. . . . Gilbert and Mulkay portray the historian as a methodological naif, but they offer no evidence to support this view."[35] Ironically enough, Shapin pointed out, Mulkay and Gilbert were proposing discourse analysis "as a form of theory- and interpretation-free historical and sociological practice," and that was itself naive.[36]

The whole posture of "discourse analysis" was misguided. Thomas Gieryn diagnosed it as early as 1982: "[T]hat scientists offer different accounts of behaviour depending on the context of discourse . . . leads Gilbert and Mulkay to a non-sequitur: *because* scientists construct and reconstruct the meaning of their behaviour in a flexible manner, *therefore* sociologists are incapable of providing adequate explanations of the behaviour itself, and can do no better than explain why it was described in this or that way."[37] Gieryn had as little patience as Shapin with their posture: "by dismissing others' questions as unanswerable, by claiming for their own data an exclusive validity for insights into what scientists do (say), by burdening other accounts with pretentious claims to definitiveness, Mulkay and Gilbert do little to help their [project] survive."[38] Gieryn astutely dismissed the projected standard of "definitive" accounting as irrelevant to any empirical inquiry.[39] "Little is to be gained by Mulkay and Gilbert's timid assumption that since the goal is unattainable, the pursuit must be abandoned."[40] This was the view, as well, of Ellsworth Fuhrman and Kay Oehler in a review of discourse analysis in 1986. They commented that Mulkay and Gilbert "dismiss as inadequate, or as an illegitimate form of sociological inquiry, studies which use scientists' discourse as a resource (rather than as a topic of study)," but they found this posture unjustified.[41] This paper drew a defense of discourse analysis from Jonathan Potter, who denied that Mulkay and Gilbert proposed "to wipe out 'traditional' science studies." Rather, he argued, "the concern is with the way analytic categories are dealt with . . . in particular, the incorporation of *participants'* categories . . . as if they were unproblematically usable as *analysts'* categories."[42] In their response, Fuhrman and Oehler pointed out that "our concern is with [discourse analysis's] lack of inspection of its own (the analyst's) categories," in other words, that Mulkay and Gilbert were not reflexive enough.[43] That betokened the drift of the whole debate. Reflexivity loomed as the inevitable outcome of a fixation upon discourse.

Peter Halfpenny picked up this drift in a review of the two major works of discourse analysis. He took up a phrase from Mulkay and Gilbert and turned it two ways. The phrase they used was "lack of inferential nerve," and they used it to express their unwillingness to extend sociological inquiry past the discursive forms of the accounts scientists offered about their practices to the character of those practices themselves. Halfpenny seriously faulted them for this lack of inferential nerve: there was no justification for such timidity. "Even when faced with a variety of reports, scientists themselves have no great difficulty in establishing what is going on."[44] Halfpenny discerned the trick of invoking a "definitive" standard to despair of any contingent and fallible empirical endeavor. And that brought him to the second turn of the phrase, "lack of inferential nerve." He called it "only a pretense, for they offer definitive descriptions of the actions that the scientists achieve through their reports."[45] More generally, Halfpenny concluded that Gilbert and Mulkay "have deluded themselves into thinking that they are making a more profound and novel contribution than they are."[46] Similarly, Fuhrman and Oehler could not get past a "nagging feeling that [Gilbert and Mulkay] are reinventing the wheel."[47] Jonathan Potter, with a colleague, again tried "wielding . . . rhetorical cudgels on behalf of [discourse analysis]" in rebuttal to Halfpenny, but the upshot was unimpressive.[48] Halfpenny bluntly dismissed the first part of their rebuttal as "merely rhetorical," then went on to restate his essential argument: "Gilbert and Mulkay provide reports of the actions of their scientists as if they were definitive descriptions of those actions—exactly what they castigate other sociologists of science for doing."[49]

In "Epistemological Chicken," Harry Collins, writing with an apostate from discourse analysis, Steven Yearley, summed up the program in a couple of paragraphs. "Discourse analysis refused to accept the evidence marshaled by SSK case studies as an unproblematic representation of scientists' activities. . . . The exercise would have been unexceptionable but for its proponents' insistence that discourse analysis itself was epistemologically foundational. They thought that discourse analysis comprised a critique of SSK while itself being invulnerable to the same critique."[50] Ironically, Collins and Yearley observed, "discourse analysis paved the way for more radical deconstruction which goes under the title of 'reflexivity.'"[51] In other words, it was, all too unwittingly, just another move in the game of epistemological chicken.

"Since the mid-1970s," Collins and Yearley tell us, "each new variant of SSK has tended to be a little more radical than the one before it. Each new variant has stood longer on the relativist road."[52] Being more radical, being more relativist: that is what epistemological chicken boiled down to. "In sum . . . each new fashion in SSK has been more epistemologically daring. . . . Each group has made the same mistake at first; they have become so enamored of the power of their negative levers on the existing structures as to believe they rest on bedrock. But this is not the case. Though each level can prick misplaced epistemological pretensions, they stand in the same relationship to each other as parallel cultures; no level has priority and each is a flimsy building on the plain."[53] Dick Pels has diagnosed this plunge into epistemological chicken as a consequence of the "politics of symmetry." The "successive extensions and radicalizations of the symmetry principle" have led to "the broad dissemination of moralizing psycho-talk about a 'failure of nerve,' 'lack of determination,' 'chicken-heartedness,' or 'self-betrayal,' which each new radicalizer of the symmetry principle hatched against his supposedly more timid and faint-hearted predecessor."[54] Pels astutely places Collins and Yearley in this charade: "With a perspicacity typically granted to losers in the epistemological chase, Harry Collins and Steven Yearley begin to see that there is no definite limit to this proliferation."[55] Thus "not only the 'Epistemological Chicken' debate itself, but the entire two-decade history of S&TS [science and technology studies] positively bristles with such debunking attributions."[56]

Somewhat plaintively, Collins and Yearley close their side of the "Epistemological Chicken" debate with the question, "where should we stop, and what good is it?" Recognizing that "the possibility of moving to another level of analysis exists," they question why it should be explored, citing Bruno Latour himself as cautioning against "whirling helplessly in our efforts to outdo and outwit each other in proving that the other is a naive believer."[57] For all their differences, Collins and Latour share a deep antipathy to the radical reflexivity of Latour's one-time collaborator Steve Woolgar. In his crucial intervention, "One More Turn after the Social Turn," Latour gave voice to a sense of crisis in the field: "After years of swift progress, social studies of science are at a standstill." Fearing "a reactionary mood" which would opine "the field has suffered enough from extremism" and hunger for a "happy medium," Latour opted instead for a radical new departure, ANT, as we have seen. But en route, he offered the following telling commentary: "A few, who call themselves reflexivists, are delighted at being in a blind alley; for fifteen years they had said that social studies of science could not go anywhere if it did not apply its own tool to itself; now that it goes nowhere and is threatened by sterility, they feel vindicated."[58] Latour, by contrast, preferred "being a little *more* radical" as the way out of the blind alley, "being one of those ants, accused of being not only frantic but also French."[59] In his view, "what explains . . . the present limitations of our field is that we have not yet reconciled our discoveries with our philosophical framework."[60] But the reflexivists, as it seems, had as little interest in metaphysics as they had in methodology: they had "jumped frame."

Radical Reflexivity

The figure who carried off the "reflexive" revolution in SSK is Steve Woolgar. At the climactic moment of the narrative he emerged, in Steve Fuller's striking terms, as "an intimate of Descartes' Evil Genius."[61] The reflexive posture—elaborated by Woolgar and Malcolm Ashmore—drew upon both the crisis of sociological theory (especially under the auspices of "ethnomethodology") and the proliferation of poststructuralism across all human science disciplines to propose abandonment of any empiricism for an ever purer form of metadiscourse.[62] *Reflexivity,* not discourse analysis, posed the decisive disruption of SSK.

Steve Woolgar began in a posture much like Mulkay's, rejecting the possibility of objective historical interpretation. His account of his own initial intentions and practice portrayed him as (naively) taking up the Strong Program agenda, only to be brought up short by its inadequacies. The tale begins with an essay Woolgar published in 1976: "Writing an Intellectual History of Scientific Development: The Use of Discovery Accounts." Woolgar wrote in that essay that he had unexpectedly discovered "methodological problems . . . in the analysis of the intellectual development of [a disciplinary community which] prevent . . . providing a straightforward chronological history."[63] The methodological problems, it appears, arose from *variations in the accounts* left by his subjects in their documentation of the "discovery" of the first pulsar. Such documental reconstructions were *partial* in both senses of that word, and hence "reliance on [such] accounts would necessarily introduce an element of distortion" into the account Woolgar himself proposed to develop.[64] Thus, he concluded, the prospects of "constructing a comparatively accurate history [were] . . . very discouraging."[65] His "first attempts to provide a definitive version of 'what actually happened'" did not satisfy him.[66] This led Woolgar to "argue for a modified approach to the construction of an intellectual history."[67] "A concern with historical accuracy and the reliability of data" beset by the methodological problems introduced by variations led him to propose that intellectual history *change its subject.* "Perhaps the very difference between the accounts can tell us about the essential process of the development of ideas."[68]

This was a radical beginning. Yet Woolgar followed it with a section entitled "A 'Working Account' of the Discovery." I want to benchmark the account of his trajectory by highlighting this pragmatic achievement which seemed theoretically insignificant to Woolgar himself. At the outset, he had professed to worry about a "straightforward" account. As he developed his presentation, that modulated into a "comparatively accurate" one. But in his most important formulation he used the following phrasing: "a definitive version of 'what actually happened.'" No historian would miss—even had Woolgar left out the scare-quotes—that the phrase "what actually happened" is a gesture to Ranke's *wie es eigentlich gewesen.* But it is not that—at least not *first* that—which we must regard, but rather the modifier *definitive.* Somehow, Woolgar *could* bring about a "working account." I would contend this account can plausibly be construed as "straightforward" and—especially in light of his extensive comparison of that account with other versions from those troublingly "varying accounts" in his sources—even "comparatively accurate." What Woolgar seems to be casting forth as his apple of discord is the notion of a *definitive* account. He seems to be suggesting that *wie es eigentlich gewesen* seems to require that of intellectual history. And he implies that intellectual historians are (surprise, surprise) *naive* not to recognize that this is an unattainable goal. (He softens the insult by including this early version of himself among the naive, and for that I suppose we should feel grateful.)

Woolgar first introduced the question of reflexivity in 1981, when he took on Barry Barnes and Donald MacKenzie to debunk the "interests model" of explanation. But Woolgar was attacking not merely *this* form of causal explanation, but the very idea of causal explanation itself. Invoking Harold Garfinkel and ethnomethodology, Woolgar argued that "the social sciences have an unfortunate habit of making out actors to be cultural or judgmental 'dopes.' "[69] Woolgar discerned in sociological explanation the same "back-and-forth process" that Garfinkel had noted in his "documentary method" as between "observed appearances" and "underlying patterns." Woolgar invoked the key ethnomethodological term for this "circularity": *reflexivity*, "the general property of discourse whereby the invocation of an ostensibly 'underlying' pattern is inevitably part and parcel of the scene which it purports to 'explain'; the sense and meaning of the scene depends on, and is inextricably tied to, the invocation of underlying pattern."[70] He turned that circularity into a fundamental epistemological issue: "what counts as legitimate construction in practical argument?"[71]

Woolgar asks why in some accounts this circularity is not challenged, while in others it is denounced, since in principle it is present in all. "[W]e should develop an understanding of the practical management of correspondence between actions and underlying patterns, rather than simply engage in unreflective attempts to advance that correspondence."[72] He seeks to problematize how "sophisticated applications . . . make the suspicion of circularity or, to be more precise, the accomplished avoidance of suspicion, much less immediately obvious."[73] He postulates that "the 'facticity' of the interpretation is enhanced by minimizing or backgrounding the active involvement of the interpreter."[74]

The empirical historian whom Woolgar chose for exemplary deconstruction, Donald MacKenzie, was mystified by the apparent *prohibition* of explanatory accounts in Woolgar's critique. "Only if one holds to a strongly positivist position on the writing of history (which I presume Steve does not) are there grounds for a prohibition on such studies."[75] Historical interpretation, MacKenzie willingly acknowledged, was contingent; so what? Why the invocation of an absolutist standard? In this response, MacKenzie echoed the lines of response that Shapin, writing as a historian, had made to Gilbert and Mulkay. They were each reacting from within the disciplinary matrix of historical practice. But Woolgar had "jumped frame."[76]

Barnes understood what Woolgar was about: "Woolgar's quarrel is with causal historical explanation as such."[77] He complained that Woolgar "offers *general* criticisms of constructed theoretical discourse . . . tantamount to a rejection of discourse itself."[78] That is overstated, to be sure, but not by much! Woolgar is rejecting not all discourse, but all explanatory discourse using an empirical approach. In response, Barnes insisted that "there is no best method of assessing theories, and no way of avoiding their construction."[79] His point, and one which deserves to be underscored, is that *empirical inquiry* is not properly met with *blanket skepticism*. Woolgar, as Steve Fuller would later spell out, was bringing back the Cartesian ghost of the absolute.[80]

In response to Woolgar, Barnes challenged the ethnomethodological rationale. He noted that ethnomethodology was itself an insurrection within a disciplinary paradigm (orthodox sociology under the aegis of Talcott Parsons) which sought to undermine the adequacy of all the traditional core concepts of sociological thinking (e.g., class, institution, interest, and especially causal explanation), dismissing them as mere "folk-reasoning."[81] It was part of a general reaction against what Dennis Wrong called the "oversocialized conception of man," that precipitated a general crisis of sociological theory.[82] Barnes is a defender of a soberly fallibilist empirical sociology. For him, *radical* ethnomethodology is *so* radical as to be absurd.

Woolgar expressed "disappointment" at these responses: his targets took him too *personally*. He claimed he was trying to make the general point, implicit in a radical reception of Garfinkel's ethnomethodology, that the "problem of descriptions" is "insoluble in principle."[83] "The artful concealment to which I refer is to be understood as symptomatic of *all* explanatory practice . . . *all* such work is *essentially flawed*."[84] He explained: "all attempts at explanation involve the use of descriptions. . . . The account of the explanandum vitally informs both the warrant for and the character of the explanatory process . . . [but] explanatory work inevitably proceeds by ignoring the problem of descriptions. . . . [T]he characters of entities are regarded as fixed for the practical purposes of explanatory argument."[85] At first, Woolgar avers, "my deconstruction of the explanatory structure of interests explanations has no implications for its legitimacy." Then he admits this is "clearly disingenuous" and continues, "It is not so much that interests explanations are wrong as that they are misguided . . . [because] they fail seriously to address the consequences of the radical perspective."[86] Woolgar's disputation of explanation launched the challenge to orthodoxy by what Chris Doran has felicitously labeled the "Reflexive Faction."[87] It proved, in my view, a most instructive dead end in science studies.

In order to make his reflexive discourse analysis clearer, Woolgar has offered a categorization of "accounts" constructed in discourse. There is what he calls the "reflective" account, which is that of the naive realist: external reality reflected fully in the representation of discourse. Then there is what he calls the "mediative" account, which postulates the decisive intervention of social construction, occasioning variation in the relation between accounts and their ostensible referents. Finally, there is the "constitutive" account, which postulates that, in Woolgar's words, "accounts *are* the reality"; in other words, there is only (discursive, not social) construction.[88] Woolgar does not feel obligated to *prove* the validity of this third mode. "We do not have to worry about the validity of this epistemological position to see that it provides a powerfully analytic handle."[89] That is, its *mere possibility* permits the insinuation that the other modes of account are evasions of this one. The challenge posed by the constitutive account can elicit two responses.

The first, which Woolgar terms the reformist reaction, turns the *in principle* "problem of description" into a "technical difficulty" to be ameliorated by more assiduous methodology. But "attempts to improve upon the accuracy of descriptions are doomed; descriptions are only more or less reliable by virtue of their being treated that way for the practical purposes at hand."[90] Hence the other response, a "more radical reaction" which implies "deep scepticism about the possibility of ever producing definitive descriptions."[91] This is the first articulation of what Woolgar eventually calls "the Problem" and which he describes as a "methodological horror" or a "monster."

Invoking his three classes of "accounting," Woolgar identifies the social constructivism essential to SSK with the second, or mediative form, and sees this as a disciplinary "craft skill" for coping with the Problem. Woolgar's Problem discredits "methodological adequacy: what grounds provide the warrant for the relationship between the objects of study and statements made about those objects?"[92] Woolgar adopts an uncompromising—absolute?—vantage: "The problem is a general and unresolvable problem of epistemology, which requires artful management whenever it makes its appearance, lest it entirely disrupt research practice."[93] Woolgar is clear that to deal *definitively* with this Problem would entail "establishing *ultimate* grounds for proposed interpretations and explanations," hence it is "unresolvable." Yet he professes astonishment that while the Problem is "generally acknowledged . . . research practitioners continue to do interpretation and explanation." He accounts for this with the hypothesis that within a given disciplinary community "the acquisition of craft skills . . . involves learning how to manage a problem which is irresolvable."[94]

He writes in another essay, "I investigate how scientists accomplish connections between specific research documents and their underlying realities . . . despite philosophical and methodological arguments that such connections are in principle inadequate, indefensible, and even impossible."[95] That is, he is interested in their "artful management of possible methodological horrors." Scientists (including sociologists) "skirt around, avoid or ignore [methodological dread] in the course of their practical actions," holding accounts " 'good enough' or 'practically adequate.' "[96] "[I]n the mood of interpretivist scepticism which pervades much current scholarship, instances of 'straightforward' interpretation are rare. For even to speak of (or draw attention to) interpretation, far less scrutinize the nature of the interpretive act, is to begin to question the congruence of the image with its underlying reality."[97] Woolgar wishes to expose these rhetorical strategies. That "might appear to claim a privileged ontological status for [his analysis, but] no such claim is intended."[98] Instead, it is merely "an analyst's gambit, a methodological heuristic" to "enable us to highlight the interpretive work of our subjects."[99] Thus, there is a problematic sense in which such work is "critical," for while it sets out to "reveal deficiencies in representational practice in science" it must acknowledge that its "own sociological [?] argument would be open to the same charge of deficiency." Still, Woolgar admits that it *is* "being critical, if only by innuendo."[100]

What Woolgar faults in sociological explanation is the technique he labels "ontological gerrymandering," that is, "making problematic the truth status of certain states of affairs selected for analysis and explanation, while backgrounding or minimizing the possibility that the same problems apply to assumptions upon which the analysis depends."[101] Woolgar queries why some topics are "ripe for ontological doubt" while others are held "immune to doubt" or maintained as resources. "While the claims of the claim makers are depicted as socio-historical constructions (definitions) that require explanation, the claims and the constructive work of the authors remain hidden and are to be taken as given."[102] Woolgar and his coauthor find that "in *all* the empirical work we examined, authors assume the existence and (objective) character of underlying conditions around which definitional claims have been made."[103] For Woolgar the question is whether "the kind of inconsistencies we identify are inevitable features . . . of sociological argument" or, instead, whether "forms of argument which go beyond the current impasse" might be found, "a form of discourse which is free from the tension engendered by espousals of relativism within the conventions of an objectivist form of presentation."[104]

Woolgar enriches his account of the tension between the "reformative" and the "radical" responses to the Problem of validity by introducing rhetorical analysis via *irony*. "The mediative position in constructivist social studies of science sustains itself by the use of a kind of irony."[105] That is, it always sets out by an act of debunking juxtaposition (a "perspective by incongruity," to borrow Kenneth Burke's classic phrase) whereby it exposes one interpretation, typically the rationalist philosophy of science, to a transformative and undermining treatment in order to instantiate an alternative one, the socially constructed interpretation.[106] But Woolgar observes that this is carefully controlled irony, "stable irony" in the sense of Wayne Booth: "stable ironies are *finite* in application in the sense that the reconstructed meanings (that is, the alternative accounts provided by sociologists) are local and limited to the particular occasion chosen for analysis."[107] Woolgar insists that we "recognize the fragility of the ironicists' own account, how it can be undermined."[108] This is "dynamic irony," and it can be "generally recognized that if constructivist irony is critical, *and* it can be turned back on itself, then we are in a situation of infinite regress."[109] Thus, the authority of the analyst and the constructed identity of interest between the analyst and his or her readership as a disciplinary community (the "craft skills" for coping) get disrupted and "dynamic irony has the effect of making the reader take more seriously the deep flexibility of accounting procedures."[110]

Surveying the currents in the social study of science in 1982, Woolgar offered yet another characterization of the tension between the reformative and the radical postures. This time, he articulated two different approaches to ethnography: the instrumental and the reflexive. Instrumental ethnography seeks to offer an account of "science *as it happens,*" thereby (ironically) "finding things to be other than you supposed they were."[111] But its "routine application of a preconceived theoretical perspective [i.e., *social* constructionism] means that we neglect attention to the fundamentals of the explanatory process itself."[112] Woolgar argues that SSK is "doggedly nonreflexive in its practice."[113] Therefore he proposes an alternative, *reflexive* ethnography: "My suggestion is that a fully developed reflexive ethnography of science might provide a way of better coming to terms with what I have called the problem of fallibility . . . [it would] retain and constantly draw attention to the problem. . . . We need to explore forms of literary expression whereby the monster can be simultaneously kept at bay and allowed a position at the heart of our enterprise."[114] What Woolgar is after, he insists, is "insight into general processes of reasoning practices."[115] For Woolgar, science is of interest not so much for itself as for its exposure of our modes of knowledge-making. A reflexive approach retains this in the consciousness of its analyst and of the readers considering that analysis. "As both analysts and members we are inevitably tied to a structure of discourse which includes the use of modalities, the achievement of objectivity, the assessment of what really is the case and so on. To the extent that we are constrained in our use of available language resources, we will inevitably reproduce the rhetoric of realism. To us, this suggests the importance of a venture which attempts critically to examine the nature of this rhetoric."[116] Students of knowledge can ill afford "the way in which the conventions of realism constrain our exploration of knowledge practices and inhibit the development of reflexive practice."[117] "The conventions of the realist genre encourage the unproblematic and unhesitant singular interpretation of text, the unreflexive perception of a reported reality (subject/object) and the essentially uninteresting character of the agency involved in the report's generation. In the realist genre, the text is a neutral medium for conveying pre-existing facts about the world . . . subjects of study appear exotic in virtue of their inherent qualities rather than as a result of their construction in the text."[118] The challenge of postmodernism has been to undermine the authority, the claim to "transparency" of this realist genre of representation. "The modern predicament about the adequacy of representation arises because the observer, the agent of representation, has once again become part of the picture."[119] Woolgar insists on viewing matters from this radical "constitutive" vantage. "[N]ow that it has been demonstrated that the constructivist perspective can indeed be applied to science, it is time to ask 'What Next?' It is perhaps especially timely to ask whether social studies of science are now in danger of becoming bogged down by constructivism."[120] The rhetorical force of Woolgar's strategy turns on this avant-garde posture.[121]

Woolgar justifies this undertaking ultimately in terms of "the possibility of a form of philosophical idealism applied relentlessly to each and every instance of interpretation . . . an exploration of the potency of epistemological scepticism."[122] As he puts it in another text, "the deconstruction of deconstruction does not just 'cancel things out' but enables us to grasp the enormity of the consequences of scepticism relentlessly and authentically pursued."[123] As Steve Fuller rightly points out, this is to revive the "Big Questions of classical epistemology."[124] Woolgar is practicing traditional absolute skepticism in the manner of Descartes. "Woolgar's overall strategy is to instill doubts about the legitimacy of the representational practices of a successively wider range of science studies inquiries."[125] And the way in which he does so "comes dangerously close to an *a priori* argument for the impossibility of *any* legitimate sociological models."[126] It can be observed that, just like foundationalism, this *anti*foundationalism is a "view from nowhere."[127] But Woolgar denies this: "[I]t is simply incorrect to assume that the radical perspective entails a search for pure (or complete) description, an epistemologically privileged vantage point and a style of analysis immune from the 'strictures' of the radical perspective. There is then no basis for rejecting the radical perspective because of unattainable goals."[128] Yet Woolgar does admit "it is not possible to specify the kind of research which would most adequately meet the constraints of the radical perspective."[129]

What is one to make of this? Reflecting on Woolgar's exchange with Barnes, Harry Collins wrote: "We can use this fact [reflexivity] either to maintain a philosophical awareness of the nature of our own scientific activities and to maintain a methodological sharpness, or we can use it as a rather undiscriminating critical lever. Because of the universal applicability of this type of analysis, it ought to be clear that the mere demonstrability of the socially analyzable nature of any explanatory category should not be allowed to count, by itself, as a criticism of the use of that category. It amounts to no more than being 'more reflexive than thou.' A criticism that can be so readily applied to everything, is not really a criticism of anything."[130] But Woolgar carries on in the same volume, undeterred by Collins's doubts, invoking the category of "irony" to develop his case. Reflexivity, in this ironic formulation, becomes a *mise en abîme*.

With this appropriately poststructuralist phrase, we take up the "French connection" in Woolgar's reflexivity theory. Responding to a criticism which intimated that discourse analysis and reflexivity failed to take adequate account of *praxis*, Woolgar revealed the full intellectual arsenal behind his rhetorical approach.[131] He insisted that his idea of discourse was informed by French poststructuralism, and that Anglo-Saxon theory was woefully naive in comparison. "To speak crudely, the social study of science trades upon the essential ambiguity of the connection between language and culture."[132] That is, "this epistemological commitment finds expression in programmatic talk about getting the real picture, finding out what actually goes on in science . . . and so on. . . . [T]his central working assumption of the social studies of science stresses the availability of 'what actually goes on in science.'"[133] But "there is no reality independent of the words . . . used to apprehend it."[134] That is what the French poststructuralist analysis ostensibly establishes.[135] Woolgar cites a line from Hayden White's interpretation of Foucault to characterize the essential insight of the whole approach: "'the distinction between language on the one side and human thought and action on the other *must be dissolved* if human phenomena are to be understood as what they truly are, that is to say, elements of a communication system.'"[136] From that vantage, "the notion of '*praxis per se*' becomes meaningless. Under the continental rubric, *praxis* cannot exist outside discourse."[137] Woolgar clearly affiliates his project with French poststructuralism and postmodernism generally: "The grounds for knowledge have come under increasing challenge within a wide range of disciplines—anthropology, psychology, sociology, philosophy—and more recently in a number of intellectual movements which share a concern for the 'problem' of representation and which cut across traditionally defined disciplinary boundaries—poststructuralism, postmodernism, literary theory, and so on."[138] Under the aegis of this battery of movements, "the word-object relationship, once the paradigm of representation, is displaced by lateral, syntagmatic, and reflexive relations between communicational 'elements' in seemingly anarchistic fields."[139] Woolgar finds this fully "consistent with the position of the idealist wing of ethnomethodology" which he already endorsed.[140] As Nigel Pleasants observes, this version of reflexivity "is closer to Cartesian 'reflection' than ethnomethodological 'reflexivity.'"[141]

How does one engage in such "reflexive" practice? "The strategy is to sustain and explore the paradoxes which arise when we attempt to escape the inescapable, not to attempt their resolution."[142] In the words of his most enthusiastic follower, Malcolm Ashmore, "Can one be a participant in analytic practices and an analyst of these practices at the same time?"[143] For Ashmore, the answer is yes, and he attempts to show how. Woolgar offers at least a theory of how one might do so. The project is "to disrupt the apprehension of texts as 'objective' accounts."[144] Whose apprehension, exactly?

Can the author really forget? Can the audience be so naive? "[W]e should attempt to inject some instability into textual organization."[145] A reflexive author "juxtapos[es] textual elements such that no single (comfortable) interpretation is readily available."[146] Why? The "artfulness" of this may be questioned; so, too, can its pretense to rigor. First, if deconstruction is correct, it is not the author but *language* that fissures texts. It happens notwithstanding the author's uttermost assiduity. Second, the necessary "perspective by incongruity" can just as well be introduced by the juxtaposition of rival texts within the disciplinary community, in whose critical assessment every seam gets exposed.

Bruno Latour questions "rendering the text unfit for normal consumption (which often means unreadable)."[147] He diagnoses precisely the misguided preoccupation behind this "strategy": "This horror, the fear of contamination with empiricism is amusing, because it is exactly the counterpart of the empiricist position. They all think that objects, things-in-themselves, are somehow out of reach. As if any access to the world was forever in the hands of the empiricist programme. As if the world in which we live was the property of scientistic accounts of science."[148] He questions the soundness of a strategy of writing which would "in effect abolish the distinction between science and fiction."[149] Chris Doran takes up the theme: "[T]he Reflexive Faction itself does not understand its own contradictory stance. It wants to write in a fictional style, but wants to proclaim a truth value for its work."[150] Steve Fuller elaborates: "[T]he 'truly' reflexive ethnographer cannot write in a way that suggests her own privileged access to the truth. . . . [She must] adopt 'new literary forms,' Borgesian genres that studiously occupy the uncertain middle ground between fact and fiction, which as a result enable an author to highlight the ungroundedness of any appeal to truth, *especially* her own."[151] As Peter Halfpenny notes, "one cannot relativize one's own knowledge claims while making them without undermining the point of one's investigation."[152] Hence his conclusion: "what gets lost during the forced playfulness of self-reflexive writing is the point of it all."[153] The preciosity of such enterprises clearly provokes the irritation of the reader. Radically reflexive texts using "new literary forms" appear frivolous or narcissistic.[154]

There is certainly scope enough for the poetic or literary evocation of the paradox of form-giving and the irony of authorship, even as there is scope for philosophical exploration of the epistemological abysses of skepticism. Aristotle had scope for philosophy and poetry, appraised them more highly than history, but he *did not proscribe* the latter. Perhaps in their excessive early enthusiasm, proponents of the Strong Program made gestures toward displacing philosophy, but that has certainly not transpired. More to the point, it simply is not part of the empirical agenda of sociology or of history to proscribe or to replace poetry or philosophy. The question that we must raise pointedly is by what lights will the new "reflexivity" have it that "new literary forms" *should* proscribe and preempt empirical inquiry? One is certainly entitled to wonder whether there might not be other ideals worthy of sociological or historical pursuit, such as the traditional ones of describing and/or explaining human actions and beliefs. As Fuhrman and Oehler write, "While we admit that the use of texts is an important element in how scientific beliefs become established, the final question must be how the *beliefs* are established, not how the texts are written."[155] Rhetoric is important, but rhetoric is not everything.

Moreover, genuine *literary* talent is required if one is to play these literary games. As Peter Burke has noted in his appraisals of experiments in historiography that introduce "new literary forms," the success of the venture is entirely proportional to the literary gift of its practitioner.[156] Put bluntly, two pages from Jorge Luis Borges on Pierre Menard are worth a hundred and more from Ashmore or even Mulkay.[157] While Woolgar might find Ashmore's *The Reflexive Thesis* "witty and clever throughout," its "highly unusual format" strikes this reader as egregiously self-indulgent and, more essentially, tedious in its tergiversations. What we learn from laboring through his rhetorical gymnastics is paltry.[158]

There is about Woolgar's reflexivity that familiar trope of hyperbole that mars virtually all the "radical" postures of contemporary postmodernism. Leaping to the outermost limit of skepticism creates an *unreal* criticism. Woolgar acknowledges that "criticism necessarily involves comparison between representations, not between representations and an 'actual object,'" but he fails to see that invoking absolute skepticism misrepresents the tasks and the standards of appraisal involved in empirical research projects.[159] The question is, what does he hope to *achieve* in the reflexive regress? As Collins and Yearley put it: "Woolgar's identification of the Problem is very persuasive. But what are we to make of it; how does it benefit us to know of it? Clearly the Problem cannot be solved. Neither can it be bypassed by adopting a phenomenological policy of introspection."[160] They reject Woolgar's hyperbole of "permanent revolution" since that spiraling "meta" regress into theory is in fact an evasion of the risk and uncertainty of empirical interpretation. They endorse empirical inquiry as a program of permanent insecurity, the inevitable anxiety of offering up contingent empirical claims to the appraisal of a disciplinary community.

In several publications Paul Roth has attacked Collins and Yearley for this stance. "Faced with a demand to substantiate the legitimacy of their approach, Collins and Yearley turn from bold challengers into reactionaries . . . they can neither legitimate the status given descriptions in the mode of social realism nor give them up."[161] Again, the key is his insistence upon a warranted theory of causal explanation. He faults history and sociology *generically* with offering "no more than narratives of the post hoc, ergo propter hoc variety."[162] He dismisses any effort at a historicist or naturalist methodology as begging the question of the fact/value divide.[163] Mere description can never warrant any normative claim, in Roth's view, and so, with this categorial standard, but with inverted motives, Roth, like the traditional logicists, bludgeons empirical inquiry with the demand for absolute compliance or condemnation as mere artifice. "Appeal to interests, to close description, and to recovering how things actually were, raise questions about the scientific status of the studies that have yet to be answered."[164] Hence, "one might say that the time has come for the SSK to stop worrying about Karl Popper et al. and start thinking about their relation to Hayden White."[165] As he puts it in another essay, "I am quite skeptical about uncovering a fact of the matter with regard to the correct interpretation of historical incidents."[166] Accordingly, radical skepticism of the Rortyan variety is all that remains, and Woolgar's radical reflexivity fits this mold well because it takes the semiotic or rhetorical turn Roth affirms with his gesture toward Hayden White. "The point here is that the historical data too are underdetermined and so can be read to many different effects."[167] The effort of figures like Shapin to look to history and local specificity for a measure of objectivity seems to Roth hopeless. Invoking Peter Novick's *That Noble Dream,* he comments: "attempts to found their account of scientific objectivity on historicist notions is building on quicksand."[168] Where does that leave us, for Roth? "The live issue, I contend, is how to analyze disciplinary practices once one acknowledges that we possess no notion of what makes a particular practice a science, in some philosophically interesting sense of 'science.'"[169] For Roth, "the term 'science' has outlived the uses for which it was mainly introduced."[170] Certainly any sense for the "unity of science" or a hard "demarcation" between science and nonscience has become fruitless for post-positivism. The "thriving subdisciplines" in place of the monolith, Roth avers, indicate that "the philosophy of science has gone postmodern."[171] In this "postmodern light," Roth continues, "the normative questions that permit of answers will be internal to particular disciplines."[172] Yet at the very same time, he denies *from a transdisciplinary vantage* that history or ethnography—more generally, that *any* empirical human science—*can* establish sufficient *internal* warrant to adjudicate its practices. That is what he means in rejecting Shapin's "hopes of methodological justification on a form of historicism."[173] Roth ends up with a radical, Rortyan relativism: "I argue that there is no way to distinguish between claims that one has discovered what someone else actually meant and the charge that one is imposing a translation."[174] While Roth invokes Rudolf Carnap's distinction of internal and external questions, he persists in posing an external standard to denigrate the very possibility of an internal one.

As Halfpenny puts it in a manner meriting my heartfelt concurrence, "[M]y suggestion is that investigators should make inferences and selections by the tacit skills absorbed from their discipline's traditions, and allow the community of scientists who make up the discipline to sort acceptable from unacceptable knowledge claims. This is a practical solution that has (so far?) proved impossible to capture in a general (part descriptive, part prescriptive) epistemological rule."[175] We are all in Quine's "web of belief" in which we can only problematize a given belief by assessing it in terms of the others held stable. There is no view from nowhere. Objectivity is a contingent, fallible achievement of the disciplinary community. Global skepticism has no point in empirical inquiry beyond a moral monitory recognition of fallibility. The "reflexivists' global skepticism towards representation per se" has a place in formal philosophy, but it is ultimately a "fruitless regress on further 're-flections' on previous 'reflections'" which contributes nothing to empirical inquiry.[176] Woolgar's absolute skepticism, like the poststructuralist rhetorical *mise en abîme* on which it is modeled, simply makes no sense for a resolutely antifoundationalist and equally resolutely *naturalist* epistemology.

With the crisis of sociological explanation, with the abandonment of empirical accounts by "discourse analysis," and, most importantly, with the articulation of "new literary forms" and radical reflexivity, the whole enterprise of science studies reached what I would characterize as a fatal impasse. Here we are not considering better accounts of science but a metatheoretical intervention that preempts empirical inquiry. That is surely a *reductio ad absurdum*, a nice derangement of epistemes. When science studies slips all its moorings in the fashion of Woolgar, folly is not long to appear. It did, with hilarious (or dire, depending on one's investments) repercussions, in the "Sokal affair" of 1996.[177]

The Sokal Affair and the "Science Wars"

If Woolgar's radical reflexivity is, for me, the internal *reductio ad absurdum* of postmodernism in science studies, I see the Sokal affair as the external—public and political—*reductio* of that same misguided impulse. The Sokal affair exposed the hyperbolic character of postmodernist "theory" in science studies that is my overarching concern. That does not mean that I have no reservations about the "science wars." Quite the contrary. But, as Callon and Law said of the dispute between Barnes and Woolgar, they could affirm neither side without qualification, but they stood closer to Barnes, so I stand closer to Sokal.[178]

The "science wars" were, as Ian Hacking phrased it with piercing precision, "about metaphysics and rage."[179] That is, "the science wars are fueled by the rage against reason-masquerading-as-innocence."[180] But they also involved perennial questions about metaphysics—perennial, in Hacking's view, because ultimately irresolvable. Hacking has harsh things to say about both sides in the science wars. "What is true is that many science-haters and know-nothings latch on to constructionism as vindicating their impotent hostility to the sciences."[181] But what is also true is that the constructivists' proper "target is not the truth of propositions received in the sciences, but an exalted image of what science is up to, or the authority claimed by scientists for the work that they do."[182] In short, substance in these controversies cannot be disseevered from emotion and personal values—politics, for short.[183]

At the base of it all, however, Hacking discerns three substantive "sticking points" with rather extended lineages: inevitability versus contingency, realism versus nominalism, and internal versus external accounts of the stability of knowledge. Scientists typically maintain that the current content of science is the *inevitable* outcome of its historical emergence, and that the contingencies of that emergence are irrelevant to the outcome, like slag burned off from the ore.[184] They believe that this inevitability is in turn the consequence of *how the world really is,* and many believe that the content of science not only converges on the real but that in some crucial areas has conclusively converged, in other words, that certain elements of scientific knowledge are now so definitive that radical revision is highly dubious.[185] Finally, they believe that the process of scientific inquiry has had this inevitable outcome of convergence upon reality by virtue of the *rationality of the methods* employed. Constructivists, by contrast, hold that what has emerged as successful science (even its *content,* not just its institutions) is only a *contingent* outcome, that other successful scientific paradigms might well have arisen, concerned with *other entities,* and hence the *path* to the outcome has had a determinate impact on the *character* of that outcome.[186] Second, they generally adopt a highly skeptical view about the correspondence of any given scientific theory with the way the world really is or even a convergence toward it. That is, they are strong *nominalists* in their ontology. Finally, they believe that the path to stabilization required substantial external intervention from society and politics and that the maintenance of that stability continues to do so. For Hacking, these are "sticking points" because there are reasonable grounds for holding stronger or weaker positions along the spectra constituted by each of the three issues, and there is no prospect for a consensual (or indeed any) stabilization.

In the dispute, a central charge thrown back and forth has to do with whether there is an objective world "out there" or whether all that we as humans can intelligibly debate is a representation in discourse. Thus, for example, Roth and Barrett assailed Pickering, in a paradigm instance of the polemic, for failing to discriminate between the object, *quark,* and the *idea of a quark.* They insisted that the object quark operated as a real cause of the emergence of the idea of a quark.[187] Pickering, in turn, insisted that he meant just what his critics objected to, for *quark,* as a well-attested object in the physical world, was a *contingent,* not an inevitable outcome of a process of inquiry which might well have been pursued in other ways and have resulted, accordingly, in the discrimination of other entities as theoretically noteworthy.[188] Hacking would prefer to leave this standoff in open conflict.[189] Let us see if we can find any more coherent resolution.

Hacking and others have raised a contextual conjecture about the timing of the outbreak of the science wars. That is best phrased by Bruno Latour: the attack of physicists like Alan Sokal on science studies in the early 1990s represented "the last acrobatics of Cold War science," in which "a very small number of theoretical physicists, deprived of their hefty Cold War budgets, are seeking a new menace."[190] One of the contributors to the special issue on the science wars, George Levine, offered a similar conjecture.[191] Hacking, while he thinks this is hardly a sufficient account, does see a connection. He discerns high-energy physicists as having enjoyed enormous cultural prestige in the era of the Cold War; their field was "the queen of the sciences." But "with the end of the Cold War, the financing of high-energy physics was abruptly curtailed" and "the new queen of the sciences became molecular biology." The impact on high-energy physicists was a significant loss of "cultural authority"—"not just the ability to command vast resources of money and talent, but also the conviction that their life work is deemed to be profoundly significant not only by their peers but also by their culture, or world culture at large."[192] Hence, he implies, their motivation to launch the science wars. While the shifts in relative prestige may well have occurred, I do not believe that they are nearly so significant as Latour alleges and Hacking seeks to make plausible. Neither Paul Gross nor Norman Levitt were high-energy physicists. And the very idea that the science wars— the *thing,* not the term—began with the attacks of the scientists in the early 1990s is already a very one-sided historical perception of the event. Science studies had been systematically "demystifying" the prestige of science for twenty years before the scientists struck back. If we want to understand the origins of the science wars, the proper place is not the end of the Cold War but rather the rise of postmodernism. That was what galvanized Gross and Levitt to publish their polemic, *The Higher Superstition: The Academic Left and Its Quarrels with Science,* in 1994, bringing the matter glaringly into the public arena.[193]

The phrase "science wars" was conjured up with reference to Gross and Levitt by Andrew Ross, who linked it with the wider *culture wars* as part of a conservative counterattack on a besieged academic left—the pursuit, as it were, of the culture wars *by other means*.[194] This view inspired the ill-fated "Science Wars" special issue of the journal *Social Text*, of which Ross was an editor.[195] That issue, as it happened, carried an undetected virus that contaminated the whole venture: Alan Sokal's hoax essay. Sokal explained that he embarked upon his hoax after reading Gross and Levitt's book. Thus, the point of departure for the whole consideration of the science wars must be *Higher Superstition*.

Professors Gross and Levitt gave voice to a backlash of natural scientists against what they perceived as "humanists of the academic left who have transgressed the boundaries" by impugning scientific practice and especially scientific knowledge.[196] They expressed shock that there could be "open hostility toward the *actual content* of scientific knowledge."[197] Clearly, science studies from the Strong Program forward saw its mission as bringing scientific *content* under scrutiny. It approached this content with *skepticism*, however, not hostility. Bloor, as we have noted, was unequivocally scientistic. But for much of the "cultural studies" approach prominent among postmodernists and feminists, *hostility* is an accurate term. The content of science, in that view, had nefarious consequences.[198] Gross and Levitt were concerned to defend not the *consequences* of scientific ideas: they admitted these had some evil outcomes. They insisted rather upon the *soundness* of those ideas. Quite simply, they thought them true. As they put it, with avowed bluntness, "science *works*."[199] As to the social context of science, the two authors observed: "Modern science is seen, by virtually all of its critics, to be both a powerful instrument of the reigning order and an ideological guarantor of its legitimacy."[200] Such a formulation equivocates on whether Gross and Levitt accepted this characterization. It would appear that they should have. Modern science simply *is* a powerful instrument, and it serves, as science always has, whoever provides the funding. Furthermore, Gross and Levitt's own work, which complacently aligned the history of science with progressive impulses in the Enlightenment tradition, is evidence of the way in which the self-conception of science serves as guarantor of the legitimacy of the current order. All the same, however, unlike those who find this a basis for unequivocal condemnation of science because they are unequivocally hostile to the Enlightenment tradition, I do not find the alignment so compromising. Certainly the historical record is problematic: there is little warrant for *complacency* about the success of progressive causes in Western society or science's role in them, but against the hyperbolic radicalism of postmodernism, it has by no means been such an unremitting calamity either.

Gross and Levitt expressed considerable concern about the lumping that went into their labeling of an "academic left." They feared that there were such disparate postures involved that "to characterize them as products of one ideological subculture . . . [was to] lay ourselves open to the charge of taking an illegitimate polemical shortcut."[201] Indeed, they did not escape that failing, and their occasional gestures of recognition for serious scholarship did not quite make up for it.[202] But they did not invent what they were confuting. There is a postmodern intellectual culture, and it has many of the features they have ascribed to it.[203] Gross and Levitt identified the core of this postmodern posture with rejection of the heritage of the Enlightenment. That is indeed one of the clearest ways to identify what is in other respects, as they noted, "a congeries of different doctrines, with no well-defined center."[204] They were apt, as well, in discerning that postmodernism "is more a matter of attitude and emotional tonality than of rigorous axiomatics."[205] In addition, Gross and Levitt discerned—correctly—"an emphatically *totalizing* component" in postmodernism.[206] Generally, postmodernists appeared ready to "pronounce with supreme confidence on all aspects of human history, politics, and culture."[207] They cited anthropologist Robin Fox's observation: "English literature departments are reconstituting themselves as cultural studies departments and are trying to take over the intellectual world."[208] Gross and Levitt highlighted the presumption "that literary scholars trained in deconstruction or some other related methodology are capable of a 'deep reading' of scientific texts."[209] While the natural sciences proved thoroughly impervious to such pretense, they continued, "in the social sciences . . . the effects have been drastic."[210] Anthropology suffered massive incursions. The "New Historicism" and "cultural studies" have made inroads as well into the disciplines of history and sociology.[211] What Gross and Levitt wrote is not without some grounds. Indeed, from the vantage of the discipline of history, the "interdisciplinary" overtures from "cultural studies" sometimes have the appearance of marching orders as arrogant as any that natural scientists might discern, but with considerably greater influence.[212]

However, there are serious flaws in *Higher Superstition*. One core area prompting dissatisfaction lies in the treatment of philosophy of science. Gross and Levitt did not demonstrate the least attentiveness to the enormous amount of controversy and revision that has transpired in philosophy of science over the twentieth century. They appeared to think that, for example, Helen Longino's careful summary of the key developments in post-positivist philosophy of science represented merely her idiosyncratic conjectures: her "claimed demise of logical positivism."[213] The name of Willard Quine is conspicuously absent from their account. Carnap's complexities are lost on them. Karl Popper made it into the text twice, but without serious scrutiny. Thomas Kuhn and Paul Feyerabend hardly got more than capsule disparagement. These scientists *did not need* to think about philosophy of science, because they *already knew*: "When it comes to the core of scientific substance . . . and the deep methodological and epistemological questions—above all, the incredibly difficult *ontological* questions—that arise in scientific contexts, perspectivism can make at best a trivial contribution."[214] I find this sort of pronunciamento "transgressive" in all the ways that Gross and Levitt have protested that postmodernism is transgressive. They themselves claimed that "countering arguments must be made with reason and patience."[215] Their posturings fail miserably by that standard. Philosophically, Gross and Levitt made assertions throughout, not arguments. There *are* arguments to be made, and they can indeed beat back rash relativism of the sort that Gross and Levitt find appalling. But substantial chunks of their logical positivist pieties will have to be cast aside as well. These are not philosophers; they should defer to those who are for support of their cause.[216]

At least they did admit: "neither of us is a professional historian."[217] I am, and what they wrote about history does not inspire me with much confidence in their judgment. When they claimed to find it "doubtful that historians have yet come to grips" with the dramatic shift in the relative power of the Western world vis-à-vis the non-West from 1800 to 1900, one could only wish of them what they urged their humanist colleagues to do in the converse situation: attend a freshman *history* course![218] Their account of the cultural background of the antiscience tendencies of the twentieth century, to take another example, omits any mention of the culturally decisive First World War. More germane to this study, their commentary on Shapin and Schaffer's *Leviathan and the Air Pump*, discussed earlier, suggests an a priori approach to historical interpretation that is profoundly irksome to empirical historians. Frankly, what these two scientists found "hard to believe" weighs as little with a professional historian as they claim "metaphors" weigh in quantum mechanics.[219] We expect to go to the sources and find out as best we can what actually happened. If all it took were to project what made sense looking from here, there would be no need for a profession of history. These scholars made disciplinary propriety the essence of their case. I submit that they showed no more respect for disciplinary propriety (regarding philosophy of science or history) than they accused the postmodern academic left of showing for the natural sciences. My point is that many of us would bring *support*—reasoned and patient support—for their larger agenda. It did not help their case to have been so cavalier with other disciplines.

Gross and Levitt had no tolerance at all for the idea of "historically and socially situated" knowledge.[220] They expressed confidence in "claims to universality and timeless, uncontextual validity."[221] What could that mean? I take it that by "historically and socially situated" they meant the radical relativist notion that *only* local political agendas signify, or, as they cited one of their targets, that falsifiability is self-referential for a given community.[222] Stanley Aronowitz, they averred, holds such a view. Perhaps he does. It is not a view that has any monopoly claim on "situated knowledge," however, or on contextualization of knowledge-claims. I take Gross and Levitt at their word that they "accede in principle to what might be called the 'weak' version of cultural constructivism."[223] There are only some simple points to be made, then. First, even within the modern Western era of science, there have been dramatically different positions held by groups of scientific specialists and sometimes whole disciplines, all of them working diligently with the "twin tests of internal logical consistency and empirical verification," but also clearly in controversy.[224] Some of these controversies appear to have been resolved, and a given scientific community accepts a certain theory as—for the moment—well-attested.[225] But all of us, I am sure, stand committed to the principle of the contingency and fallibility of even the best scientific theory. Then, clearly, even what we today with good reason find to be the best available theory cannot but be a contingent and fallible (read: historically and socially situated) claim. It is historically situated, because we have confidence that scientists have not always accredited it and it is highly unlikely that they will evermore. It is socially situated because, by the end of the twentieth century, there is virtually nothing in science that does not have massive social and political connections—both at the point of origin in funding and problem formulation and at the point of delivery in technology and policy. In addition, it is a fact, pleasant or not, that most of the human race is not in a position to judge what the issues are or the evidence, and therefore a scientific claim is for most cognitive purposes the restricted purview of a disciplinary community, hence contingent upon its training, standards, and practices. Does that mean the content of science does not accord with nature? That it is not "real"? First, apart from the extreme wing of science studies, which Gross and Levitt rightly disparaged, most scholars of the history of Western science not only believe science seeks evidential warrant from nature, but carry such a belief in evidential warrant into their own interpretive practices vis-à-vis science as an object of study. But let us be very careful with the word *real*. Let us take seriously what Gross and Levitt merely said, that there are "incredibly difficult *ontological* questions . . . in scientific contexts." Philosophers have debated realism and antirealism for at least the better part of two hundred years. It may well be that it is a normal and fruitful operating assumption of professional scientists to be "realists," or to adopt what Arthur Fine has called the "natural ontological attitude."[226] But science does not require a commitment to *any* ontology. An instrumentalist ontology has proven perfectly compatible with natural science.[227] What is incompatible is an "irrealist" *epistemology,* one where "empirical verification is dismissed as a species of bluster."[228] That is where philosophy of science, empirical history of science, and natural scientists themselves all share a common cause to resist. Gross and Levitt have not invented straw men for such a posture. But they proved extremely clumsy in the way they articulated the *philosophical* and the *methodological* principles that have governed science. It does not help to write such lines as "So much, then, for three thousand years of struggle to develop a systematic method for getting reliable information about the world!"[229] There is no evidence for a historically continuous and determinate endeavor along those lines; it is a piety projected upon the past with ideological intent. Nor is it helpful to wax grandiloquent about "science as an activity of the autonomous and unfettered intellect."[230] This is especially so when these same authors wrote: "No serious thinker about science, least of all scientists themselves, doubts that personal and social factors influence problem choice and the acceptance of results by the scientific community."[231]

What balance can be drawn, then, about this controversial book? Gross and Levitt were correct in maintaining: "postmodern skepticism rejects the possibility of enduring universal knowledge in any area. It holds that all knowledge is local, or 'situated' . . . all knowledge projects are, like war, politics by other means."[232] They were wrong in thinking this exhausted the significance of perspectivism or contextualization in understanding knowledge-claims or the history of science. They were right, too, that: "The central ambition of the cultural constructivist program—to explain the deepest and most enduring insights of science as a corollary of social assumptions and ideological agenda—is futile and perverse."[233] They were wrong in presuming that "enduring insights" meant timeless truths; they overlooked too cavalierly the fallibility and contingency of knowledge. And they were right, finally, that: "A serious investigation of the interplay of cultural and social factors with the working of scientific research in a given field is an enterprise that requires patience, subtlety, erudition, and a knowledge of human nature. Above all, however, it requires an intimate appreciation of the science in question, of its inner logic and of the state of data on which it relies, of its intellectual and experimental tools."[234] Norman Levitt in a later essay observed: "A recurrent theme among the Young Turks of the field [of science studies] is that a close familiarity with the methodology and conceptual content of science is not only unnecessary but positively detrimental."[235] It is the case that some proponents of science studies, perhaps most notoriously Bruno Latour, have made a big deal about not knowing any of the science they are studying "ethnographically."[236] I think that Latour knows more science than he lets on. Be that as it may, Andrew Pickering certainly knows a lot of high-energy physics. And most scholars in science studies know and believe they *ought* to know a lot about science. That is certainly the way that the new generation of historians and philosophers of science is being trained. To this extent, Gross and Levitt were correct. But there was, beyond the sensible demand for understanding, an implicit demand for *deference*. They were wrong in presuming that scientists have an ultimate authority over the social, historical, or political meaning of their practices or ideas. A careful history of post-positivism in science studies confirms all these points.

Prior to the outbreak of the Sokal affair, the campaign launched by Gross and Levitt was carried forward in a conference sponsored by the New York Academy of Sciences, held in New York City, May 31–June 2, 1995. Its proceedings were published as *The Flight from Science and Reason* (1996), edited by Gross and Levitt in collaboration with Martin Lewis. The title was as vituperative as *Higher Superstition*, and the contents widened the scope to span the entire academy. Among others, Susan Haack tried to explain what made science more than "*simply* a social institution." Her argument was that scientific inquiry cared about evidence. She urged "a distinction between questions of warrant—how good is the evidence for the theory?—and questions of acceptance—what is the standing of the theory in the relevant scientific community?"[237] This would suggest not only that somehow scientific communities *accept* theories for largely nonevidential reasons but also that somehow it is possible, apart from their deliberations, to discriminate the evidence and "how good it is." What we have here, it would appear, is a new articulation of the idea of a "context of justification" and the implicit suggestion that its unequivocal standards are externally administered—by disinterested epistemologists from outside the relevant scientific community. Haack wrote: "how justified they are in accepting [a scientific claim] does not depend on how justified they *think* they are, but on how good their evidence is."[238] By whose lights and by what rights? Haack notwithstanding, the relation between "warrant" and "acceptance" is considerably more intricate. It is essentially negotiated *within* the given scientific community, engendering precisely "acceptance." We have conclusive historical evidence that it is not *administered to* that scientific community by logical legislators from philosophy. In "Science As Social?—Yes and No" (1996), Haack explains that by *warrant* she means a normative notion and by *acceptance* a strictly descriptive one.[239] By that light, to conflate them, and especially to assimilate the normative element in warrant into a matter of mere agreement would miss something important about inquiry. Yet I think that this misses precisely what Haack's essay is ostensibly about: the social element in science, and indeed, the social element in warrant. Warrant is not, any more than objectivity, something that any given individual—not even a philosopher—can attain or even adjudicate autonomously.[240] The only even proximal recourse we have is that of collective judgments, and *acceptance* may be a descriptive term, but it refers to a normative process.

Haack ultimately makes this point herself: "the social character of science [is] one of the (of course, very fallible and imperfect) factors which help to *keep acceptance appropriately correlated with warrant.*"[241]

Alan Sokal began putting together his pastiche in 1994, upon reading *Higher Superstition.* His text was submitted to *Social Text* in that year. The editorial board of this "tendency journal" did not employ peer review. Instead, they appraised submissions internally, including this one, even though it patently exceeded their expertise. The most the editors endeavored was to send Sokal a suggestion that he purge many of the citations and footnotes, but since both were central to his hoax-strategy, Sokal declined. Eventually, the editors decided to publish it, unrevised, in a special issue they were commissioning from noted "academic left" critics of science and technology. As it happened, the special issue generated little interest and sold few copies.[242] Instead, Sokal's virtually simultaneous exposé in *Lingua Franca* made his hoax and the debate over it far more famous than anything in the original issue.

The ambition of the special issue needs to be retrieved to situate the hoax and its reception. John Brenkman observed that "*Social Text* was trying to do a very difficult thing in the 'Science Wars' issue . . . on the one hand, . . . a critique of the epistemology of Western science, and on the other hand . . . talk about the political implications of the funding of scientific research, the application of science to industry and the military and to the inner workings of a capitalist economy."[243] Brenkman discerned "an unexamined tendency to believe that these two sets of problems are intimately related, if not identical." He termed this "the sort of Foucault paradigm run amok," which was nonetheless "a very deep part of the common sense of cultural studies."[244] Ellen Willis comments: "Those who actually read *Social Text*'s 'Science Wars' issue will find it mostly devoted to analyzing—in English, not poststructuralese—how the questions scientists ask, the research they choose to pursue, the ways they collect and interpret their data, and the technological applications of their work both reflect and shape certain social and political ends."[245] Setting apart the question of jargon, the contributions of figures like Sandra Harding, Steve Fuller, Dorothy Nelkin, Sharon Traweek, and others are hardly without a strong postmodernist intellectual agenda. And in the contribution from Ruth Hubbard claims were made to equal any that Sokal was trying to discredit by satire and Gross and Levitt by blunt polemic.[246] In short, one cannot plead political or theoretical innocence for these authors and the issue for which they wrote. It was a clearly tendentious volume. It just got lost in the wake of the hoax controversy.

In a *Lingua Franca* roundtable in 1997, John Brenkman, one of the founders of *Social Text*, made an observation that seems a good jumping off point for consideration of the Sokal affair: "The parody itself, it seems to me, was brilliant, but Sokal's explanation in *Lingua Franca* of what he'd done makes two massive claims, neither of which I think is true."[247] First, Sokal lumped all of cultural studies into "a morass of relativism and confused logic," and second, he advocated "a very narrow position on the nature of scientific inquiry."[248] What was so "brilliant" about the parody? And did Sokal make these two errors in his revelation of the hoax in *Lingua Franca*? In his own comments on the parody, Sokal emphasized the importance of his citations from dominant figures in the postmodern pantheon, all of them inept, yet nonetheless invoked with sycophantic praise.[249] Such gestures of affiliation with a privileged "way of knowing" have, indeed, been all-too-characteristic of postmodern writing, where appeals to authority and the invocation of what "everyone knows or else should know" (to pirate William Butler Yeats to profane ends) are commonplace. Here is a wonderful sentence from the opening of Sokal's "Transgressing the Boundaries": "It has thus become increasingly apparent that physical 'reality,' no less than social 'reality,' is at bottom a social and linguistic construct; that scientific 'knowledge,' far from being objective, reflects and encodes the dominant ideologies and power relations of the culture that produced it; that the truth claims of science are inherently theory-laden and self-referential; and consequently, that the discourse of the scientific community, for all its undeniable value, cannot assert a privileged epistemological status with respect to counter-hegemonic narratives emanating from dissident or marginalized communities."[250] He packed about as many dogmas of postmodernism into that one sentence as it could carry! In it we can discern the elements of postpositivist philosophy that we have traced over the last half-century, admixed with the political and cultural studies agendas of postmodernism which emerged most clearly in connection with the feminist critique of science. In short, Sokal got the argot of the antiscience literature perfectly.[251] To call it behind the times, as Andrew Ross and Bruce Robbins attempt to do, is a piece of unequivocal bad faith.[252] "Poststructuralese" (to use Willis's neologism) continues alive and well, both linguistically and conceptually.[253]

In his revelation in *Lingua Franca*, Sokal charged postmodernism with "the proliferation of a particular kind of nonsense . . . one that denies the existence of objective realities." That is, he asked rhetorically: "Is it now dogma in cultural studies that there exists no external world?" Against such a posture, Sokal asserted: "There *is* a real world; its properties are *not* merely social constructions; facts and evidence do matter."[254] It would appear that Brenkman's description of Sokal's statement captures what Sokal wrote. It remains to ask, were these statements grossly in error, as Brenkman alleges?

The controversy focused on just these issues, and the best way to assess Brenkman's claims (and Sokal's behind them) is to consider the controversy.

First, it is important to ask what Sokal and his opponents thought the hoax meant. Sokal viewed it as a decisive exposé of the incompetence of cultural studies of science. The editors of *Social Text* and their defenders sought to maintain that it proved no such thing, but rather the ethical impropriety of Sokal.[255] Of the two views, the latter has not come off well in the debate. As David Albert comments, "what the hoax should be construed as having pointed up is a failure of science studies, to the extent that science studies is represented by *Social Text*, to have done its job right."[256] Brenkman put it precisely: "I think cultural studies is suffering from a kind of crisis of over-confidence, and it shows."[257] Paul Boghossian observed: "the conclusion is inescapable that the editors of *Social Text* didn't know what many of the sentences in Sokal's essay actually meant; and that they just didn't care."[258] He went on, "what's at the heart of the issue [is] allowing ideological criteria to displace standards of scholarship."[259] Evelyn Fox Keller called the defense that Ross and Robbins offered for their gullibility "embarrassing."[260] Brenkman found "the editors' response to the hoax disturbing and misguided." As the debate went on, he commented, "their own credibility crumbles with each new statement . . . far worse, they are undermining the critical study of science itself."[261] In particular, the recourse to *ad hominem* ethical reproaches appeared lame, especially when figures like Stanley Fish and Stanley Aronowitz could find no better rebuttals.[262] Kathy Pollitt observed: "It's hard not to enjoy the way this incident has made certain humanities profs look self-infatuated and silly—most recently Stanley Fish."[263]

What intellectual defense did the partisans of cultural studies of science actually present in the aftermath of the hoax? Evelyn Fox Keller made a first effort: "Scholars in science studies who have turned to postmodernism have done so out of a real need: Truth and objectivity turn out to be vastly more problematic concepts than we used to think, and neither can be measured simply by the weight of scientific authority, nor even by demonstrations of efficacy."[264] Here, at least, there are philosophical issues about science that can be discriminated and debated, and the question whether recourse to postmodernism can avail scholars of science in their need can be put to the test. I am frankly skeptical that postmodernism avails at all to resolve the issues that trouble Keller, but that at least is an issue of intellectual substance, not of personal invective or rhetorical evasion.

Stanley Fish, before he resorted to *ad hominem* self-righteousness, also offered a defense of the postmodernist stance. He wrote: "What sociologists of science say is that of course the world is real and independent of our observations but that accounts of the world are produced by observers and are therefore relative to their capacities, education, training, etc. It is not the world or its properties but the vocabularies in whose terms we know them that are socially constructed—fashioned by human beings—which is why our understanding of those properties is continually changing."[265] This is a remarkable piece of writing. If, indeed, science studies took the stance that Fish represented, there would be nothing radical whatever about it. That is, in fact, why Fish is unbelievable, for science studies does seek to be radical. Indeed, a careful study of Collins, of Pickering, and above all of Latour—to say nothing of Harding and Haraway—suggests not only that they would repudiate Fish's intervention but recognize it for what it is—disingenuous rhetoric. There is a characteristic move here, one which features in much postmodernist posturing. Extreme positions are taken; when challenged, authors deny the extremity and affirm they really meant a far more modest posture.[266] At least in Collins, Pickering, and Latour we have authors strong enough in their convictions, whatever others think of their claims, to refuse to water them down to escape criticism. Fish makes a mockery of what radical science studies wants to say about the epistemology of science. Furthermore, when he elaborates his analogy to baseball, he makes a fundamental category error, as Ian Hacking has mercilessly demonstrated. "Fish wanted to aid his allies, but did nothing but harm. Balls and strikes are real *and* socially constructed, he wrote. Analogously, he was arguing, quarks are real *and* socially constructed. . . . Unfortunately for Fish, the situation with quarks is fundamentally different from that for strikes. Strikes are quite self-evidently ontologically subjective. Without human rules and practices, no balls, no strikes, no errors. Quarks are not self-evidently ontologically subjective. The shortlived quarks (if there are any) are all over the place, quite independently of any human rules or institutions."[267] Fish is equivocating, in short, between epistemology and ontology. "Perhaps it is the idea of quarks, rather than quarks, which is the social construction," Hacking suggests as a way to rescue Fish's position. Still, "quarks, the objects themselves, are not constructs, are not social, and are not historical."[268] That is, *if* they exist. We can be epistemologically skeptical about whether we can know this, but what the dispute is *about* is the ontologically objective character of quarks, not the mere representation. In a letter on this very matter, Andrew Pickering made that entirely clear: "I would never say that *Constructing Quarks* is about 'the idea of quarks.' That may be your take on constructionism re the natural sciences, but it is not mine."[269] Pickering has no interest in blurring the issue in a cloud of rhetoric.

Bruno Latour, too, stands by his guns. "Relativism is an asset, not a weakness," he writes. "It is the ability to change one's point of view."[270] For Sokal, Latour's ANT program "mixes up ontology and epistemology," deliberately playing with "a profound confusion between 'Nature's representation' and 'Nature,' that is, between our theories about the world and the world itself."[271] Where Fish wants to preserve the distinction and pretends science studies is in the innocuous position of addressing only epistemology,

Latour wants to make the very discrimination between epistemology and ontology yet another of the dualisms his approach "transcends." That is radical. In short, Sokal has a far better sense of what *radical* science studies is about than Fish (pretends to).

Postmodernism *does* adopt a radical stance. The Italian commentator Marco D'Eramo put it bluntly: "I am incapable of distinguishing between reality and perception (and theories) of physical reality"; therefore, "we must resign ourselves to an inescapable epistemological uncertainty."[272] What D'Eramo is claiming is simply that epistemology washes away ontology: we cannot even say what we are unsure about! For D'Eramo, Sokal is guilty of "nineteenth-century scientism."[273] Presumably, any distinction between ontology and epistemology is fruitless. That, I submit, is radical postmodernism. But is it viable? I believe that a moderate historicism is a far better stance than this radical postmodernism, precisely because, as Paul Boghossian perceived, it "doesn't entail that there is no such thing as objective truth." He contrasted historicism explicitly with postmodernism in a very useful way:

> Historicism leaves intact, then, both the claim that one's aim should be to arrive at conclusions that are objectively true and justified, independent of any particular perspective, and that science is the best idea anyone has had about how to satisfy it. Postmodernism, in seeking to demote science from the privileged epistemic position it has come to occupy, and thereby to blur the distinction between it and "other ways of knowing"—myth, and superstition, for example—needs to go much further than historicism, all the way to the denial that objective truth is a coherent aim that inquiry may have. Indeed, according to postmodernism, the very development and use of the rhetoric of objectivity, far from embodying a serious metaphysics and epistemology of truth and evidence, represents a mere play for power, a way of silencing those "other ways of knowing."[274]

The representation of postmodernism is clearly polemical here, but the substance is essentially correct. What is at issue is whether science has any right to a "privileged epistemic position"—is that not simply "scientism"?

The question of "scientism" became acute in connection with Steven Weinberg's important intervention in the Sokal affair in the *New York Review of Books*.[275] Weinberg's essay carried forward his views on the social studies of science which had been articulated earlier in his popular work, *Dream of a Final Theory*.[276] What Weinberg expounded was the stance of a practicing physicist who found the interventions of "many sociologists, historians, and philosophers as well as postmodern literary theorists" beside the point. "[O]f course we know very well how complicated the relation is between theory and experiment, and how much the work of science depends on an appropriate social and economic setting."[277] Scientists know "the choice of scientific questions and the method of approach may depend on all sorts of extrascientific influences, but the correct answer when we find it is what it is because that is the way the world is."[278] That is, Weinberg maintains that laws of physics (when we find them) are as real as rocks. But, he goes on, "The objective nature of scientific knowledge has been denied" by a variety of figures, including Ross, Latour, Rorty and Kuhn, even though "it is taken for granted by most natural scientists."[279] That is, Weinberg adopts a realist ontology as the natural concomitant of the epistemic credibility laws of physics enjoy (at least among scientists). "For those who have not lived with the laws of physics," he writes, "I can offer the obvious argument that the laws of physics as we know them work, and there is no other known way of looking at nature that works in anything like that sense."[280]

Weinberg drew a series of rejoinders to his essay. The one from Michael Holquist and Robert Shulman proved full of the self-befuddling rhetoric that Sokal tried to shake awake by his satire. If they professed that the hoax was "rapidly ceasing to be funny," I would agree: with their response it had become very sad indeed. They claimed that Weinberg was guilty of maintaining that science was "fundamentally different from other human activities . . . because of its relation to a reality that is ultimate." They protested this "radical dualism" between scientific knowledge-claims and the rest of culture, implying that Weinberg was carrying on like some high priest in "the innermost sanctum of his temple." Thus "the particle physicist stands ready in his role as pre-Kantian shaman to declare taboo all that which falls outside his narrow definitions." Weinberg was accused of making "privative claims for an immaculate conception of science." By contrast, these authors celebrated "the greatest minds" who had "striven to understand the relation of the incommensurable to our situatedness at a particular moment in a specific time."[281] Please! As Bruno Latour said in another context, "it is indeed this absurd position that has made the whole field of SSK look ridiculous."[282]

Norton Wise tries to make the point in an intellectually responsible manner: "Weinberg presents us with an ideology of science, an ideology which radically separates science from culture."[283] That was what Holquist and Shulman were getting at, but what Wise does that they do not is offer a series of historically pertinent examples of the disputes which the best science at a given historical time struggled over, which problematizes Weinberg's claims about the clarity and cumulativity of scientific knowledge. Wise makes the utterly apt point that Weinberg's stance "will never do for comprehending the history of science." As a matter of fact, Weinberg couldn't care less. In his response to Wise he writes: "Just as anyone may get inspiration from scientific discoveries, scientists in their work may be inspired by virtually anything in their cultural background, but that does not make these cultural influences a permanent part of scientific theories."[284] More precisely: "Whatever cultural influences went into the discovery of Maxwell's equations and other laws of nature have been refined away, like slag from ore. . . . The cultural backgrounds of the scientists who discovered such theories have thus become irrelevant to the lessons that we should draw from the theories."[285]

This response deserves serious consideration. Physics is *not* the history of physics.[286] Physicists have every right to be interested in the current content of physical theory and to presume that getting that right is their primary disciplinary task. But *historians* of physics—and of science more generally—believe that knowing how science made its creative breakthroughs in the past illuminates how science happens and therefore has relevance to current scientific practice and, in the measure that it discerns the multiplicity and controversy of scientific theories over time, even to current scientific content. This is, however, a *monitory,* not a *constitutive* relevance. Physicists *must* take the latest, most critically scrutinized theories as their basis of operations. They can postulate their fallibility and contingency; more, they can probe their anomalies to seek to revise them. But it is rather silly to believe that historians or sociologists of science are making some fundamental contribution to ongoing natural science by pointing out merely that it is contingent and fallible.[287] That, I believe, is what Weinberg was trying to argue. Physicists learn little about how to manipulate or transform current laws of physics by becoming informed of the history of their emergence. Where Weinberg may be wrong is in his belief that many of the laws of physics currently in place are likely never to be significantly displaced. But historical and sociological insights can only problematize, not refute, this belief. We have no claim to know the future, either. Similarly, we may find fault with Weinberg's presumption that a metaphysical commitment to ontological realism is incumbent upon scientists. But that would hardly imply that it has not been commonplace in the practice of science or that it is illegitimate. Agnosticism about ontology may be more appropriate for science studies, but evidence is an *epistemic* category, not an ontological one. We have enough to do wrestling with epistemic questions.

An interesting exploration of these issues came in an email exchange between English Professor Michael Bérubé and Alan Sokal. Bérubé challenged Sokal about the implicit claim that "the humanities don't give a shit [this is email, remember] about evidence."[288] He presented a few grand theses from figures like Marx and Freud and continued: "in the humanities, to our shame [?], we give these kinds of statements a great deal of leeway. We're not that concerned with whether they're true. We're more concerned with whether they have interesting, productive, illuminating, destructive, or earth-shattering results, and we'll gauge those results by the evidence of the *effects* of those claims rather than by their pre-existing evidentiary basis."[289] This is a remarkable statement of the methodology of the human sciences, and I, for one, view it with considerable reservation. Bérubé went on to raise John Searle's distinction between "social reality" and "brute fact."[290] He asked: "is the distinction between 'brute fact' and 'social reality' *itself* a brute fact or a social reality?"[291] (Thus, improperly in my view, Bérubé presumed this was an *exhaustive* disjunct.) He believed that Searle's distinction—a conceptual distinction—had to be registered as a social, not a brute, fact and therefore that its ontological status was overwhelmed by its epistemological one. This is to replicate, not to clarify, the issue. Sokal professed not to understand what Bérubé was getting at. Bérubé explained in a follow-up: "What I meant is this: Searle's brute fact/social reality distinction is one of the more sophisticated versions of the fact/value distinction in our time."[292] I think this is philosophically misguided: it blurs epistemology with value theory and makes ontology hopelessly obscure between them. But what Bérubé is after comes clear as he tries to elaborate this idea: "Let's say we agree that there's a distinction between facts and values . . . and that it's hard to derive the one from the other. Which, then, has 'priority' over the other, and what might 'priority' mean?"[293] As a Kant scholar, I could try to gloss this as a question of the "primacy" of practical versus theoretical reason, which is an important question. But it appears to me eminently misplaced because what is really at issue is not a question of *value* at all, as Bérubé manages to work himself through to in the end: "So there must be a strong, perhaps unassailable, sense in which physical facts precede human deliberation about facts."[294] That is, ontology can and philosophically *must* be distinguished from epistemology.

In *A House Built on Sand: Exposing Postmodernist Myths about Science* (1998), Sokal presented his considered reflections on the meaning of his hoax and the ensuing controversy.[295] His text is largely a replicant of the text included in *Fashionable Nonsense,* jointly authored with Jean Bricmont, and I will take both versions as largely expressing Sokal's personal views. In both contexts, Sokal presented evidence relevant to the two criticisms that Brenkman had made of his original account. He delimited the significance of his hoax to the failure of one journal in cultural studies to show due diligence in its appraisal of an essay. No larger claim about the status of postmodernism in general could be made *on that basis alone.* But Sokal insisted that the content of his hoax could demonstrate something considerably more general, namely a pervasive disregard for epistemic standards in the use and appraisal of natural scientific ideas among the most prominent figures in postmodernism, both in America and in France. It was Brenkman's second charge, however, that Sokal answered more fully, namely, that his epistemology of science amounted to a crude realism—positivism, in short. Sokal demonstrated that he had a far more measured sense of the issues in post-positivist philosophy of science and that the position he was taking showed considerably more sophistication than Brenkman alleged. Sokal observed: "Much contemporary skepticism claims to find support in the writings of philosophers such as Quine, Kuhn, or Feyerabend who have called into question the epistemology of the first half of the twentieth century. This epistemology is indeed in crisis."[296] Sokal identified the three key issues: the theory-ladenness of observation, the underdetermination of theories, and incommensurability, but he also registered that these were ideas with a considerable elasticity of claims, and that "the radical interpretation is often taken not only as the 'correct' interpretation of the original text but also as a well-established fact ('X has shown that . . . ')."[297] What Sokal proceeded to do was to criticize the radical reception of post-positivism as unwarranted in its inferences. In particular he criticized the Strong Program and Bruno Latour, primarily for their apparent unwillingness to provide a role for evidence from nature in the explanation of scientific developments. His essay concluded by supporting strongly the position of Philip Kitcher.[298] With that, Sokal came to a position which I find thoroughly compatible.

"Something has gone badly wrong in contemporary science studies," Kitcher begins his important survey of the field.[299] But Kitcher, unlike Gross and Levitt and others, does not wish to "dismiss the entire field as a mess." His is "A Plea *for* Science Studies" (my emphasis).[300] The form the science wars have taken suggests that extremists of both camps have shared too much of the limelight. The endeavor to study science in a more open and complicated form cannot be abandoned. I concur. Science is too important to be left to scientists alone, especially when "science" has once again become the name for any form of disciplined inquiry, as in the old German sense of *Wissenschaft*. What Kitcher wants, then, is to offer constructive suggestions for getting science studies *out* of the mess in which it finds itself. In his view, this can only occur by taking up both sides of a fruitless polarization into what he dubs the "realist-rationalist" and the "socio-historical" clusters. Kitcher hopes that certain cognitive claims should be "uncontroversial"—basically, certain general maxims about science which encourage "rational confidence in the whole," even while recognizing fallibility.[301] Equally, Kitcher believes that certain social perspectives on science ought to be accepted as appropriate, for example, the social influence upon research practices, the choice of problems, and so on.[302] "The challenge for science studies," he writes, "is to do justice to both clusters."[303] Whereas the "Received View" systematically neglected the socio-historical cluster up to about 1960, "since the 1970s, Science Studies has sometimes ignored the first cluster entirely."[304] Kitcher diagnoses the problem along the same lines I have tried to develop in this monograph: "The root of the problem is some bad philosophy that has been strikingly influential in contemporary history and sociology of science (and occasionally in some contemporary philosophy of science)."[305] He identifies the theory-ladenness of observation, the underdetermination of theory by evidence, "variety of belief" (i.e., "cultural relativism"), and radical historicism (the exclusive attention to "actors' categories"). He reasons, in my view aptly, that "overinterpretation of the theory-ladenness of observation leads to a kind of global skepticism that makes it impossible to say anything at all."[306] Ironically, this should have silenced even science studies, but its advocates routinely refuse to take seriously *any* suggestion of the aptness of the idea of performative self-contradiction. Similarly, underdetermination "has been dramatically overblown."[307] It, too, is a "reformulation of a form of skepticism" that throws back to Descartes.[308] Neither of these core dogmas of antiempirical epistemology warrant such extreme readings, and the more moderate ones, again as I have been trying to demonstrate, simply do not justify the extravagances of radical postmodernism in science studies. Under the heading of "variety of belief," Kitcher discusses Bloor's famous "principle of symmetry."[309] His argument is that in fact not all claimants to knowledge are equally creditworthy. We might supplement Kitcher's case in this area with the reflections that Paul Boghossian offered: "To get the desired relativistic result, a postmodernist would have to claim that the two views are equally justified, given their respective rules of evidence, and add that there is no objective fact of the matter which set of rules is to be preferred. . . . [S]ince there are no perspective-independent rules of evidence that could adjudicate between the two sets of rules, both claims would be equally justified."[310] Yet it is by no means clear that this impasse is, apart from a politically condescending or guilt-ridden unwillingness to face cultural value-conflicts, insuperable. Satya Mohanty has argued eloquently against the self-defeating elements in this postmodern politics for those who really wish to be "liberatory" or "egalitarian" about marginalized voices, and Clifford Geertz has made a powerful critique of Richard Rorty's complacent ethnocentrism.[311] As for the critique of radical historicism, Kitcher argues that "later knowledge can be employed in history to fulfill an explanatory function, different from that of immersing us in the world of the protagonists."[312] More generally, "any account of past people will involve assumptions about motivation and action, the character of the public world and human responses to it, and we rightly make those assumptions using the best information we have."[313] As a practicing contextual intellectual historian, I can fully embrace these claims, observing merely that the problem arises when we seek to impute to historical actors *as conscious rational choices* concepts or arguments that would have been inconceivable to them. That sort of "rational reconstruction" is insufferably anachronistic. The rest of Kitcher's argument can be accommodated.

As Kitcher puts it, "the road to relativism is paved with the best of intentions and the worst of arguments," but the resulting "Four Dogmas" have now become such conventional wisdom that students simply cannot be brought to question them: "Anyone who has tried to talk to people who have recently been trained in Science Studies will know that the conclusions of the four arguments I have criticized are treated as axiomatic. There is just no questioning them."[314] Rachel Laudan has made similar observations regarding graduate students in history of science.[315] It is precisely to contravene these tendencies that my book was written.

Because there has been so much speedy propaganda to the contrary, let me assert that *of course* there is a rough and ready distinction between theory and observation, and *of course* we often look and see what is true. *Of course* some theories are just false and, after diligent attempts at patching, have to be abandoned.

— IAN HACKING, "IMRE LAKATOS'S PHILOSOPHY OF SCIENCE"

Old-guard Diltheyans, their shoulders hunched from years-long resistance against the encroaching pressure of positivist natural science, suddenly pitch forward on their faces as all opposition ceases to the reign of universal hermeneutics.

— CHARLES TAYLOR, "UNDERSTANDING IN HUMAN SCIENCE"

Conclusion: The Hyperbolic Derangement of Epistemes

Three hyperbolic dogmas of *anti*empiricism have dominated "theory" over the past fifty years of post-positivist thought: theory-ladenness, underdetermination, and incommensurability. *None* is justified in the radical form which alone empowers the extravagances of postmodernism. Radical "theorists" have made bold to deny categorically that there could be *any* distinction of theory from observation, indeed, to deny that language has *any* referential feature, such that there could be evidential verification. They are disposed to hold all theories of a constraining external world equally implausible (radical "underdetermination"), the better to pursue other agendas in domains of "rhetoric" where they feel themselves at home. It appears "liberatory" to them to find that theories, languages, and cultures are separated by radical "incommensurability" and, hence, are impervious to objective interpretation and evidentiary adjudication of arguments. The upshot has not been salutary—not for an understanding of the natural sciences, and especially not for a proper self-understanding of empirical inquiry in the human sciences.

Something has gone awry in the linguistic turn. It seemed quite *de rigueur* to take it, but we now have good grounds to suspect it has led us into a dead end. In this suspicion I am happy to invoke Ian Hacking as at least something of a forerunner: "whatever be the interest in the philosophy of language, it has very little value for understanding science."[1] More specifically, "Incommensurability, transcendental nominalism, surrogates for truth, and style of reasoning are the jargon of philosophers. They arise from contemplating the connection between theory and the world. All lead to an idealist cul-de-sac."[2] Along similar lines, and for similar reasons, Dudley Shapere has "protest[ed] against . . . the view . . . that what science must aim at can be established by an examination of the nature of language," that "by an analysis of the features of language, we should be able to lay down inviolable conditions on the knowledge-seeking enterprise."[3] More extensively, Shapere spells out what it has been the burden of this study to illuminate:

> [T]he technical concepts of meaning and reference stemming from the philosophy of language have failed to clarify the scientific enterprise. On the contrary, they have only succeeded either in introducing hopeless confusion or in contradicting some of the most fundamental aspects and achievements of that enterprise. Their vagaries, confusions, and paradoxes, their arbitrary presuppositions and apriorisms, their epistemological relativisms and metaphysical absolutes, must all be avoided. The only way of doing this is to abandon those technical concepts themselves, as philosophers and others have understood them, and to exorcise completely the error of supposing that scientific reasoning is subservient to certain alleged necessities of language, and that the study of the latter is therefore deeper than the study of the former. The situation . . . is rather the reverse.[4]

Philosophy of language ceases to illuminate any of the concrete concerns that drive actual research. Rather than elucidate and concretize decisive concerns of such inquiry—for example, experience, evidence, and knowledge—it has pronounced such terms pointless. The situation is precisely the reverse; philosophy of language has argued *itself* into irrelevancy for empirical inquiry.

The basis for the authoritative posture of philosophy of science was the confidence that what constituted the context of justification was precisely a priori rules, and that philosophy had access to them.[5] But that pretense shattered in the era of post-positivism, and the result was a fundamental deflation of the claims of philosophy. Ronald Giere came to this conclusion by the mid-1980s: "[M]ethodological foundationalism is a hopeless program and thus . . . naturalism, in spite of the circle argument, is our only alternative."[6] Some, like Richard Rorty, considered the business of philosophy as a separate and authoritative discipline to be concluded: philosophy—or at least epistemology—was *over*.[7] Others, like Giere, realized that philosophy had to take up a more modest place among the empirical sciences, no longer in judgment over them.[8] It had to renounce its claims to a priori rules.[9] If it had a role, it could only be as a naturalized epistemology. As Bonnie Paller notes, "The perceived failure of philosophers to provide an account of rules for rational theory choice and theory justification has paralleled an increased philosophical interest in the history of science," for if there are no "unchanging, essential, and transcontextual rules which are discoverable a priori," the only source for methodological and epistemological standards must be immanent in the history of science itself.[10] Giere pointedly sums it up, "In arguing a 'role for history,' Kuhn was proposing a naturalized philosophy of science."[11] That is, "the breakdown of the demand for a priori rules for theory justification has precipitated a breakdown in the distinction between (the context of) justification and (the context of) discovery."[12] Mainstream philosophy of science has increasingly abandoned logicism and the a priori, either deflating itself to another descriptive empirical science engaged in characterizing how science works, or reconstructing a minimal normativity out of its descriptions in a circular but not vicious "bootstrap" or "reticulated" inference from past success to future fertility.

In contemporary methodological discourse, *radical* post-positivism has entered into a still-fulminating compound with poststructuralism: their global admixture is the essence of "postmodernism." Until quite recently, the two more specific rubrics—post-positivism and poststructuralism—had largely distinct trajectories, attached to two longstanding traditions of philosophy conventionally termed "analytic" and "continental," respectively. Yet as post-positivism has reached the apparent impasses documented in this study, it has become fashionable to have recourse to poststructuralism as therapeutic or at least ludic escape. Perhaps the most important American figure in this (con)fusion of post-positivism with poststructuralism has been Richard Rorty. Indeed, his *Philosophy and the Mirror of Nature* (1979) carried out the decisive maneuver of summing up the course of post-positivism as the dissolution of epistemology (if not the "end of philosophy"), opening the way to its displacement by "continental" discourses—initially of "hermeneutics" but ultimately of poststructuralism. If there is a trope which has become inveterate in current methodological discourse, it is that of "turns." Rorty, of course, became famous in offering us the "linguistic turn." Thereafter have come a whole host of such turns: the "rhetorical turn," the "interpretive turn," the "cultural turn," the "practice turn," and turns away from those turns.[13] It is a quick study, this business of turns, and it takes a "conversationalist" of Rorty's caliber to change subjects so swiftly. His wondrous passage from partner to partner across the postmodern dance floor offers a suitable coda to this history of post-positivism run "all the way down." Setting out from Quine, Kuhn, and Davidson, Rorty has executed several elegant turns through Gadamer and Heidegger to come more and more to partner with Derrida.

Epistemology, Rorty confidently asserts, stands as one undoubted casualty of post-positivist progress. He has been waging a grand campaign to dispel forever the notion of "correspondence," the impression that natural science has some privileged access to reality which makes it paradigmatic for knowledge.[14] Herbert Dreyfus has called this "theoretical holism."[15] It entails, in Rorty's own words, "thinking of the entire culture, from physics to poetry, as a single, continuous, seamless activity in which the divisions are merely institutional and pedagogical."[16] In another context, he puts it this way: "we shall say that *all* inquiry is interpretation, that *all* thought is re-contextualization."[17] Or, in yet a third formulation, "To say that something is better 'understood' in one vocabulary than another is always an ellipsis for the claim that a description in the preferred vocabulary is more useful for a certain purpose."[18] While Rorty is perfectly comfortable with the notion that different vocabularies are "irreducible," he insists that it is wrong to think "that the irreducibility of one vocabulary to another implies something ontological."[19] He concludes: "I have been arguing . . . that the notion that we know *a priori* that nature and man are distinct sorts of objects is a mistake. It is a confusion between ontology and morals."[20] Rorty acknowledges that "There *seems* to be a difference between the hard objects with which chemists deal and the soft ones with which literary critics deal. This apparent difference is the occasion for all the neo-Diltheyan theories which insist on a distinction between explanation and understanding, and all the neo-Saussurean theories which insist upon a distinction between lumps and texts."[21] Instead, Rorty insists strategies for interpreting "texts" and "lumps" are for the most part assimilable to each other.[22] "My holistic strategy . . . is to reinterpret every such dualism as a momentarily convenient blocking-out of regions along a spectrum, rather than as recognition of an ontological, or methodological, or epistemological divide."[23]

The dissolution of boundaries that Rorty carries out between the *Naturwissenschaften* and the *Geisteswissenschaften* has a profoundly ironic dimension, perhaps best caught in the image Charles Taylor conjured in the epigraph cited above: "Old-guard Diltheyans, their shoulders hunched from years-long resistance against the encroaching pressure of positivist natural science, suddenly pitch forward on their faces as all opposition ceases to the reign of universal hermeneutics."[24] Taylor and many others find this collapse into indiscriminate "pragmatism" implausible, insisting not only on the uniqueness of humanistic inquiry, but also on a distinctive epistemological claim for scientific knowledge. In insisting that only *moral* stipulation animates any of the discriminations that traditionally appeared *epistemic,* Rorty wraps himself in the final dogma of positivism, the fact-value distinction. For him there are only arbitrary value judgments over against an "ontology" which he deliberately relegates to epistemic inaccessibility. This represents an "idealism," in Hacking's phrase, radical as anything Kant had in mind with "things in themselves." What is left is language and the arbitrary "poetics" of conversation. Rorty dissolves too many distinctions; his new "pragmatism" entails a cavalier disdain for rational adjudication of dispute.

There has been a derangement of epistemes. Philosophy of science pursued "semantic ascent" into a philosophy of language so "holistic" as to deny determinate purchase on the world of which we speak. History and sociology of science has become so "reflexive" that it has plunged "all the way down" into the *abîme* of an almost absolute skepticism. In that light, my fears are for empirical inquiry not in the natural sciences, whose practitioners brush all this off as impertinence, but in the human sciences. Hyperbolic "theory" threatens especially the prospect of learning anything from others that we did not already presume. It is time for a hard reckoning, for a rigorous deflation. Willard Quine put it with uncharacteristic bluntness: "To disavow the very core of common sense, to require evidence for that which both the physicist and the man in the street accept as platitudinous, is no laudable perfectionism; it is a pompous confusion."[25]

Notes

Introduction

1. John Greenwood and Mary Hesse have written of a "new empiricism" and its dogmas: Greenwood, "Two Dogmas of Neo-Empiricism: The 'Theory Informity' of Observation and the Quine-Duhem Thesis," *Philosophy of Science* 57 (1970): 553–74; and Hesse, "Duhem, Quine and a New Empiricism" (1970), reprinted in *Can Theories Be Refuted?* ed. Sandra Harding (Dordrecht: Reidel, 1976), 184–204.

2. Thomas Haskell has made the very important distinction here. See *Objectivity Is Not Neutrality: Explanatory Schemes in History* (Baltimore: Johns Hopkins University Press, 1998).

3. No less a figure than Bruno Latour has stated, this once without any apparent irony, "postmodernism rejects all empirical work as illusory and deceptively scientific" (*We Have Never Been Modern* [Cambridge, Mass.: Harvard University Press, 1993], 46).

4. My formulation explicitly holds open the possibility that the original thinkers may not have indulged in the extravagances that have followed from their ideas: that is to be established empirically.

5. Chris Doran, "Jumping Frames: Reflexivity and Recursion in the Sociology of Science," *Social Studies of Science* 19 (1989): 515–31.

6. Malcolm Ashmore, *The Reflexive Thesis: Wrighting the Sociology of Knowledge* (Chicago: University of Chicago Press, 1989), 103.

7. As Paul Roth has put it, "When Kuhn entered the scene, what happened was that the sort of skepticism which had been raised concerning the viability of efforts to specify the logic of contemporary science were extended to include the history of Western science" ("Resolving the *Rationalitätsstreit*," *Archives Européenes de Sociologie* 26 (1985): 142–57, citing 143).

8. For the extended use of this metaphor, see the literature cited in chapter 4 below.

9. That is to say, Donald Davidson and Richard Rorty represent elaborations of Quine, as both acknowledge. As for "Hilary Putnams," which refer causally to the philosopher of that name, several will become important in this narrative as both instantiating and querying the impetus associated with Quine.

10. Most prominently under the rubric of the "Duhem-Quine thesis." But it is also commonplace to find the "indeterminacy of translation" invoked as preemptively established. These are dogmas indeed of a new (anti)empiricism.

11. John Passmore has noted this: "in such writers as David Bloor or Barry Barnes we encounter large slabs of epistemology, often of the sort which philosophers have long ago rejected as indefensible" ("The Relevance of History to the Philosophy of Science," in *Scientific Explanation and Understanding,* ed. Nicholas Rescher [Lanham, Md.: University Press of America, 1983], 83–105, citing 91).

12. The upshot of that debate has raised the question whether there is any meaning left to a demarcation between them, or between their objects—what earlier German treatment discriminated as the *Naturwissenschaften* and *Geisteswissenschaften,* respectively. These terms were developed by Wilhelm Dilthey and Heinrich Rickert in the late nineteenth century as a response from within the human sciences to the hegemonic claims of "positivism" for natural science.

13. Alan Sokal's article appeared in the constructivist journal *Social Text* in 1996:

"Transgressing the Boundaries: Toward a Transformative Hermeneutics of Quantum Gravity," *Social Text* 46/47 (1996): 217–52. His exposure of the hoax marked a moment of consternation rarely encountered by the ebullient constructivists, and their response has been hilariously old-fashioned. That alone suffices to mark the absurdism of the moment.

Chapter One

1. Positivism has a complex conceptual history, from its coinage by Auguste Comte to designate the culminating phase in a historical theory of scientific growth, through its spread across European culture in the late nineteenth century as a program for the relentless banishment of all "metaphysics" from scientific inquiry, to its most sophisticated articulation, by the Vienna Circle, as a theory about the "unity of science" under the auspices of physics, preserved in its mutated form of "logical empiricism." See Leszek Kolakowski, *Positivist Philosophy* (Harmondsworth, U.K.: Penguin, 1972).

2. On "scientism," see the recent overview by Tom Sorell, *Scientism: Philosophy and the Infatuation with Science* (London: Routledge, 1991).

3. Hilary Putnam, *Reason, Truth and History* (Cambridge: Cambridge University Press, 1981), 126.

4. A striking example is the guru of American deconstruction, Paul de Man. The number of times de Man invokes historical work with a preceding, disparaging "pseudo" or "positivistic" cannot escape any careful reader. It is important to ascertain (as clearly as de Man's global derision permits) exactly what he means by historical "positivism." Perhaps the best indication of what de Man disdains as "positivist" emerges from the following assertion: "few historians still believe that a work of the past can be understood by reconstructing, on the basis of recorded evidence, the set of conventions, expectations, and beliefs that existed at the time of its elaboration" (*The Resistance to Theory* [Minneapolis: University of Minnesota Press, 1986], 58). By this formulation de Man not only misrepresents what current historians believe, but he crudely collapses *empirical inquiry* into "positivism."

5. Anthony Giddens, "Positivism and Its Critics," in *A History of Sociological Analysis,* ed. Tom Bottomore and Robert Nisbet (New York: Basic Books, 1978), 237–86, citing 238.

6. Giddens stresses the "importance of the line of connection from Comte to Durkheim" for the genesis of "positivistic sociology" (ibid., 243). "Durkheim's writings have been more influential than those of any other author in academic social science in the spread of 'positivistic sociology'" (245). For a substantial literature on positivist sociology, see Peter Halfpenny and Peter McMylor, eds., *Positivist Sociology and Its Critics,* 3 vols. (Brookfield, Vt.: Elgar Publishing, 1994); Halfpenny, *Positivism and Sociology: Explaining Social Life* (London: Allen & Unwin, 1982); Jeffrey Alexander, *Positivism, Presuppositions, and Current Controversies* (Berkeley: University of California Press, 1982).

7. On neo-Kantians and Dilthey and the distinction of *Naturwissenschaften* from *Geisteswissenschaften,* see Charles Bambach, *Heidegger, Dilthey, and the Crisis of Historicism* (Ithaca, N.Y.: Cornell University Press, 1995).

8. "Neurath and Carnap developed their physicalist thesis in some part in direct opposition to the tradition of the *Geisteswissenschaften*" (Giddens, "Positivism and Its Critics," 251). That is, logical positivism rejected any insulation of the "human sciences" from the canons of the natural sciences. That was the thrust of the idea of the "unity of science." Ironically post-positivism exploded this unity, but at the same time dissolved the distinction between the human and the natural sciences by undermining the hegemonic claims of the philosophy of natural science associated with positivism.

9. Giddens, "Positivism and Its Critics," 240. John Locke's famous profession of service as "underlaborer" comes from his *Essay Concerning Human Understanding* (1689; Oxford: Clarendon, 1975), 10.

10. Giddens, "Positivism and Its Critics," 248.

11. Michael Friedman, "The Re-evaluation of Logical Positivism," *Journal of Philosophy* 88 (1991): 505–23; Friedman, "Philosophy and the Exact Sciences: Logical Positivism As a Case-Study," in *Inference, Explanation, and Other Frustrations: Essays in the Philosophy of Science,* ed. John Earman (Berkeley: University of California Press, 1992), 84–98; Friedman, "Remarks on the History of Science and the History of Philosophy," in *World Changes: Thomas Kuhn and the Nature of Science,* ed. Paul Horwich (Cambridge, Mass.: MIT Press, 1993), 37–54; Thomas Uebel, *Overcoming Logical Positivism from Within: The Emergence of Otto Neurath's Naturalism in the Vienna Circle's Protocol Sentence Debate* (Amsterdam: Rodopi, 1992).

12. George Reisch, "Did Kuhn Kill Logical Empiricism?" *Philosophy of Science* 58 (1991): 264–77; Guy Axtell, "In the Tracks of the Historicist Movement: Re-assessing the Carnap-Kuhn Connection," *Studies in History and Philosophy of Science* 24 (1993): 119–46; John Earman, "Carnap, Kuhn, and the Philosophy of Scientific Methodology," in *World Changes,* ed. Horwich, 9–38; Gürol Irzik and Teo Grünberg, "Carnap and Kuhn: Arch Enemies or Close Allies?" *British Journal for the Philosophy of Science* 46 (1995): 285–307; Mark A. Notturno, "Thomas Kuhn and the Legacy of Logical Positivism," *History of the Human Sciences* 10 (1997): 131–34.

13. Thomas S. Kuhn, "Afterwords," in *World Changes,* ed. Horwich, 311–42.

14. That Kuhn's revisions moved back toward logical positivism was remarked by Frederic Suppe (*Structure of Scientific Theories,* 2d, rev. ed. [Urbana: University of Illinois Press, 1977]); and also by Trevor Pinch ("Paradigm Lost? A Review Symposium," *Isis* 70 [1979]: 429–40). A good instance of this is Kuhn's welcoming attitude toward the Sneed-Stegmüller formalism. See Wolfgang Stegmüller, "Structures and Dynamics of Theories: Some Reflections on J. D. Sneed and T. S. Kuhn," *Erkenntnis* 9 (1975): 75–100; Thomas S. Kuhn, "Theory-Change As Structure-Change: Comments on the Sneed Formalism," *Erkenntnis* 10 (1976): 179–99; Paul Feyerabend, "Changing Patterns of Reconstruction," *British Journal for the Philosophy of Science* 28 (1977): 351–82; David Pearce, "Stegmüller on Kuhn and Incommensurability," *British Journal for the Philosophy of Science* 33 (1983): 389–96; Walter Van der Veken, "Incommensurability and the Structuralist View," *Philosophica* 32 (1983): 43–56; Wolfgang Balzer, "Incommensurability, Reduction, and Translation," *Erkenntnis* 23 (1985): 255–67; David Pearce, "Incommensurability and Reduction Reconsidered," *Erkenntnis* 24 (1986): 293–308; David Pearce, "The Problem of Incommensurability: A Critique of Two Instrumentalist Approaches," in *Scientific Knowledge Socialized,* ed. Imre Hronsky, Marta Feher, and Balazs Dajka (Dordrecht: Kluwer, 1988), 385–98.

15. Michael Friedman certainly holds that the contrast is overdrawn: "the currently popular diagnosis of the failure of logical positivism (a diagnosis due largely to the work of Kuhn and his followers) is fundamentally misleading" ("Remarks on the History of Science and the History of Philosophy," 54). John Earman contends that what is "askew" in the notion of post-positivism "is the notion that a philosophical revolution as opposed to an evolution has taken place" ("Carnap, Kuhn, and the Philosophy of Scientific Methodology," 9). Larry Laudan's important essay "Sins of the Fathers" would have us collapse post-positivism back into positivism ("'The Sins of the Fathers . . .': Positivist Origins of Postpositivist Relativism," reprinted in *Beyond Positivism and Relativism* [Boulder, Colo.: Westview, 1996], 3–25). While he makes some good points, that overstates the matter. Similarly, David Weissman has made a very interesting case for the persistence of "transcendental idealism" all the way from Kant through logical positivism to Quine and Putnam ("Logical Positivism: A Retrospective," *Journal of Philosophy* 88 [1991]: 520–21; Weissman, "Positivism Reconsidered," *Journal of Speculative Philosophy* 8 [1994]: 1–19). The continuity claims deserve careful assessment below.

16. "It is well known that Kuhn quite shocked the philosophy of science. . . . Kuhn

asserts that the separation of the context of discovery from the context of justification is not justified. It is well known that this separation is the starting point of both logical positivism and critical rationalism" (Paul Hoyningen-Huehne, "The Interrelations between the Philosophy, History, and Sociology of Science in Thomas Kuhn's Theory of Scientific Development," *British Journal for the Philosophy of Science* 43 [1992]: 487–501, citing 491).

17. Evandro Agazzi, "Commensurability, Incommensurability, and Cumulativity in Scientific Knowledge," *Erkenntnis* 22 (1985): 51–77, citing 52.

18. On the importance of Karl Popper, see Gerard Radnitzky, "Analytic Philosophy As the Confrontation between Wittgensteinians and Popper," in *Scientific Philosophy Today*, ed. Joseph Agassi and Robert S. Cohen, Boston Studies in the Philosophy of Science 67 (Dordrecht: Reidel, 1981), 239–86.

19. Keekok Lee, "Kuhn—a Re-appraisal," *Explorations in Knowledge: An International Journal in the Philosophy of Science* 1 (1984): 33–88, esp. 35, 42.

20. See esp. Imre Lakatos, "Falsification and the Methodology of Scientific Research Programmes," in *Criticism and the Growth of Knowledge,* ed. Imre Lakatos and Alan Musgrave (Cambridge: Cambridge University Press, 1970), 91–196.

21. Agazzi, "Commensurability, Incommensurability, and Cumulativity in Scientific Knowledge," 60.

22. Dudley Shapere, "Notes toward a Post-positivistic Interpretation of Science, Part I," in *Reason and the Search for Knowledge: Investigations in the Philosophy of Science*, Boston Studies in the Philosophy of Science 78 (Dordrecht: Reidel, 1984), 102–19, citing 104.

23. Ibid., 106.

24. Friedrich Nietzsche, *The Will to Power* (New York: Vintage, 1968), 267.

25. Shapere, "Notes toward a Post-positivistic Interpretation of Science, Part I," 112.

26. Ibid., 113.

27. Grover Maxwell, "The Ontological Status of Theoretical Entities," in *Scientific Explanation, Space, and Time*, ed. Herbert Feigl and Grover Maxwell, Minnesota Studies in the Philosophy of Science 3 (Minneapolis: University of Minnesota Press, 1962), 3–27, citing 13.

28. Robert Nola, "Interpretation of 'The Facts' in the Light of Theory," *Philosophica* 31 (1983): 25–44.

29. Hilary Putnam, "Meaning and Reference," in *Naming, Necessity, and Natural Kinds,* ed. Stephen Schwartz (Ithaca, N.Y.: Cornell University Press, 1977), 119–32, citing 130.

30. Robert Nola, "'Paradigms Lost, or the World Regained'—an Excursion into Realism and Idealism in Science," *Synthese* 45 (1980): 317–50, citing 329.

31. See Michael Friedman, *The Dynamics of Reason* (Stanford, Calif.: CSLI Publications, 2001).

32. "Popper, Reichenbach and other members of the Vienna Circle all agreed that there is a sharp distinction between the 'context of discovery' and the 'context of justification.' Only the latter lies within the domain of methodology. . . . As for the context of discovery, it belongs to the psychology of invention" (Elie Zahar, "Logic of Discovery or Psychology of Invention?" *British Journal for the Philosophy of Science* 34 [1983]: 243–61, citing 243).

33. Harvey Siegel wrote: "Reichenbach is not concerned to demonstrate anything like a 'logical gulf' between the two contexts; nor is he concerned to demonstrate any irrelevance of principles of justification to discovery. . . . Justification *is* relevant to discovery. It is only the converse—that discovery can be relevant to justification—that Reichenbach is concerned to deny" ("Justification, Discovery and the Naturalizing of Epistemology," *Philosophy of Science* 47 [1980]: 297–321, citing 300). If, indeed, this is true of Reichenbach personally, it is *not* true of the tradition of usage of the contrast.

34. Ibid., 302.

35. Clifford Hooker, "Surface Dazzle, Ghostly Depths: An Exposition and Critical Evaluation of van Fraassen's Vindication of Empiricism against Realism," in *Images of Science: Essays on Realism and Empiricism,* with a reply from Bas C. van Fraassen, ed. Paul M. Churchland and Clifford A. Hooker (Chicago: University of Chicago Press, 1985), 153–96, citing 156–57.

Chapter Two

1. James Bohman, *New Philosophy of Social Science: Problems of Indeterminacy* (Cambridge, U.K.: Polity, 1991).

2. One of the most impressive efforts in this vein, which I will frequently cite, is Paul Roth, *Meaning and Method in the Social Sciences* (Ithaca, N.Y.: Cornell University Press, 1987).

3. Murray G. Murphey, *Philosophical Foundations of Historical Knowledge* (Albany: SUNY Press, 1994).

4. William Dray, review of *Philosophical Foundations of Historical Knowledge,* by Murray G. Murphey, *Clio* 24 (1995): 434–36, citing 434–35.

5. Gary Hardcastle, "Presentism and the Indeterminacy of Translation," *Studies in History and Philosophy of Science* 22 (1991): 321–45.

6. Ibid., 331–33.

7. Raymond Weiss, "Historicism and Science: Thoughts on Quine," *Dialectica* 29 (1975): 157–65. The term *historicism* has been used by many authors in many ways. Weiss is simply using it as a rubric for postmodernism.

8. Ibid., 158.

9. Ibid., 159.

10. Ibid., 162–63.

11. Ibid.

12. Ibid., 159. Weiss draws for this last observation upon Quine's statement: "For my part I do, qua lay physicist, believe in physical objects and not in the gods of Homer; and I consider it a scientific error to do otherwise. But in point of epistemological footing the physical objects and the gods differ only in degree and not in kind. Both sorts of entities enter our conceptions only as cultural posits" (Willard van Orman Quine, "Two Dogmas of Empiricism," in *From a Logical Point of View,* 2d, rev. ed. [Cambridge, Mass.: Harvard University Press, 1980], 20–46, citing 44). Of course, one should be a bit skeptical of a reading which assimilates Quine to Nietzsche. Nelson Goodman, perhaps, but not Quine.

13. Philip Kitcher, "A Plea for Science Studies," in *A House Built on Sand: Exposing Postmodernist Myths about Science,* ed. Noretta Koertge (Oxford: Oxford University Press, 1998), 32–56, esp. 38–39. Kitcher's assessment tallies with the overall argument developed in this study.

14. Willard van Orman Quine, "Three Indeterminacies," in *Perspectives on Quine,* ed. Robert B. Barrett and Roger F. Gibson (London: Blackwell, 1990), 1–16.

15. Quine, "Epistemology Naturalized," in *Ontological Relativity and Other Essays* (New York: Columbia University Press, 1969), 69–90, citing 74.

16. Willard van Orman Quine, "Two Dogmas in Retrospect," *Canadian Journal of Philosophy* 21 (1991): 265–74, citing 272. Quine insisted upon a "moderate" notion of reductionism because "extreme reductionism, the notion that every scientific sentence should have a full translation in sense-datum language, is by now a straw man" (272). One wonders whether it was such a "straw man" in 1950.

17. Quine, "Epistemology Naturalized," 82.

18. Ibid., 79.

19. Ibid., 74.

20. Quine, "Ontological Relativity," in *Ontological Relativity and Other Essays*, 26–68.

21. Mary Hesse, "Duhem, Quine and a New Empiricism" (1970), reprinted in *Can Theories Be Refuted?* ed. Sandra Harding (Dordrecht: Reidel, 1976), 184–204.

22. Quine, "Two Dogmas of Empiricism."

23. Thomas Uebel, "From the Duhem Thesis to the Neurath Principle," in *Austrian Philosophy Past and Present*, ed. Keith Lehrer and Johann Christian Marek (Dordrecht: Kluwer, 1987), 87–100.

24. See Uebel's major study, *Overcoming Logical Positivism from Within: The Emergence of Otto Neurath's Naturalism in the Vienna Circle's Protocol Sentence Debate* (Amsterdam: Rodopi, 1992).

25. John Greenwood, "Two Dogmas of Neo-Empiricism: The 'Theory Informity' of Observation and the Quine-Duhem Thesis," *Philosophy of Science* 57 (1970): 553–74.

26. Ibid. It should be noted that Quine is not comfortable being associated with Hanson, Kuhn, and Feyerabend, the prime advocates of the notion of the "theory-ladenness of facts." See Quine, "In Praise of Observation Sentences," *Journal of Philosophy* 90 (1993): 107–18.

27. Quine, "Two Dogmas of Empiricism," 41.

28. Ibid., 64 n. 17.

29. Quine, "Two Dogmas in Retrospect," 269.

30. Adolph Grünbaum, "The Duhemian Argument" (1960), reprinted in *Can Theories Be Refuted?* ed. Harding.

31. Adolph Grünbaum, "The Falsifiability of Theories: Total or Partial? A Contemporary Evaluation of the Duhem-Quine Thesis," in *Proceedings of the Boston Colloquium for the Philosophy of Science, 1961–1962*, ed. Marx W. Wartofsky, Boston Studies in the Philosophy of Science 1 (Dordrecht: Reidel, 1963). Quine responded to this second effort in a personal letter to Grünbaum (June 1, 1962), which is reprinted in *Can Theories Be Refuted?* ed. Harding, 132.

32. Larry Laudan, "Grünbaum on the Duhemian Argument" (1965), reprinted in *Can Theories Be Refuted?* ed. Harding; Imre Lakatos, "Popper, Falsificationism and the 'Duhem-Quine Thesis,'" appendix to "Falsification and the Methodology of Scientific Research Programmes," in *Criticism and the Growth of Knowledge*, ed. Imre Lakatos and Alan Musgrave (Cambridge: Cambridge University Press, 1970), 180–89.

33. That can be construed as a question of *mere* history of ideas, but the point I hope to develop is that probing the historical coherence of the thesis will pay dividends in assessing its current status. Indeed, one reason to try to ascertain what Duhem actually argued is that it has been so very difficult to get straight what Quine himself means in any of his most famous positionings.

34. Richard Rorty, "Historiography of Philosophy: Four Genres," in *Philosophy in History*, ed. Richard Rorty, J. B. Schneewind, and Quentin Skinner (Cambridge: Cambridge University Press, 1984), 49–76.

35. Laudan, "Grünbaum on 'The Duhemian Argument,'" 155.

36. Laudan, "Demystifying Underdetermination," in *Scientific Theories*, ed. C. Wade Savage, Minnesota Studies in the Philosophy of Science 14 (Minneapolis: University of Minnesota Press, 1990), 267–97, citing 274. Laudan is one of those who identified the strong form of the Duhem-Quine thesis with the underdetermination thesis, with some deleterious consequences (see below, pp. 27–29).

37. Laudan, "Grünbaum on 'The Duhemian Argument,'" 160.

38. Pierre Duhem, [selections from] *The Aim and Structure of Physical Theory*, reprinted in *Can Theories Be Refuted?* ed. Harding, 4.

39. Ibid., 6.

40. Ian Hacking, "Philosophers of Experiment," *PSA 1988*, vol. 2, 147–56, citing 150.

41. Philip Quinn, "What Duhem Really Meant," in *Methodological and Historical Essays in the Natural and Social Sciences*, ed. Robert S. Cohen and Marx W. Wartofsky, Boston Studies in the Philosophy of Science 14 (Dordrecht: Reidel, 1973), 33–56. See the ensuing discussion: Nancy Tuana, "Quinn on Duhem: An Emendation," *Philosophy of Science* 45 (1978): 456–62; and Philip Quinn, "Rejoinder to Tuana," 463–65.

42. Quinn, "What Duhem Really Meant," 43.

43. Ibid.

44. Ibid. Quinn takes Quine to hold this extreme holist view.

45. Ibid., 41.

46. Tuana, "Quinn on Duhem," 460.

47. Duhem, *The Aim and Structure of Physical Theory;* cited in Tuana, "Quinn on Duhem," 461.

48. Tuana, "Quinn on Duhem," 461.

49. Ibid.

50. Gary Wedeking, "Duhem, Quine and Grünbaum on Falsification" (1969), reprinted in *Can Theories Be Refuted?* ed. Harding, 176–83, citing 178; the Quine passage is from "Two Dogmas of Empiricism."

51. Carlo Giannoni, "Quine, Grünbaum, and the Duhemian Thesis" (1967), reprinted in *Can Theories Be Refuted?* ed. Harding, 162–75, citing 162.

52. Ibid., 163. The cite from Quine is from "Two Dogmas of Empiricism."

53. Giannoni, "Quine, Grünbaum, and the Duhemian Thesis," 162.

54. Ibid., 163.

55. Jules Vuillemin, "On Duhem's and Quine's Theses" (1979), reprinted in *The Philosophy of W. V. Quine*, ed. Lewis Edwin Hahn and Paul Arthur Schilpp (LaSalle, Ill.: Open Court, 1986), 595–618; with a reply by Quine, 619–22.

56. Ibid., 599.

57. Ibid., 598.

58. Ibid., 599.

59. Ibid., 610.

60. Ibid., 611.

61. Henry Krips, "Epistemological Holism: Duhem or Quine?" *Studies in History and Philosophy of Science* 13 (1982): 251–64, citing 261.

62. Ibid., 253.

63. Quine, "Two Dogmas of Empiricism," 43.

64. Krips, "Epistemological Holism," 256.

65. See esp. Andrew Pickering, ed., *Science As Practice and Culture* (Chicago: University of Chicago Press, 1992).

66. Krips, "Epistemological Holism," 259.

67. Roger Ariew, "The Duhem Thesis," *British Journal of the Philosophy of Science* 35 (1984): 313–25, citing 313.

68. Ibid., 313–14.

69. Ibid., 320.

70. Ibid., 321.

71. Duhem, *The Aim and Structure of Physical Theory;* cited in Ariew, "The Duhem Thesis," 322.

72. Ulrich Gähde and Wolfgang Stegmüller, "An Argument in Favor of the Duhem-Quine Thesis: From the Structuralist Point of View," in *The Philosophy of W. V. Quine*, ed. Hahn and Schilpp, 117–36, citing 117.

73. Ibid., 133.

74. Willard van Orman Quine, "Reply to Ulrich Gähde and Wolfgang Stegmüller," in *The Philosophy of W. V. Quine*, ed. Hahn and Schilpp, 137-38, citing 138. In "Three Indeterminacies," Quine articulates his stance in terms of a "maxim of minimum mutilation." He argues that this generally requires that one "safeguard any purely mathematical truth; for mathematics infiltrates all branches of our system of the world, and its disruption would reverberate intolerably." In addition to this "conservatism," Quine also emphasizes simplicity as "another guiding consideration." But he emphasizes that "the ultimate objective is so to choose the revision as to maximize future success in prediction." Conservatism and simplicity are merely proxies for this goal ("Three Indeterminacies," 11).

75. E.g., in "Two Dogmas in Retrospect," 270.

76. Yuri Balashov, "Duhem, Quine, and the Multiplicity of Scientific Tests," *Philosophy of Science* 61 (1994): 608-28, citing 613.

77. For Lakatos's model, see "Falsification and the Methodology of Scientific Research Programmes."

78. Willard van Orman Quine, *Pursuit of Truth*, rev. ed. (Cambridge, Mass.: Harvard University Press, 1992), 15.

79. Greenwood, "Two Dogmas of Neo-Empiricism," 565.

80. Ibid., 566.

81. Ibid., 567.

82. Ibid., 568.

83. Ian Hacking, "The Self-Vindication of the Laboratory Sciences," in *Science As Practice and Culture*, ed. Pickering, 29-30.

84. The phrase "semantic ascent" is taken from Willard van Orman Quine, *Word and Object* (Cambridge, Mass.: MIT Press, 1960), 270-76.

85. Quine, "Two Dogmas of Empiricism," 42.

86. Quine, "Two Dogmas in Retrospect," 268.

87. Ibid., 270; Quine, *Pursuit of Truth*, 15.

88. Quine, *Pursuit of Truth*.

89. Ibid., 17.

90. Quine's most explicit articulations of the underdetermination thesis came in three essays, "On the Reasons for Indeterminacy of Translation," *Journal of Philosophy* 67 (1970): 178-83; "On Empirically Equivalent Systems of the World," *Erkenntnis* 9 (1975): 313-28; and "Things and Their Place in Theories," in *Theories and Things* (Cambridge, Mass.: Harvard University Press, 1981), 1-23. Quine's various formulations provoked an incisive formulation of a dilemma for Quine by Roger Gibson, "Translation, Physics, and Facts of the Matter," in *The Philosophy of W. V. Quine*, ed. Hahn and Schilpp, 139-54, here 153 n. 2. Quine acknowledged the dilemma (156-57). He has twisted upon its horns for some twenty years, only to end up in a cul-de-sac, in Gibson's view: "Quine's thinking about underdetermination over the last twenty-five years has landed him in a contradiction" ("Quine's Dilemma," *Synthese* 69 [1986]: 27-39, citing 27). See also Gibson, "More on Quine's Dilemma of Underdetermination," *Dialectica* 45 (1991): 59-68. For a related argument, see A. C. Genova, "Quine's Dilemma of Underdetermination," *Dialectica* 42 (1988): 283-93. For earlier recognitions of incoherency, see Paul Roth, "Semantics without Foundations," in *The Philosophy of W. V. Quine*, ed. Hahn and Schilpp, 433-58; Roth, "Paradox and Indeterminacy," *Journal of Philosophy* 75 (1978): 347-67; Roth, "Reconstructing Quine: The Troubles with a Tradition," *Metaphilosophy* 14 (1983): 249-66; P. William Bechtel, "Indeterminacy and Underdetermination: Are Quine's Two Theses Consistent?" *Philosophical Studies* 38 (1980): 309-20; Murray G. Murphey, "The Underdetermination Thesis," in *Frontiers in American Philosophy*, ed. Robert W. Burch and Herman J. Saatkamp, Jr., vol. 1 (College Station: Texas A&M University Press, 1992), 157-65; and Dorit Bar-On, "Seman-

tic Indeterminacy and Scientific Underdetermination," *Pacific Philosophical Quarterly* 67 (1986): 245–63.

91. Laudan, "Demystifying Underdetermination," 271.

92. Roth, "Semantics without Foundations," 434.

93. "This is a point on which I expect wide agreement, if only because the observational criteria for theoretical terms are commonly so flexible and fragmentary" (Quine, "On the Reasons for Indeterminacy of Translation," 179).

94. Ibid.

95. Actually, Quine's whole point makes more sense as a claim about *language* and its limited tangency with the world. There are myriad sentences—the bulk of the language—that have no unique observable referent.

96. On the distinction of *in fact* versus *in principle* argumentation in Quine, see Gibson, "Quine's Dilemma," 33.

97. Quine, "On the Reasons for Indeterminacy of Translation," 179. I have cited this passage so extensively because it can serve as the benchmark for the entire problematic of the multiplication of underdetermination theses.

98. Quine, "On Empirically Equivalent Systems of the World," 313.

99. In the terms of Louise Antony, does "confirmational holism" entail "semantic holism"? Quine's own concessions about the limits of the Duhem thesis on confirmational holism only reinforce Antony's contention that there is *no* entailment sufficient to Quine's purposes here (Louise Antony, "Naturalized Epistemology and the Study of Language," in *Naturalistic Epistemology,* ed. Abner Shimony and Debra Nails, Boston Studies in the Philosophy of Science 100 [Boston: Reidel, 1987], 235–57, esp. 244).

100. Roth, *Meaning and Method in the Social Sciences,* 7, esp. 23: "The Duhemian thesis does not require that theories be underdetermined. . . . Nor does the Duhem thesis of itself entail that theories will be underdetermined, for the Duhem thesis is a thesis about the truth conditions for a single theoretical sentence, whereas the underdetermination thesis is one about the ability to formulate entire theories . . . that . . . account for all available evidence in logically incompatible fashions."

101. Roth, "Semantics without Foundations," 457.

102. Roth cites from that text, ibid.

103. Laudan, "Demystifying Underdetermination," 294 n. 14.

104. Quine, "On Empirically Equivalent Systems of the World," 313–28.

105. Ibid., 313.

106. Laudan, "Demystifying Underdetermination," 293 n. 9.

107. Ibid., 296 n. 37, where Laudan cites Quine's line from "On Empirically Equivalent Systems of the World," cited above, and interpolates as follows: " 'the holism thesis [egalitarianism] lends credence to the underdetermination thesis [nonuniqueness].' "

108. Roth, "Semantics without Foundations," 457.

109. Ibid.

110. Ibid., 457–58.

111. Carl Hoefer and Arthur Rosenberg, "Empirical Equivalence, Underdetermination, and Systems of the World," *Philosophy of Science* 61 (1994): 592–607, esp. abstract, 592: "The underdetermination of theory by evidence must be distinguished from holism . . . a doctrine about the testing of scientific hypotheses"; and 593ff.

112. Willard van Orman Quine, "Reply to Roth," in *The Philosophy of W. V. Quine,* ed. Hahn and Schilpp, 459.

113. Quine, *Theories and Things,* 71.

114. Quine, "On Empirically Equivalent Systems of the World," 314. Robert Nozick helps us to understand what is so simultaneously frustrating and lucid about Quine: "Quine

is the theorist of slack. Data underdetermine theory, there is leeway about which component to modify when data conflict with a theory, and translation is indeterminate. Also, theory underdetermines the world (the doctrine of ontological relativity)" ("Experience, Theory and Language," in *The Philosophy of W. V. Quine,* ed. Hahn and Schilpp, 339–63, citing 339).

115. See, e.g., Roth, "Reconstructing Quine," 250, for a recognition of this.

116. Laudan, "Demystifying Underdetermination," 294 n. 19.

117. Laudan makes reference to this explicit sense of holism in Quine (appendix to "Demystifying Underdetermination," 292).

118. Quine concedes this explicitly in "Two Dogmas in Retrospect," 268.

119. Murphey, "The Underdetermination Thesis," 163–64.

120. Genova, "Quine's Dilemma of Underdetermination," 285.

121. Murphey, "The Underdetermination Thesis," 160.

122. Quine, "On Empirically Equivalent Systems of the World," 326.

123. Quine, *The Pursuit of Truth,* 100.

124. Quine, "On Empirically Equivalent Systems of the World," 326.

125. Gibson, "Quine's Dilemma," 34. See also Bar-On, "Semantic Indeterminacy and Scientific Underdetermination," 249.

126. Larry Laudan and Jarrett Leplin, "Empirical Equivalence and Underdetermination," *Journal of Philosophy* 88 (1991): 449–72, citing 451.

127. Laudan, "Demystifying Underdetermination," 286.

128. Ibid., 267.

129. Ibid., 268. Philip Kitcher makes this same case in "A Plea for Science Studies," 44: "When the Four Dogmas [Kitcher's summary of the false inferences from post-positivist philosophy] have been thoroughly absorbed, so that younger scholars start from their conclusions as if they were gospel, then enterprises of real peculiarity can be launched." See Rachel Laudan on this preemptive mind-set among many younger "postmodernist" scholars: "The 'New' History of Science: Implications for Philosophy of Science," *PSA 2002,* vol. 2, 478–79.

130. Laudan, "Demystifying Underdetermination," 271.

131. "Quine nowhere explicitly expresses the egalitarian thesis in precisely this form. . . . What follows is not meant to be an exegesis of Quine's intentions; it is meant, rather, as an exploration of whether Quine's position on this issue will sustain the broad implications that many writers (sometimes including Quine himself) draw from it" (ibid., 271). And: "In 'Two Dogmas,' Quine propounded a thesis of normative, ampliative, egalitarian underdetermination" (281).

132. Ibid., 271: "Quine's numerous pronouncements on the retainability of theories, in the face of virtually any evidence, presuppose the egalitarian thesis, and make no sense without it."

133. Here, again, see Kitcher, "A Plea for Science Studies," for one corroboration.

134. Quine admits that he "give[s] logic a special status," but he goes on: "However, we remain free here to adjust and to vary the limits of what to count as logic" ("Two Dogmas in Retrospect," 268). Quine is here elaborating the idea that all sentences are subject to revision, even the "law of the excluded middle." But the converse of that argument is that inferences warranted by looser forms of logic—probabilistic accounts, say—would have a place in real-time scientific reasoning, which is what Laudan is trying to establish.

135. Willard van Orman Quine and J. S. Ullian, *The Web of Belief,* 2d ed. (New York: Random House, 1978), 64–82; see Quine's rather equivocal comment on this issue in *Pursuit of Truth,* 20.

136. Quine writes: "No general calibration of either conservatism or simplicity is known, much less any comparative scale of the one against the other. For this reason alone—and it is not alone—there is no hope of a mechanical procedure for optimum hypothesizing. Creating

good hypotheses is an imaginative art, not a science" (*From Stimulus to Science* [Cambridge, Mass.: Harvard University Press, 1995], 49).

137. Mary Hesse has formulated something along this line in *Revolutions and Reconstructions in the Philosophy of Science* (Bloomington: Indiana University Press, 1980), ix.

138. On Bayesianism, see Wesley Salmon, "Rationality and Objectivity in Science, or Tom Kuhn Meets Tom Bayes," in *Scientific Theories,* ed. Savage, 175–204; Colin Howson and Peter Urback, *Scientific Reasoning: The Bayesian Approach,* 2d ed. (Chicago: Open Court, 1993). For a specific application to the Duhem problem, see Jon Dorling, "Bayesian Personalism, the Methodology of Scientific Research Programmes, and Duhem's Approach," *Studies in History and Philosophy of Science* 10 (1979): 177–87.

139. Samir Okasa, "Laudan and Leplin on Empirical Equivalence," *British Journal for Philosophy of Science* 48 (1997): 251–56.

140. Laudan and Leplin, "Empirical Equivalence," 463.

141. Hoefer and Rosenberg, "Empirical Equivalence," 598; but see John Earman, "Underdetermination, Realism, and Reason," *Midwest Studies in Philosophy* 18 (1993): 19–38, for a rejoinder defending the underdetermination thesis.

142. Laudan, "Demystifying Underdetermination," 277.

143. Ibid.

144. Laudan and Leplin, "Empirical Equivalence," 463 n. See also Jarrett Leplin, "Surrealism," *Mind* 96 (1987): 519–24.

145. André Kukla, "Laudan, Leplin, Empirical Equivalence and Underdetermination," *Analysis* 53 (1993): 1–7, citing 5.

146. Laudan and Leplin, "Determination Undeterred," *Analysis* 53 (1993): 8–16, citing 12.

147. Laudan and Leplin, "Empirical Equivalence," 459.

148. Ibid., 466 n.

149. Quine, "Epistemology Naturalized," 89.

150. On many of these issues see Paul Horwich, "How to Choose between Empirically Indistinguishable Theories," *Journal of Philosophy* 79 (1982): 61–77. See esp. 68 n. 5: "Although Quine, more than any other philosopher, is responsible for the demise of traditional empiricism, he is nonetheless one of its diehard exponents. . . . [H]e retains the idea that the content of a total theory may be identified with its observational consequences."

151. See Quine, *Word and Object,* 270–76, for his classic statement of this.

152. Laudan and Leplin, "Empirical Equivalence," 468–69, 460.

153. Ian Hacking, *Representing and Intervening* (Cambridge: Cambridge University Press, 1983), esp. 101–11, 167–85.

154. Laudan makes this point clearly in "Demystifying Underdetermination," 271.

155. Hoefer and Rosenberg, "Empirical Equivalence," 593.

156. Laudan, "Demystifying Underdetermination," 281.

157. See, e.g., Hacking, "The Self-Vindication of the Laboratory Sciences," 55: "It is extraordinarily difficult to make one coherent account, and it is perhaps beyond our powers to make several. The philosophical task is less to understand an indeterminacy that we can imagine but almost never experience than to explain the sheer determinateness of mature laboratory science."

158. This agenda is most forthrightly encompassed in his two latest books, *The Pursuit of Truth* and *From Stimulus to Science.*

159. Ian Hacking, "A Tradition of Natural Kinds," *Philosophical Studies* 61 (1991): 109–26, citing 111; and see 124 n. 3 for a reference to Plato's *Statesman* 287c.

160. On Quine and "Natural Kinds," see esp. Ian Hacking, "Natural Kinds," in *Perspectives on Quine,* ed. Barrett and Gibson, 129–41; and Hacking, "A Tradition of Natural Kinds."

161. Quine, "Three Indeterminacies," 6-7.

162. Willard van Orman Quine, "Natural Kinds," in *Ontological Relativity and Other Essays*, 114-38, citing 114.

163. Ibid., 126.

164. Ibid., 123.

165. See Hesse, *Revolutions and Reconstructions;* and Barry Barnes, *Scientific Knowledge and Sociological Theory* (London: Routledge & Kegan Paul, 1974) on "natural rationality."

166. Quine, "Natural Kinds," 116-17.

167. Ibid., 128-29.

168. Ibid., 121.

169. Quine, *Pursuit of Truth,* 34.

170. Quine, *Theories and Things,* 9.

171. Ibid., 17.

172. Quine, "Three Indeterminacies," 7.

173. Ibid.

174. Quine, *Pursuit of Truth,* 31.

175. Ibid., 26, 31.

176. Ibid., 33.

177. Quine introduced the notion "gavagai" in *Word and Object* in the context of his theory of "radical translation" and the "indeterminacy of translation," but he has insisted that its proper bearing is essentially on the "inscrutability of reference." See Quine, "Indeterminacy of Translation Again," *Journal of Philosophy* 84 (1987): 5-10, citing 8; "On the Reasons for Indeterminacy of Translation," 178; "Three Indeterminacies," 6.

178. Quine, "Speaking of Objects," in *Ontological Relativity and Other Essays,* 1-25, citing 15.

179. Quine, "Ontological Relativity," 41.

180. Ibid., 46.

181. Roth, "Paradox and Indeterminacy," 264.

182. Quine, *Pursuit of Truth,* 27.

183. Quine, "Three Indeterminacies," 6.

184. Ibid., 20.

185. Quine, "Ontological Relativity," 51.

186. Ibid., 49.

187. Ibid., 21.

188. Roth, "Reconstructing Quine," 253.

189. Roth, "Paradox and Indeterminacy," 366.

190. Dagfin Föllesdal, "Indeterminacy of Translation and Under-determination of the Theory of Nature," *Dialectica* 27 (1973): 289-301, citing 289. Michael Friedman says the same: "both the content of the thesis and the arguments for it remain relatively unclear" ("Physicalism and the Indeterminacy of Translation," *Nous* 9 [1975]: 353-74, citing 353).

191. Miriam Solomon, "Quine's Point of View," *Journal of Philosophy* 86 (1989): 113-36, citing 113.

192. Ibid.

193. Ibid., 114.

194. For a rich treatment of actual translation problems as a context for a rigorous examination of Quine's thesis, see Bar-On, "Semantic Indeterminacy and Scientific Underdetermination."

195. Christopher Boorse, "The Origins of the Indeterminacy Thesis," *Journal of Philosophy* 72 (1975): 369-87, citing 380.

196. Richard Rorty, "Indeterminacy of Translation and of Truth," *Synthese* 23 (1972): 443–62, citing 453; Solomon, "Quine's Point of View," 116–17.

197. On the "principle of charity" in Quine and in Davidson, see David Henderson, "The Principle of Charity and the Problem of Irrationality (Translation and the Problem of Irrationality)," *Synthese* 73 (1987): 225–52; Henderson, "The Importance of Explanation in Quine's Principle of Charity," *Philosophy of the Social Sciences* 18 (1988): 355–69; Jeffrey Malpas, "The Nature of Interpretive Charity," *Dialectica* 42 (1988): 17–36; Christopher Gauker, "The Principle of Charity," *Synthese* 69 (1986): 1–25; Paul Thagard and Richard Nisbett, "Rationality and Charity," *Philosophy of Science* 50 (1983): 250–67.

198. Solomon, "Quine's Point of View," 120 n. But she adds, correctly: "Indeterminacy is a stronger thesis than the rejection of the propositional model" (119).

199. Christopher Hookway develops this case quite convincingly in *Quine: Language, Experience, Reality* (Stanford, Calif.: Stanford University Press, 1988).

200. Föllesdal, "Indeterminacy of Translation and Under-determination," 290–91.

201. Joseph Levine, "Quine's Psychology," in *Naturalistic Epistemology*, ed. Shimony and Nails, 259–90, citing 262.

202. Ibid., 281.

203. Willard van Orman Quine, "Comment on Hintikka," in *Perspectives on Quine*, ed. Barrett and Gibson, 176.

204. Willard van Orman Quine, "Facts of the Matter," in *American Philosophy from Edwards to Quine,* ed. Robert W. Shahan and Kenneth R. Merrill (Norman: University of Oklahoma Press, 1977), 193.

205. Friedman, "Physicalism," 360.

206. Gibson, "Translation, Physics, and Facts of the Matter," in *The Philosophy of W. V. Quine,* ed. Hahn and Schilpp, 139–54; and Quine's appreciation, 155–58.

207. Indeed, Quine persists in a measure of this conviction: "Taken analytically, the indeterminacy of translation is trivial and indisputable. . . . [O]ne can scarcely question the holophrastic indeterminacy thesis," he wrote in 1992 (*Pursuit of Truth,* 50–51), as if thirty years of contestation amounted to scarcely a question!

208. Quine, "On the Reasons for Indeterminacy of Translation," 183.

209. "The serious and controversial thesis of indeterminacy of translation is not [inscrutability of reference]; it is rather the holophrastic thesis, which is stronger. It declares for divergences that remain unreconciled even at the level of the whole sentence, and are compensated for only by divergences in the translations of other whole sentences" (Quine, *Pursuit of Truth,* 50).

210. Willard van Orman Quine, "Reply to Chomsky," in *Words and Objections,* ed. Donald Davidson and Jaako Hintikka, 2d ed. (Dordrecht: Reidel, 1975), 302–11, citing 303.

211. Quine, "On the Reasons for Indeterminacy of Translation," 179–80.

212. Michael Dummett, "The Significance of Quine's Indeterminacy Thesis," *Synthese* 27 (1974): 351–97, citing 383.

213. Friedman, "Physicalism," 356.

214. Föllesdal, "Indeterminacy of Translation and Under-determination of the Theory of Nature," 289–301.

215. Quine, "On the Reasons for Indeterminacy of Translation," 178.

216. Solomon, "Quine's Point of View," 113–36.

217. Quine, "Epistemology Naturalized," 80–81.

218. Föllesdal, "Indeterminacy of Translation and Under-determination of the Theory of Nature," 289–301.

219. Roth, "Paradox and Indeterminacy," 351.

220. Quine claims that he is not a reductionist, but that claim is in stark conflict with his practices.

221. Quine, *From Stimulus to Science,* 49.

222. Quine, *Word and Object,* 73.

223. Antony, "Naturalized Epistemology," 249.

224. Friedman, "Physicalism," 354–59.

225. Antony, "Naturalized Epistemology," 249.

226. Ibid., 245.

227. Hookway, *Quine,* chap. 12.

228. Hilary Putnam puts it well: "Today we tend to be too realistic about physics and too subjectivistic about ethics, and these are connected tendencies. It is *because* we are too realistic about physics, because we see physics (or some hypothetical future physics) as the One True Theory, and not simply as a rationally acceptable description suited for certain problems and purposes, that we tend to be subjectivistic about descriptions we cannot 'reduce' to physics" (*Reason, Truth and History* [Cambridge: Cambridge University Press, 1981], 143).

229. Friedman, "Physicalism," 369.

230. Quine, "Epistemology Naturalized," 75.

231. Quine, *Pursuit of Truth,* 19.

232. Donald Davidson, "Meaning, Truth, and Evidence," in *Perspectives on Quine,* ed. Barrett and Gibson, 68–79, citing 70.

233. Ibid., 71.

234. Ibid.

235. Ibid., 74.

236. Ibid., 76.

237. Quine, "In Praise of Observation Sentences," 107.

238. Ibid., 108.

239. Ibid., 109.

240. Ibid., 111.

241. Quine, *Pursuit of Truth,* 7.

242. Quine, "Epistemology Naturalized," 89.

243. Ibid., 80–81.

244. Davidson, "Meaning, Truth, and Evidence," 74.

245. Roth, "Paradox and Indeterminacy," 366–67.

246. Bjørn T. Ramberg, *Donald Davidson's Philosophy of Language* (Oxford: Blackwell, 1989), 16.

247. Roth, "Reconstructing Quine," 263.

248. Quine, "Things and Their Place in Theories," 23; cited by Gibson, with his emphasis, in "Quine's Dilemma," 29.

249. Gibson, "Quine's Dilemma," 29.

250. Ibid., 149.

251. Hesse, "Is There an Independent Observation Language?" in *Revolutions and Reconstructions,* 63–110.

252. Genova, "Quine's Dilemma of Underdetermination," 283.

253. Ibid., 291.

254. Ibid., 292.

255. Gibson, "Translation, Physics, and Facts of the Matter," 152.

256. Quine, "Ontological Relativity," 34.

257. Gibson, "Quine's Dilemma," 27.

258. Bar-on, "Semantic Indeterminacy," 255.

259. Gibson, "Quine's Dilemma," 27. What would it mean for someone who professed

to be an empiricist in epistemology to contend that two theories which are equally adequate to *all possible* evidence could nevertheless be logically incompatible and thus pose the dilemma that only one could be true? Quine himself recognizes that this poses acute problems. He has vacillated between what he terms his "sectarian" and his "ecumenical" stances on this question for some two decades now.

260. On this see Roth, "Paradox and Indeterminacy," 354; and Murphey, *Philosophical Foundations of Historical Knowledge*, 38: "Quine's strategy is to show that a natural language can be treated as if it were an abstract formal theory, so that it too will admit of interpretation by multiple models and the model actually referred to in any case can only be stipulated in a metatheory. . . . But can natural languages in fact be treated as formal theories?"

261. Quine, *Roots of Reference*, 38.

262. Quine, "Epistemology Naturalized," 87. But by the same token, when these special languages enter into conflict, or simply become so specialized that they split away from one another, the ultimate recourse is always descent into natural language, slack and sloppy as it may be. See Nobuharu Tanji, "Quine on Theory and Language," *British Journal of the Philosophy of Science* 40 (1989): 233–47.

263. Gibson, "Translation, Physics, and Facts of the Matter," 151.

264. Ibid.

265. Rorty, *Philosophy and the Mirror of Nature* (Princeton, N.J.: Princeton University Press, 1979).

266. Rorty, "Indeterminacy," 461 n. 28.

267. Quine, *From Stimulus to Science*, 49. I am at a loss to know whether I should be grateful for the ascription of "glory" or indignant at Quine's condescension; at any rate, it is clear that history hardly has claim to status as an "empirical" inquiry in his view.

268. Bar-On, "Semantic Indeterminacy and Scientific Underdetermination," 254.

269. Ibid., 258.

270. Indeed, in later writings, e.g., *Pursuit of Truth* and *From Stimulus to Science*, it would appear that Quine is willing to take far more seriously the complexities of psychology required to get his theory of the relation between reception and perception and between perception and utterance off the ground. Thus we find a considerable invocation of "empathy" in these writings and a firmer acknowledgment that in language teaching or learning one cannot proceed far at all without making judgments not only about assent or dissent to sentences but about what a person is in fact perceiving—a mental state if ever there was one!

271. Quine, "Speaking of Objects," 11.

272. Föllesdal, "Indeterminacy of Translation and Under-determination of the Theory of Nature," 299.

273. Quine, "Indeterminacy of Translation Again," 5.

274. Quine, "Epistemology Naturalized," 81.

275. Quine, "Speaking of Objects," 5.

276. Quine, "Three Indeterminacies," 4.

277. Quine, *Pursuit of Truth*, 37–38.

278. John Searle, "Indeterminacy, Empiricism, and the First Person," *Journal of Philosophy* 84 (1987): 123–46.

279. Quine is stunningly cavalier about empirical linguistics, especially for a naturalized epistemologist. He dismisses any claims they might make about linguistics with the following condescension: "It may well be the efficient terminology for working empirical linguists, and I wish them Godspeed; but it is precisely the contrary for philosophers or psychologists of language" (*From Stimulus to Science*, 94).

280. Noam Chomsky, "Quine's Empirical Assumptions," in *Words and Objections*, ed.

Davidson and Hintikka, 53–68; with Quine's vigorous response, 302–11. For background on the issue of "behaviorism" here, see Noam Chomsky, "A Review of B. F. Skinner's *Verbal Behavior,*" in *The Structure of Language,* ed. Jerry A. Fodor and Jerrold J. Katz (Englewood Cliffs, N.J.: Prentice Hall, 1964), 547–78; and see the debate about the cogency of Chomsky's critique of Skinner between Peter Slezak and Steve Fuller: Slezak, "Scientific Discovery by Computer As Empirical Refutation of the Strong Programme," *Social Studies of Science* 19 (1989): 563–600, esp. 593–94; Fuller, "Of Conceptual Intersections and Verbal Collisions: Towards the Routing of Slezak," *Social Studies of Science* 19 (1989): 625–38, esp. 631–33; and Slezak, "Computers, Contents and Causes: Replies to My Respondents," *Social Studies of Science* 19 (1989): 671–95, esp. 689–92.

281. For a consideration of much of this empirical evidence, see Murphey, "Meaning and Reference," chap. 1 of *Philosophical Foundations of Historical Knowledge,* 1–61.

282. Tanji, "Quine on Theory and Language," 245.

283. Davidson, "The Structure and Content of Truth," *Journal of Philosophy* 87 (1990): 279–328, citing 325. Yet Davidson endorses the idea of indeterminacy—for interpretation, if not translation—even as he dismisses Quine's empiricism and his behaviorism. What grounds does Davidson offer for *his* version of indeterminacy, and does he rescue the essence of Quine's position on language with that shifted form of the argument?

Chapter Three

1. Thomas S. Kuhn, *The Structure of Scientific Revolutions,* 2d ed. (Chicago: University of Chicago Press, 1970), 1.

2. On the lust for paradigm status in social sciences, see, e.g., Douglas Lee Eckberg and Lester Hill, Jr., "The Paradigm Concept and Sociology: A Critical Review," in *Paradigms and Revolutions: Appraisals and Applications of Thomas Kuhn's Philosophy of Science,* ed. Gary Gutting (Notre Dame, Ind.: University of Notre Dame Press, 1980), 117–36; and esp. Albert Hirschman, "The Search for Paradigms As a Hindrance to Understanding," in *Interpretive Social Science: A Second Look,* ed. Paul Rabinow and William Sullivan (Berkeley: University of California Press, 1987), 177–94.

3. On the antiscience attitude, see Stephen Toulmin, "The Historical Background to the Antiscience Movement," in *Civilization and Science in Conflict or Collaboration?* Ciba Foundation Symposium (Amsterdam: Elsevier, 1972), 23–32. There is something right about the concern with antiscience, but it can easily get overstated in a reaction as excessive as the hyperbole it condemns.

4. "Kuhn certainly believed, at one time, that a history department was the right place for him, even if, as he admitted, historians were far less susceptible to his blandishments than scientists or philosophers" (John Passmore, "The Relevance of History to the Philosophy of Science," in *Scientific Explanation and Understanding,* ed. Nicholas Rescher [Lanham, Md.: University Press of America, 1983], 84). Historians of science have indeed been sharply suspicious of Kuhn, identifying and belaboring problems in Kuhn's theory of scientific development as a historical model from the moment he set it forth. See, e.g., Maurice Mandelbaum, "A Note on Thomas S. Kuhn's *The Structure of Scientific Revolutions,*" *Monist* 60 (1977): 445–52; David Hollinger, "T. S. Kuhn's Theory of Science and Its Implications for History," in *Paradigms and Revolutions,* ed. Gary Gutting, 195–222; Nathan Reingold, "Through Paradigm-Land to a Normal History of Science," *Social Studies of Science* 10 (1980): 475–96. There is thus a whole school of historical critics to match Kuhn's philosophical critics. One reason for this is the recognition of a difference between history of science and what Kuhn was up to, which is the mining of history for a theory of scientific development, what

Wolf Lepenies has distinguished as "the historical study of science" ("Problems of a Historical Study of Science," in *The Social Production of Scientific Knowledge: Sociology of the Sciences*, vol. 1 [Dordrecht: Reidel, 1977], 55–67). That is an endeavor in social theory or
philosophy; it was to the latter that Kuhn instinctively drifted. But the point is, "he is unlikely
to feel completely at home anywhere, as Kuhn's later apotheosis as a philosopher rather suggests" (Passmore, "Relevance of History to the Philosophy of Science," 85).

5. Sergio Sismondo, "The *Structure* Thirty Years Later: Refashioning a Constructivist
Metaphysical Program," *PSA 1992*, vol. 1, 300–312, citing 305.

6. Thomas S. Kuhn, "The Relations between the History and the Philosophy of Science," in *The Essential Tension* (Chicago: University of Chicago Press, 1977), 3–20, citing 3.

7. "A Discussion with Thomas S. Kuhn," in *The Road since Structure: Philosophical Essays, 1970–1993, with an Autobiographical Interview,* by Thomas S. Kuhn, ed. James Conant and John Haugland (Chicago: University of Chicago Press, 2000), 276.

8. Ibid.

9. Ibid., 314.

10. Ibid., 322.

11. "The view that the history of science has relevance for the philosophy of science was,
and still is, anathema to most practitioners of the latter discipline" (Daniel G. Cedarbaum,
"Paradigms," *Studies in History and Philosophy of Science* 14 [1983]: 173–213, citing 178).
Frederic Suppe minced no words in 1977: "Despite his sustained efforts to reply to critics
and clarify, modify, or improve his position when he feels his critics have a legitimate objection, since [1969] Kuhn's views have undergone a sharply declining influence on contemporary philosophy of science" (afterword to *Structure of Scientific Theories*, 2d, rev. ed.
[Urbana: University of Illinois Press, 1977], 647).

12. For a continuation of standard philosophy of science refutation, see Harmon Hol-
comb III, "Circularity and Inconsistency in Kuhn's Defense of His Relativism," *Southern
Journal of Philosophy* 25 (1987): 467–80; and "Interpreting Kuhn: Paradigm-Choice As
Objective Value Judgement," *Metaphilosophy* 20 (1989): 51–67.

13. Writing much earlier, John Passmore came to the same realization: "the outcome of
our discussion is rather that [the historian of science] is unlikely to feel completely at home
anywhere, as Kuhn's later apotheosis as a philosopher rather suggests" ("The Relevance of
History to the Philosophy of Science," 85).

14. Hilary Putnam put it nicely: Kuhn's *Structure* "enthralled vast numbers of readers,
and appalled most philosophers of science" (*Reason, Truth and History* [Cambridge: Cambridge University Press, 1981], 113).

15. Kuhn himself puts it in these terms in his "Reflection on My Critics," in *Criticism
and the Growth of Knowledge,* ed. Imre Lakatos and Alan Musgrave (Cambridge: Cambridge University Press, 1970), 231–78.

16. Ronald Giere, "Kuhn's Legacy for North American Philosophy of Science," *Social
Studies of Science* 27 (1997): 497.

17. Kuhn, "The Relations between the History and the Philosophy of Science," 14.

18. The locus classicus of this idea of the "Received View," and of its demise, is to be
found in Suppe, introduction and afterword to *The Structure of Scientific Theories*. For an
even earlier retrospective, see Peter Achinstein and Stephen Barker, eds., *The Legacy of Logical Positivism: Studies in the Philosophy of Science* (Baltimore: Johns Hopkins University
Press, 1969). Very insightful on this historical issue is Stephen Toulmin, "Rediscovering
History: New Directions in Philosophy of Science," *Encounter* 36 (1971): 53–64; Toulmin, "From Form to Function: Philosophy and History of Science in the 1950s and Now,"
Daedalus 106.3 (1977): 143–62.

19. Toulmin, "Rediscovering History," 56. "The central convictions were: (1) that careful scrutiny and analysis of the arguments which emerge within the scientific 'context of justification' will reveal that properly conducted natural science does indeed have a canon, 'method,' or organon; (2) that the essential procedures of that method can be captured and expressed in formal algorithms . . . and (3) that the 'rationality' of the natural sciences lies in conforming to that set of formally valid procedures" (Toulmin, "From Form to Function," 147).

20. Lakatos and Musgrave, eds., *Criticism and the Growth of Knowledge;* Suppe, ed., *The Structure of Scientific Theories.*

21. Dudley Shapere, "The Structure of Scientific Revolutions," in *Paradigms and Revolutions,* 27–38; and "Meaning and Scientific Change," in *Mind and Cosmos: Essays in Contemporary Science and Philosophy,* ed. Robert Colodny (Pittsburgh: University of Pittsburgh Press, 1966), 41–85; Israel Scheffler, *Science and Subjectivity* (Indianapolis: Bobbs-Merrill, 1967).

22. Suppe lumps these figures in terms of a *"Weltanschauungen-*approach" (introduction [1971] and afterword [1977] to *The Structure of Scientific Theories*).

23. See, e.g., Carl Kordig, "The Theory-Ladenness of Observation," *Review of Metaphysics* 24 (1971): 448–84, esp. 471: "Roughly put, essential revisions of beliefs about experience could not be made in a rational way if Feyerabend, Hanson, Toulmin, or Kuhn were correct." Kordig found Kuhn's work full of "systematically misleading expressions" as Gilbert Ryle defined this category, broadly as implying things quite different from what the case is in reality. See, as well, Gary Jones, "Kuhn, Popper, and Theory Comparison," *Dialectica* 35 (1981): 389–97, esp. 397: "in his thesis of incommensurability and his rejection of Popper's point, Kuhn has dissolved every apparent criteria [*sic*] for stating of two theories that they are competitive or even that they are two theories in the same discipline."

24. Shapere, "The Structure of Scientific Revolutions," 37.

25. Ibid., 34. In "Meaning and Scientific Change," Shapere stated this baldly: Kuhn's relativism "is not the result of an investigation of actual science and its history; rather it is the purely logical consequence of a narrow preconception about what 'meaning' is" (68).

26. See their respective contributions to *Criticism and the Growth of Knowledge.*

27. Karl Popper, "The Myth of the Framework" (1965, rev. 1972), in *Rational Changes in Science,* ed. Joseph C. Pitt and Marcello Pera (Dordrecht: Reidel, 1976), 35–62.

28. Imre Lakatos, "Falsification and the Methodology of Scientific Research Programmes," in *Criticism and the Growth of Knowledge,* ed. Imre Lakatos and Alan Musgrave (Cambridge: Cambridge University Press, 1970), 91–196.

29. Alan Musgrave, "Kuhn's Second Thoughts" (1971), reprinted in *Paradigms and Revolutions,* ed. Gutting, 39–53; Dudley Shapere, "The Paradigm Concept," *Science* 172 (1971): 706–9; Israel Scheffler, "Vision and Revolution: A Postscript on Kuhn," *Philosophy of Science* 39 (1972): 366–74; Jack W. Meiland, "Scheffler, Kuhn, and Objectivity in Science," *Philosophy of Science* 41 (1974): 179–87; Harvey Siegel, "Meiland on Scheffler, Kuhn, and Objectivity in Science," *Philosophy of Science* 43 (1976): 441–48; Abner Shimony, "Comments on Two Epistemological Theses of Thomas Kuhn," in *Essays in Memory of Imre Lakatos,* ed. Robert S. Cohen, Paul K. Feyerabend, and Marx W. Wartofsky, Boston Studies in the Philosophy of Science 39 (Dordrecht: Reidel, 1976), 569–88. For a powerful rebuttal to all these critics, see Gerald Doppelt, "Kuhn's Epistemological Relativism: An Interpretation and Defense," *Inquiry* 21 (1978): 33–86.

30. Harvey Siegel, "Objectivity, Rationality, Incommensurability and More," *British Journal for the Philosophy of Science* 31 (1980): 359–84, esp. 365: "He continually tries both to retract the more radical claims of his earlier writing, and at the same time to maintain them under different guises"; and the summary verdict on 373: "His position remains both

confusing and confused." See, too, Ian Hacking, "Essay Review of *The Essential Tension,*" *History and Theory* 18 (1979): 223–36.

31. See "Paradigm Lost? A Review Symposium [on Kuhn's *Black-Body Theory and the Quantum Discontinuity, 1894–1912*]," *Isis* 70 (1979): 429–40. All three commentators—Martin Klein, Abner Shimony, and Trevor Pinch—observe critically that there is no effort on the part of Kuhn in this work to use the theoretical constructs of *Structure* to elucidate this major moment in the history of modern physics.

32. See, e.g., Musgrave, "Kuhn's Second Thoughts"; and Trevor Pinch, "Kuhn—the Conservative and Radical Interpretations: Are Some Mertonians 'Kuhnians' and Some Kuhnians 'Mertonians'?" *Social Studies of Science* 27 (1997): 465–82.

33. This is meticulously elaborated in the most careful study of *Structure,* Paul Hoyningen-Huehne's *Reconstructing Scientific Revolutions: Thomas S. Kuhn's Philosophy of Science* (Chicago: University of Chicago Press, 1993), esp. 3ff.

34. These two senses of history (the past versus its construals) are not always carefully distinguished by Kuhn or by his commentators. Some historicists are very critical of his notion of history merely as a "repository." Thus Marga Vicedo writes: "by using the history of science only as a repository of cases, we are not dealing with the history of science, but only with past episodes in science." This fails to do justice to the "historicity of knowledge." Vicedo contends that "the evidential role that can be attributed to an isolated episode from the history of science is usually very low." Accordingly, she urges the pursuit of a "systematics" modeled after biology: "the analysis of the genealogical relationship between species" as the analogue of "the genealogical relationship between different belief systems, methodologies, and evaluative strategies" (Vicedo, "Is the History of Science Relevant to the Philosophy of Science?" *PSA 1992,* vol. 2, 490–96, citing 492).

35. Kuhn's expression of admiration for the "new internal history of science" associated with Alexandre Koyré should not mislead us into believing that Kuhn thinks a *historiographical model* could organize philosophy of science.

36. These are, in fact, the focus of his properly historical work. See Thomas S. Kuhn, *The Copernican Revolution: Planetary Astronomy in the Development of Western Thought* (Cambridge, Mass.: Harvard University Press, 1957).

37. Kuhn addresses this in "Postscript, 1969," *Structure,* 174–210, citing 180–81.

38. Stephen Toulmin's critique stresses this point already in "Does the Distinction between Normal and Revolutionary Science Hold Water?" in *Criticism and the Growth of Knowledge,* 39–48; and Jack Carloye develops it into a rigorous disputation of Kuhn's whole distinction ("Normal Science and the Extension of Theories," *British Journal for the Philosophy of Science* 36 [1985]: 241–56).

39. Margaret Masterman, "The Nature of a Paradigm," in *Criticism and the Growth of Knowledge,* 59–89.

40. While we can follow the debate over such Kuhnian terms as "incommensurability" forward into the ensuing philosophy of science, "paradigm" has been lost to the general language. Its usage has proliferated beyond any reckoning. This allowed "paradigm" to be something relevant to domains beyond the natural sciences, though Kuhn was not sure that was appropriate. See Gutting, ed., *Paradigms and Revolutions,* for an early survey of this proliferation.

41. See *Criticism and the Growth of Knowledge,* esp. Karl Popper, "Normal Science and Its Dangers," 51–58; John Watkins, "Against Normal Science," 25–38; and Toulmin, "Does the Distinction between Normal and Revolutionary Science Hold Water?"

42. Popper, "Normal Science and Its Dangers"; and Paul Feyerabend, "Consolations for the Specialist," in *Criticism and the Growth of Knowledge,* 197–230.

43. Discipline has those two senses of constraint and empowerment; they cannot be dis-

aggregated. It is a frantic sense of freedom which finds this oppressive in principle. Rather, it can be construed as the soundest vehicle for cultural creativity to which humans have attained.

44. The analogy to common law and precedent is one that deserves more consideration as a model for historical interpretation. (See Toulmin, "From Form to Function.")

45. Popperians and others were highly sensitive to the shift away from "*problem*-solving," and concerned that this trivialized the openness and the creativity of science.

46. Kuhn mentions Polanyi in *Structure*, 44 n; see Michael Polanyi, *Personal Knowledge* (Chicago: University of Chicago Press, 1958). Polanyi is very important, not just for Kuhn but for the subsequent evolution of science studies, yet he is virtually forgotten in mainstream philosophy of science.

47. See, e.g., Andrew Pickering, ed., *Science As Practice and Culture* (Chicago: University of Chicago Press, 1992).

48. Kuhn, "Postscript, 1969," *Structure*, 182.

49. "There is no neutral algorithm for theory-choice, no systematic decision procedure which, properly applied, must lead each individual in the group to the same decision" (ibid., 200).

50. For a more recent elaboration of a similar view, see Rachel Laudan and Larry Laudan, "Dominance and the Diversity of Method: Solving the Problems of Innovation and Consensus," *Philosophy of Science* 56 (1989): 221–38.

51. Kuhn made this clear in *Structure*, and he remained committed to this position throughout, as a letter to the physicist Steven Weinberg from the early 1990s confirmed. Weinberg cited that letter in response to criticisms by Norton Wise of comments Weinberg made about Kuhn in the context of the Sokal hoax. See Weinberg, "The Sokal Hoax," in *The Sokal Hoax*, 155, 169–70.

52. Lakatos, "Falsification and the Methodology of Scientific Research Programmes," 93.

53. "To my dismay, . . . my 'purple passages' led many readers of *Structure* to suppose that I was attempting to undermine the cognitive authority of science rather than to suggest a different view of its nature" (Thomas S. Kuhn, "Afterwords," in *World Changes: Thomas Kuhn and the Nature of Science*, ed. Paul Horwich (Cambridge, Mass.: MIT Press, 1993), 314).

54. Paul Hoyningen-Huehne, "Idealist Elements in Thomas Kuhn's Philosophy of Science," *History of Philosophy Quarterly* 6 (1989): 393–401. Kuhn himself averred that this Kantian analogy had some merit, that he sought a historicized Kantianism; but this was with reference *not* to ontology but to epistemology. He was interested in relativizing the Kantian categories of judgment historically, but not in taking up the Kantian conundrum of any thing-in-itself.

55. Sergio Sismondo, "The *Structure* Thirty Years Later: Refashioning a Constructivist Metaphysical Program," *PSA 1992*, vol. 1, 300–312. Ernan McMullin, "Rationality and Paradigm Change in Science," in *World Changes*, ed. Horwich, 55–78, esp. 71: "The radical challenge of *SSR* is directed not at rationality but at realism." McMullin holds "Kuhn's influence on the burgeoning antirealism of the last two decades can scarcely be overestimated" (71).d

56. Thomas S. Kuhn, "Metaphor in Science," in *Metaphor and Thought*, ed. Andrew Ortony, 2d ed. (Cambridge: Cambridge University Press, 1993), 533–42, citing 539.

57. For a sense of the issue at the very time Kuhn published *Structure*, see Grover Maxwell, "The Ontological Status of Theoretical Entities," in *Scientific Explanation, Space, and Time*, ed. Herbert Feigl and Grover Maxwell, Minnesota Studies in the Philosophy of Science 3 (Minneapolis: University of Minnesota Press, 1962): 3–27.

58. Jack W. Meiland, "Kuhn, Scheffler, and Objectivity in Science," *Philosophy of Science* 41 (1974): 179–87, citing 186.

59. Kuhn, *Structure of Scientific Revolutions;* Paul Feyerabend, "Explanation, Reduction, and Empiricism," in *Scientific Explanation, Space, and Time,* ed. Feigl and Maxwell, 28–97.

60. "If 'meaningful communication is not possible' across paradigms, or if paradigms share no common concepts or entities, how do the statements of each contradict each other? If instead we take incommensurability or contradiction in some other sense, the incommensurabilist arguments suffer a lack of precision. Specifically, in what other sense of 'contradictory' or 'incommensurable' are different paradigms contradictory or incommensurable?" (Gary Weaver and Dennis Gioia, "Paradigms Lost: Incommensurability vs. Structurationist Inquiry," *Organization Studies* 15.4 [1994]: 565–90, citing 571).

61. Norwood Russell Hanson, *Patterns of Discovery* (Cambridge: Cambridge University Press, 1958); Feyerabend, "Explanation, Reduction, and Empiricism," 28–97.

62. Kuhn, preface to *The Essential Tension,* xxii—xxiii.

63. Cedarbaum, "Paradigms," 203.

64. Ibid.

65. Keekok Lee, "Kuhn—a Re-appraisal," *Explorations in Knowledge* 1 (1984): 70.

66. Ian Hacking, *Representing and Intervening* (Cambridge: Cambridge University Press, 1983), 67.

67. Ernest Nagel, *The Structure of Science* (New York: Harcourt, Brace & World, 1961).

68. See Joseph Pitt, ed., *Theories of Explanation* (Oxford: Oxford University Press, 1988); Wesley Salmon, *Four Decades of Scientific Explanation* (Minneapolis: University of Minnesota Press, 1989).

69. If "a *change of meaning* occurs when we pass from T to T' . . . it would be a logical mistake to overlook this change of meaning in a formal deduction. In other words it is a mistake to assume that T and T' are able to explain *the same* explanandum e, because e means two *different* things when it is considered as a sentence of T and when it is considered as a sentence of T'" (Evandro Agazzi, "Commensurability, Incommensurability, and Cumulativity in Scientific Knowledge," *Erkenntnis* 22 [1985]: 60–61).

70. Hacking, *Representing and Intervening,* 68–72.

71. "Aristotle appeared not only ignorant of mechanics, but a dreadfully bad physical scientist as well. . . . These conclusions were unlikely. . . . Perhaps his words had not always meant to him and his contemporaries what they meant to me and mine. . . . Suddenly the fragments in my head sorted themselves out in a new way, and fell into place together. My jaw dropped, for all at once Aristotle seemed a very good physicist indeed, but of a sort I'd never dreamed possible" (Thomas S. Kuhn, "What Are Scientific Revolutions?" in *The Probabilistic Revolution,* vol. 1, *Ideas in History,* ed. Lorenz Krüger, Lorraine Daston, and Michael Heidelberger [Cambridge, Mass.: MIT Press, 1987], 7–20, citing 9).

72. This notion of the historicization of the inquiring subject seems central to the entire theoretical situation of postmodernity.

73. Thomas S. Kuhn, "Speaker's Reply," in *Possible Worlds in Humanities, Arts and Sciences: Proceedings of Nobel Symposium 65,* ed. Sture Allen (New York: de Gruyter, 1989), 49. Kuhn elaborated: "Historians, working backwards, regularly experience as a single conceptual shift a transposition for which the developmental process required a series of stages" (50). But Kuhn was no more consistent in his later writings than he was in *Structure,* and in his essay "What Are Scientific Revolutions?" he insisted upon using just this sudden revelation and holistic notion to characterize an essential feature of scientific revolutions: "the cen-

tral change cannot be experienced piecemeal, one step at a time. Instead, it involves some relatively sudden and unstructured transformation" ("What Are Scientific Revolutions?" 9).

74. Hacking, *Representing and Intervening,* 70.

75. Ibid., 70, 72.

76. Ibid., 74.

77. See Jarrett Leplin, ed., *Scientific Realism* (Berkeley: University of California Press, 1984).

78. Bas van Fraassen, "Empiricism in the Philosophy of Science," in *Images of Science: Essays on Realism and Empiricism,* ed. Paul M. Churchland and Clifford A. Hooker (Chicago: University of Chicago Press, 1985), 245–308; responses to van Fraassen, in *Images of Science, passim.*

79. Dale Moberg, "Are There Rival, Incommensurable Theories?" *Philosophy of Science* 46 (1979): 244–62, citing 249.

80. Thus, Richard Boyd maintained that the domain of philosophy of science was a subset of that of philosophy of language (Boyd, "Metaphor and Theory Change," in *Metaphor and Thought,* 481–532).

81. Shapere, "The Structure of Scientific Revolutions," 27–38.

82. Shapere, "Meaning and Scientific Change," 66.

83. Ibid., 67.

84. Ibid., 65.

85. Ibid., 68.

86. Ibid., 69.

87. Ibid., 67.

88. That Kuhn himself was prepared to go down that "road since *Structure*" is more a token of his longing to be part of the philosophical community than it is of the fruitfulness of that way out of the difficulties from which his text clearly suffered.

89. Scheffler, *Science and Subjectivity,* 55.

90. Ibid., 17.

91. Ibid., 39.

92. Ibid., 58.

93. Ibid., 59.

94. Ibid., 60–61.

95. Ibid., 61.

96. Carl Kordig, "The Theory-Ladenness of Observation," *Review of Metaphysics* 24 (1971): 448–84, citing 464.

97. Ibid., 479.

98. Michael Martin, "Referential Variance and Scientific Objectivity," *British Journal for the Philosophy of Science* 22 (1971): 17–26, citing 17.

99. Ibid., 19.

100. Jarrett Leplin, "Reference and Scientific Realism," *Studies in History and Philosophy of Science* 10 (1979): 265–84, citing, 279.

101. Ibid., 278.

102. Ibid., 284.

103. Robert Nola, " 'Paradigms Lost, or the World Regained'—an Excursion into Realism and Idealism in Science," *Synthese* 45 (1980): 317–50, 323.

104. Ibid., 325.

105. Michael Devitt, "Against Incommensurability," *Australasian Journal of Philosophy* 57 (1979): 29–50, citing 33, 35.

106. Ibid., 46.

107. Doppelt, "Kuhn's Epistemological Relativism."

108. For corroboration of Doppelt's construal, see Richard Grandy, "Incommensurabil-ity: Kinds and Causes," *Philosophica* 32 (1983): 7–24, esp. 15: "The day-to-day work of the scientist is as much a matter of forging a suitable new language and set of concepts to de-scribe the world as it is a matter of making adjustments of beliefs in a given language. . . . [A]ll semantic theories assume fixity of language that is incompatible with the phenomena that are at the center of our investigation. But this is a reason for being dissatisfied with our semantic theories."

109. Doppelt, "Kuhn's Epistemological Relativism," 41.

110. Ibid., 51.

111. Ibid., 52.

112. Ibid., 83 n. 31.

113. Ibid., 54.

114. Ibid., 65.

115. Ibid., 70.

116. On "causal theory of reference," see Stephen P. Schwartz, ed., *Naming, Necessity, and Natural Kinds* (Ithaca, N.Y.: Cornell University Press, 1977).

117. Gottlob Frege, "Über Sinn und Bedeutung," *Zeitschrift für Philosophie und philosophische Kritik* 100 (1892): 25–50.

118. Michael Devitt and Kim Sterelny, *Language and Reality: An Introduction to the Philosophy of Language* (Cambridge, Mass.: MIT Press, 1987), 15–35.

119. On this see, e.g., Donald Davidson, "Epistemology Externalized," *Dialectica* 45 (1991): 191–202.

120. Keith Donnellan, "Reference and Definite Description," in *Naming, Necessity, and Natural Kinds*, ed. Schwartz, 42–65.

121. Saul Kripke, "Identity and Necessity," in *Naming, Necessity, and Natural Kinds*, ed. Schwartz, 66–101; Kripke, *Naming and Necessity* (Cambridge, Mass.: Harvard Univer-sity Press, 1972).

122. Stephen Schwartz, introduction to *Naming, Necessity, and Natural Kinds*, ed. Schwartz, 13–41, citing 29.

123. Hilary Putnam, "Meaning and Reference," in *Naming, Necessity, and Natural Kinds*, ed. Schwartz, 119–32; Putnam, *Mind, Language and Reality: Philosophical Papers*, vol. 2 (Cambridge: Cambridge University Press, 1975).

124. Hacking, *Representing and Intervening*, 79.

125. There was a connection between Putnam's "causal theory of reference" and his commitment to realism; both changed significantly with his abandonment of "metaphysi-cal" for "internal" realism. See Putnam, *Reason, Truth and History*, and the comments on his changes in position, e.g., Hartry Field, "Realism and Relativism," *Journal of Philosophy* 79 (1982): 553–67; David Pearce and Veikko Rantala, "Realism and Reference," *Synthese* 52 (1982): 439–48.

126. Keith Donnellan, "Kripke and Putnam on Natural Kind Terms," in *Knowledge and Mind: Philosophical Essays*, ed. Carl Ginet and Sydney Shoemaker (Oxford: Oxford Uni-versity Press, 1983), 84–104; Donnellan, "Speaking of Nothing," in *Naming, Necessity, and Natural Kinds*, ed. Schwartz, 216–44.

127. Schwartz, introduction to *Naming, Necessity, and Natural Kinds*, 39.

128. Robert Nola, "Fixing the Reference of Theoretical Terms," *Philosophy of Science* 47 (1980): 505–31; Berent Enç, "Reference of Theoretical Terms," *Nous* 10 (1976): 261–82; John Dupré, "Natural Kinds and Biological Taxa," *Philosophical Review* 90 (1981): 66–125; Kim Sterelny, "Natural Kind Terms," *Pacific Philosophical Quarterly* 64 (1983): 110–25. For more recent discussions, see Frederick Kroon, "Theoretical Terms and the Causal View of Reference," *Australasian Journal of Philosophy* 63 (1985): 143–66; Harry Deutsch,

"Semantics for Natural Kind Terms," *Canadian Journal of Philosophy* 23 (1993): 389–412; Danielle Macbeth, "Names, Natural Kind Terms, and Rigid Designation," *Philosophical Studies* 79 (1995): 259–81; Jessica Brown, "Natural Kind Terms and Recognitional Capacities," *Mind* 107 (1998): 275–303.

129. Enç, "Reference of Theoretical Terms," 261–82.

130. Jarrett Leplin, "Meaning Variance and the Comparability of Theories," *British Journal for the Philosophy of Science* 20 (1969): 69–80; see also Leplin, "Reference and Scientific Realism," *Studies in History and Philosophy of Science* 10 (1979): 265–84.

131. Leplin, "Reference and Scientific Realism," 271.

132. Nola, "Fixing the Reference," 508.

133. Ibid., 512.

134. Dudley Shapere, "Evolution and Continuity in Scientific Change," *Philosophy of Science* 56 (1989): 419–37.

135. Another fundamental criticism takes up the question of the idea of "causality" in the theory, for it suggests that the physicalism required for causality runs up against fundamental epistemological and ontological dilemmas having to do with the relation between physiology of perception and linguistic consciousness. See, e.g., Samuel C. Wheeler, "Indeterminacy of Radical Interpretation and the Causal Theory of Reference," in *Meaning and Translation: Philosophical and Linguistic Approaches*, ed. Franz Guenther and M. Guenther-Reutter (New York: New York University Press, 1978), 83–94.

136. Arthur Fine, "How to Compare Theories: Reference and Change," *Nous* 9 (1975): 17–32, citing 17.

137. Ibid.

138. Ibid., 28.

139. Ibid., 30.

140. Philip Kitcher, "Theories, Theorists, and Theoretical Change," *Philosophical Review* 87 (1978): 519–47, citing 520.

141. Ibid.

142. Compare his framing remarks with the arguments developed in *Philosophy in History*, ed. Richard Rorty, J. B. Schneewind, and Quentin Skinner (Cambridge: Cambridge University Press, 1984).

143. Kitcher, "Theories, Theorists, and Theoretical Change," 519.

144. Ibid., 522.

145. Ibid., 523.

146. Ibid., 527.

147. Ibid., 528–29.

148. Ibid., 534.

149. Ibid., citing Richard Grandy, "Reference, Meaning, and Belief," *Journal of Philosophy* 70 (1973): 439–72.

150. Ibid., 535.

151. Ibid., 547.

152. Hilary Putnam, "Anarchism Is Self-Refuting," in *Reason, Truth and History*, 113–19; Donald Davidson, "On the Very Idea of a Conceptual Scheme" (1974), in *Inquiries into Truth and Interpretation* (Oxford: Clarendon, 1984), 183–98.

153. Davidson, "On the Very Idea of a Conceptual Scheme," 184.

154. Ibid., 192.

155. Ibid., 191–92.

156. Ibid., 190.

157. Ibid., 189.

158. See Wilfred Sellars, "Scientific Realism and Irenic Instrumentalism," *Proceedings*

of the Boston Colloquium for the Philosophy of Science, 1962–1964, ed. Robert S. Cohen and Marx W. Wartofsky, Boston Studies in the Philosophy of Science 2 (New York: Humanities, 1965), 171–204.

159. On this central problem of the indeterminacy of "theory" in modern philosophy of language, see Dudley Shapere, "Notes toward a Post-positivistic Interpretation of Science, Part I," in *Reason and the Search for Knowledge,* Boston Studies in the Philosophy of Science 78 (Dordrecht: Reidel, 1984), 112. This is hammered at from many vantages in Jerry Fodor and Ernest LePore, *Holism: A Shopper's Guide* (Oxford: Blackwell, 1990).

160. Karl Popper, "Myth of the Framework," 35–62.

161. For a rich discussion of this, see Jeffrey Malpas, "The Intertranslatability of Natural Languages," *Synthese* 78 (1989): 233–64.

162. Davidson, "On the Very Idea of a Conceptual Scheme."

163. Ibid., 194.

164. Donald Davidson, "In Defense of Convention T," in *Inquiries into Truth and Interpretation,* 65–75.

165. "If we cannot find a way to interpret the utterances and other behavior of a creature as revealing a set of beliefs largely consistent and true by our own standards, we have no reason to count that creature as rational, as having beliefs, or as saying anything" (Donald Davidson, "Radical Interpretation," in *Inquiries into Truth and Interpretation,* 125–40, citing 127).

166. Ibid., 125–40.

167. Davidson, "On the Very Idea of a Conceptual Scheme," 195, 197.

168. Ibid., 197.

169. Ibid., 196–97.

170. David Larson, "Correspondence and the Third Dogma," *Dialectica* 41 (1987): 231–37; Hilary Putnam, "Truth and Convention: On Davidson's Refutation of Conceptual Relativism," *Dialectica* 41 (1987): 69–77; Marie McGinn, "The Third Dogma of Empiricism," *Proceedings of the Aristotelian Society,* n.s., 82 (1981/82): 89–101; William Child, "On the Dualism of Scheme and Content," *Proceedings of the Aristotelian Society,* n.s., 93 (1993): 53–71.

171. Davidson, "On the Very Idea of a Conceptual Scheme," 198.

172. Nicholas Rescher, "Conceptual Schemes," *Midwest Studies in Philosophy* 5 (1980): 323–45, citing 324.

173. Ibid.

174. Ibid., 327.

175. Ibid.

176. Roberto Salinas, "Realism and Conceptual Schemes," *Southern Journal of Philosophy* 27 (1989): 101–23, citing 121 n. 22.

177. Rescher, "Conceptual Schemes," 336.

178. Ibid., 337.

179. Ibid., 331.

180. Ibid., 333.

181. Ibid., 332.

182. Ibid., 333.

183. Robert Kraut, "The Third Dogma," in *Truth and Interpretation: Perspectives on the Philosophy of Donald Davidson,* ed. Ernest LePore (Oxford: Blackwell, 1986), 398–416, citing 407.

184. Ibid., 409.

185. Ibid., 413.

186. Ibid., 415.

187. McGinn, "The Third Dogma of Empiricism," 93.

188. Rorty, "Pragmatism, Davidson and Truth," in *Truth and Interpretation*, ed. Ernest LePore (Oxford: Blackwell, 1986), 333–55.

189. David Larson, "Correspondence and the Third Dogma": "Nothing can count as a reason for holding a belief except another belief" (231). "Davidson speaks of the absurdity of trying to confront our beliefs with the world or experience. I take it that this is an epistemological point. . . . There is no possibility of achieving a vantage point of independence from all our beliefs (or, equivalently, from our language) from which our beliefs can be compared to something else. We can only compare our beliefs with other beliefs" (234).

190. Hence the vexed question whether Davidson is a "realist."

191. For one view, see Richard Rorty, "Pragmatism, Davidson, and Truth," in *Truth and Interpretation*, ed. LePore, 333–55. For another, see Frederick Stoutland, "Realism and Antirealism in Davidson's Philosophy of Language, I," *Critica* 14.41 (1982): 13–51; Stoutland, "Realism and Antirealism in Davidson's Philosophy of Language, II," *Critica* 14.42 (1982): 19–47. For a balanced view, see Jeffrey Malpas, *Donald Davidson and the Mirror of Meaning: Holism, Truth, Interpretation* (Cambridge: Cambridge University Press, 1992).

192. Thomas S. Kuhn, "The Road since *Structure*," *PSA 1990*, vol. 2, 3–13, citing 3.

193. Thomas S. Kuhn, "Dubbing and Redubbing: The Vulnerability of Rigid Designation," in *Scientific Theories*, ed. C. Wade Savage (Minneapolis: University of Minnesota Press, 1990), 298–318, citing 315 n. 4.

194. Thomas S. Kuhn, "Commensurability, Comparability, Communicability," *PSA 1982*, vol. 2, 669–88, citing 684 n. 3.

195. Thomas S. Kuhn, "Response to Commentaries," *PSA 1982*, vol. 2, 712–16, citing 715.

196. Kuhn, "Commensurability, Comparability, Communicability," 670–71.

197. Ibid., 670.

198. Ibid., 672.

199. Thomas S. Kuhn, "Theory-Change As Structure-Change: Comments on the Sneed Formalism," *Erkenntnis* 10 (1976): 191.

200. Ibid.

201. Kuhn, "Commensurability, Comparability, Communicability," 676.

202. See Dorit Bar-On, "Indeterminacy of Translation—Theory and Practice," *Philosophy and Phenomenological Research* 53 (1993): 781–810.

203. Kuhn, "Commensurability, Comparability, Communicability," 682–83.

204. Philip Kitcher, "Implications of Incommensurability," *PSA 1982*, vol. 2, 689–703, citing 699 n. 7.

205. Ibid., 691.

206. Mary Hesse, "Comment on Kuhn's 'Commensurability, Comparability, Communicability,'" *PSA 1982*, vol. 2, 704–11, citing 706.

207. Ibid., 705–6.

208. Ibid., 708.

209. Kuhn, "Dubbing and Redubbing," 299.

210. Ibid., 300.

211. Kuhn, "The Road since *Structure*," 4–5.

212. Kuhn, "Possible Worlds in the History of Science," in *Possible Worlds in Humanities, Arts, and Sciences*, 9–32, citing 15.

213. Ibid., 15 n.

214. Ibid., 11.

215. Kuhn, "Dubbing and Redubbing," 303. But Kuhn offers an important elaboration: "A difference in the language learning route, one which had had no effect while the world be-

haved as anticipated, would lead to differences of opinion when anomalies were found" ("Possible Worlds in the History of Science," 21). This could be crucial in historical reconstruction of theory change.

216. Kuhn, "Possible Worlds in the History of Science," 22.

217. Ibid., 9–10.

218. Ibid., 24.

219. Ibid.

220. Ibid., 23 n.

221. Dale Moberg, "Are There Rival, Incommensurable Theories?" *Philosophy of Science* 46 (1979): 244–62, citing 261.

222. Ibid., 249.

223. Nancy Nersessian, "Scientific Discovery and Commensurability of Meaning," in *Imre Lakatos and Theories of Scientific Change*, ed. Kostas Gavroglu, Yorgos Goudaroulis, Pantelis Nicolacopoulos (Dordrecht: Kluwer, 1989), 324–34.

224. Steve Fuller, *Thomas Kuhn: A Philosophical History for Our Times* (Chicago: University of Chicago Press, 2000), 32.

Chapter Four

1. Thomas S. Kuhn, "The Relation between the History and the Philosophy of Science," in *The Essential Tension* (Chicago: University of Chicago Press, 1977), 3–20, citing 12.

2. The sense in which knowledge itself evolves; see *Evolutionary Epistemology, Rationality, and the Sociology of Knowledge,* ed. Gerard Radnitzky and W. W. Bartley (La Salle, Ill.: Open Court, 1987). Another, not incompatible, sense is that human reason is an evolutionary emergence, accountable as part of the (neo-Darwinian) survival mechanism. See Michael Ruse, *Taking Darwin Seriously* (New York: Prometheus, 1998).

3. Gerald Doppelt, "Kuhn's Epistemological Relativism: An Interpretation and Defense," *Inquiry* 21 (1978): 33–86, citing 79. Again, Richard Grandy can be cited as offering some support to Doppelt's view: "To attack the myth of observation sentences is to attack a certain conception of the essence of the objectivity of science, but it is not to attack its objectivity. It is only if we carefully scrutinize the significant features of actual science as it develops instead of the crystalline rational reconstruction that emerges at the end that we will find the features characteristic of its objectivity" (Grandy, "Incommensurability: Kinds and Causes," *Philosophica* 32 [1983]: 7–24, citing 21).

4. In the words of Thomas Nickles, "Kuhn, more than any single person, is responsible for philosophers now taking history seriously" (quoted in Werner Callebaut, *Taking the Naturalist Turn, or How Real Philosophy of Science Is Done* [Chicago: University of Chicago Press, 1993], 13).

5. "The work of Thomas Kuhn has had a healthy long-term effect: both by example, because he prompted scholars to confront issues in the history of science which he himself stated in highly philosophical language; and also by reaction, because historians and philosophers have had to join forces in order to criticize his initial formulations effectively" (Stephen Toulmin, "From Form to Function: Philosophy and History of Science in the 1950s and Now," *Daedalus* 106 [1977]: 143–62, citing 154). For an interesting argument that a better historical approach than Kuhn's was needed in philosophy of science, see Satosi Watanabe, "Needed: A Historico-Dynamical View of Theory Change," *Synthese* 32 (1975): 113–34.

6. "Whatever one may think of Thomas Kuhn, he did a marvelous job of making history and philosophy of science (HPS)—two fledgling fields in their own home disciplines—appear to be a disciplinary blend of utmost importance to the intellectual community at large"

(Steve Fuller, "Is History and Philosophy of Science Withering on the Vine?" *Philosophy of the Social Sciences* 21 [1991]: 149–74, citing 149).

7. Thus, Michael Friedman: "Thomas Kuhn's *Structure of Scientific Revolutions* (1962) forever changed our appreciation of the philosophical importance of the history of science" ("Remarks on the History of Science and the History of Philosophy," in *World Changes: Thomas Kuhn and the Nature of Science*, ed. Paul Horwich [Cambridge, Mass.: MIT Press, 1993], 37). And Thomas Nickles, cited in note 4 above.

8. Giere, "Kuhn's Legacy," 497.

9. There is considerable support for this view. See, e.g., Ernan McMullin, "Rationality and Paradigm Change in Science," in *World Changes*, ed. Horwich, 55–78, esp. 70–71.

10. Kuhn wrote a very insightful essay on this score, which deserves attention even today: "The Relations between History and the History of Science" (1971), in *The Essential Tension*, 127–61. He observed, correctly: "Despite the universal lip service paid by historians to the special role of science in the development of Western culture during the past four centuries, the history of science is for most of them still foreign territory" (128). Kuhn offered a shrewd hunch about why: "What historians generally view as historical in the development of individual creative disciplines are those aspects which reflect its immersion in a larger society. What they all too often reject, as not quite history, are those internal features which give the discipline a history in its own right" (152). That is, the issue of history of science for historians turned on the "internal/external" controversy. That will concern us again.

11. Even here there were tensions: "history of science is historical in a rather different sense from most other forms of intellectual history, perhaps as compared with any form of intellectual history. . . . [A]s Kuhn recognizes, . . . philosophers do not write in the manner of historians. They treat their predecessors almost as if they were contemporaries" (John Passmore, "The Relevance of History to the Philosophy of Science," in *Scientific Explanation and Understanding*, ed. Nicholas Rescher [Lanham, Md.: University Press of America, 1983], 84). One of the consequences of the dispute about the "marriage" of the disciplines was a sharp revisionism in the *history* of science, away from Koyré and toward a more social-historical and politically relevant historiography. See Arnold Thackray, "Science: Has Its Present Past a Future?" in *Historical and Philosophical Perspectives of Science*, ed. Roger Stuewer (1970; reprint, New York: Gordon & Breach, 1989), 112–27; followed by an exchange with Larry Laudan, 127–33.

12. On the institutional emergence of HPS and in particular of its flagship journal, *Studies in History and Philosophy of Science*, see Larry Laudan, "Thoughts on HPS: 20 Years Later," *Studies in History and Philosophy of Science* 20 (1989): 9–13, esp. 9–11.

13. Gerd Buchdahl, "History and Philosophy of Science at Cambridge," *History of Science* 1 (1962): 62–66.

14. Passmore, "The Relevance of History to the Philosophy of Science," 83.

15. Norwood Russell Hanson, "The Irrelevance of History of Science to Philosophy of Science," in *What I Do Not Believe, and Other Essays* (Dordrecht: Reidel, 1971), 286 n.

16. "I urge that history and philosophy of science continue as separate disciplines. What is needed is less likely to be produced by marriage than by active discourse" (Kuhn, "Relation between the History and the Philosophy of Science," 20).

17. In terse terms, Carl Kordig summarizes the "Standard Account" as follows: "Discovery is for description alone, for psychology and the history of the sociology of science. Justification, however, is for the philosophy of science and epistemology. Discovery is subjective. Justification is objective. It is also normative" ("Discovery and Justification," *Philosophy of Science* 45 [1978]: 110–17, citing 110). See also Elie Zahar, "Logic of Discovery or Psychology of Invention?" *British Journal for the Philosophy of Science* 34 (1983): 243–61;

Ewa Chmielecka, "The Context of Discovery and the Context of Justification: A Reappraisal," in *Polish Essays in the Philosophy of the Natural Sciences*, ed. Wladyslaw Krajewski, Boston Studies in the Philosophy of Science 68 (Dordrecht: Reidel, 1982), 63–74; Carl Kordig, "Discovery and Justification," *Philosophy of Science* 45 (1978): 110–17; Harvey Siegel, "Justification, Discovery and the Naturalizing of Epistemology," *Philosophy of Science* 47 (1980): 297–321; Paul Hoyningen-Huehne, "Context of Discovery and Context of Justification," *Studies in History and Philosophy of Science* 18 (1987): 501–15.

18. See Thomas Nickles, "Scientific Discovery and the Future of Philosophy of Science," in *Scientific Discovery, Logic, and Rationality,* ed. Thomas Nickles, Boston Studies in the Philosophy of Science 56 (Dordrecht: Reidel, 1980), 1–59, for a very thorough problematization of this notion.

19. "[T]he old epistemology claimed to be prior to, and conceptually more fundamental than, the diverse empirical sciences" (Larry Briskman, "Historicist Relativism and Bootstrap Rationality," *Monist* 60 [1977]: 509–39, citing 509).

20. George Basalla, ed., *The Rise of Modern Science: External or Internal Factors* (Lexington, Mass.: D. C. Heath, 1968); for an excellent discussion, see Steven Shapin, "Discipline and Bounding: The History and Sociology of Science As Seen through the Externalism/Internalism Debate," *History of Science* 30 (1992): 333–69.

21. Bonnie Tamarkin Paller, "Naturalized Philosophy of Science, History of Science, and the Internal/External Debate," *PSA 1986*, vol. 1, 258–68, citing 262.

22. Robert McGlaughlin, "Invention and Induction: Laudan, Simon and the Logic of Discovery," *Philosophy of Science* 49 (1982): 198–211, citing 200.

23. That is why I find Steve Fuller's title *Philosophy of Science and Its Discontents* (Boulder, Colo.: Westview, 1989) so apt. Fuller's account and mine diverge in places, but the overall conclusions tally tightly.

24. Alan Musgrave, "Logical versus Historical Theories of Confirmation," *British Journal for the Philosophy of Science* 25 (1974): 1–23.

25. Ernan McMullin, "Philosophy of Science and Its Rational Reconstructions," in *Progress and the Rationality of Science*, ed. Gerard Radnitzky, Gunnar Andersson, Robert S. Cohen, and Marx W. Wartofsky, Boston Studies in the Philosophy of Science 58 (Dordrecht: Reidel, 1978), 221–52, citing 233.

26. Gerard Radnitzky, "Popperian Philosophy of Science As an Antidote against Relativism," in *Essays in Memory of Imre Lakatos*, ed. Robert S. Cohen, Paul K. Feyerabend, and Marx W. Wartofsky, Boston Studies in the Philosophy of Science 39 (Dordrecht: Reidel, 1976), 505–46, citing 516.

27. Siegel, "Justification, Discovery and the Naturalizing of Epistemology," 307.

28. Ibid., 310.

29. Ibid., 320.

30. "Siegel's proposed alternatives either beg the question (for they presume that we have some clear, non-scientific account of validation and justification, which is just what Siegel needs to prove) or they conflict with Siegel's own avowed rejection of first philosophy" (Paul Roth, "Siegel on Naturalized Epistemology and Natural Science," *Philosophy of Science* 50 [1983]: 482–93, citing 491). "If there is no justification, e.g., via rational reconstruction or via derivation from extra-scientific certainties, then there seems to be no 'science of justification' which the epistemologist practices but the scientist does not" (489). But then, "if epistemology cannot be the 'science of justification,' what is it?" (492). The radicalism of "naturalized epistemology" may be precisely the abandonment of epistemology as traditionally conceived. This is Richard Rorty's conclusion, and not his alone.

31. Harold Brown, "For a Modest Historicism," *Monist* 60 (1977): 540–55, citing 541.

32. Larry Laudan, "The Demise of the Demarcation Problem," in *Physics, Philosophy*

and Psychoanalysis, ed. Robert S. Cohen and Larry Laudan, Boston Studies in the Philosophy of Science 76 (Dordrecht: Reidel, 1983), 111–27, citing 111.

33. Ibid., 122.

34. W. W. Bartley, "Theories of Demarcation between Science and Metaphysics," in *Problems in the Philosophy of Science,* ed. Imre Lakatos and Alan Musgrave (Amsterdam: North Holland Publishing Co., 1968), 40–64; with ensuing, vigorous discussion, 65–119.

35. "Justificationism is unable to solve the severe problems which beset it, it provides no means by which we could actually distinguish pseudoscience from genuine science" (Briskman, "Historicist Relativism and Bootstrap Rationality," 513). "A fairly broad agreement appears to have developed over the past decade among philosophers of science . . . that the logicist monolith has been shattered" (McMullin, "Philosophy of Science and Its Rational Reconstructions," 248).

36. Karl Popper, *The Logic of Scientific Discovery* (1959; reprint, New York: Harper, 1965); Popper, *Objective Knowledge: An Evolutionary Approach,* 2d, rev. ed. (Oxford: Clarendon, 1979).

37. Imre Lakatos, "Popper on Demarcation and Induction," in *The Philosophy of Karl Popper,* ed. Paul Arthur Schilpp (Glencoe, Ill.: Free Press, 1974), 241–70, citing 256.

38. Imre Lakatos, "Falsification and the Methodology of Scientific Research Programmes," in *Criticism and the Growth of Knowledge,* ed. Imre Lakatos and Alan Musgrave (Cambridge: Cambridge University Press, 1970), 103.

39. Ibid., 108.

40. In this context, we should never forget the peculiar and vitriolic sense of "historicism" which Karl Popper introduced in his *Poverty of Historicism* (Boston: Beacon, 1957), a work certainly animating much of the Popperian response to this whole episode. Of course, Popper distinguished "historicism" from "historism," and what is here in dispute properly falls under the latter rubric, but the point holds notwithstanding.

41. Gerald Doppelt, "Relativism and Recent Pragmatic Conceptions of Scientific Rationality," in *Scientific Explanation and Understanding: Essays on Reasoning and Rationality in Science,* ed. Nicholas Rescher (Lanham, Md.: University Press of America, 1983), 107–42, citing 107.

42. Lakatos, "Falsification and the Methodology of Scientific Research Programmes," 178.

43. McMullin, "Philosophy of Science and Its Rational Reconstructions," 245.

44. ". . . there is no such thing as the logical method of having new ideas or a logical reconstruction of this process" (Popper, *Logic of Scientific Discovery,* 31–32). The whole question for post-positivism, on the other hand, is whether there is any possibility of a *rational* reconstruction.

45. Thomas S. Kuhn, "Logic of Discovery or Psychology of Research?" (1965), reprinted in *The Essential Tension,* 266–92, citing 267.

46. Karl Popper, "Normal Science and Its Dangers," in *Criticism and the Growth of Knowledge,* 57–58.

47. Ibid., 58. There are not a few philosophers even now who would claim that philosophy has no need of history, e.g., André Kukla: "I think that the importance of history of science to philosophy of science has been greatly exaggerated" ("Scientific Realism, Scientific Practice, and the Natural Ontological Attitude," *British Journal for the Philosophy of Science* 45 [1994]: 955–75, citing 969).

48. Lakatos, "Falsification and the Methodology of Scientific Research Programmes," 177–81.

49. Ibid., 116.

50. "Basically, Lakatos was trying to reconstruct Popper's theory of the growth of scien-

tific knowledge while accepting the criticisms of Agassi and Feyerabend, and some points of Kuhn's theory" (William Berkson, "Lakatos One and Lakatos Two: An Appreciation," in *Essays in Memory of Imre Lakatos,* ed. Cohen, Feyerabend, and Wartofsky, 39–54, citing 50). Popper clearly recognized the apostasy of his former disciple. "Professor Lakatos has . . . misunderstood my theory of science," the patriarch proclaimed in *The Philosophy of Karl Popper,* ed. Schilpp, 999.

51. McMullin, "Philosophy of Science and Its Rational Reconstructions," 240.

52. Roger Stuever, ed., *Historical and Philosophical Perspectives of Science* (1970; reprint, New York: Gordon & Breach, 1989).

53. I. Bernard Cohen, "History and the Philosopher of Science," in *Structure of Scientific Theories,* ed. Frederic Suppe, 2d, rev. ed. (Urbana: University of Illinois Press, 1977), 308–49; Peter Achinstein, "History and Philosophy of Science: A Reply to Cohen," *Structure of Scientific Theories,* 350–60; and the ensuing discussion, 361–73.

54. See esp. Ernan McMullin, "The History and Philosophy of Science: A Taxonomy," in *Historical and Philosophical Perspectives of Science,* ed. Stuever, 12–67. For an overview of this material that reaches roughly parallel conclusions, see Jacquelyn Kegley, "History and Philosophy of Science: Necessary Partners or Merely Roommates?" in *History and Anti-History in Philosophy,* ed. T. Z. Lavine and Victorino Tejera (Dordrecht: Kluwer, 1989), 237–55.

55. Ronald Giere, "History and Philosophy of Science: Intimate Relationship or Marriage of Convenience?" *British Journal for the Philosophy of Science* 24 (1973): 282–97.

56. Ernan McMullin, "History and Philosophy of Science: A Marriage of Convenience?" *PSA 1974,* ed. Robert S. Cohen, Clifford A. Hooker, Alex Michalos, and James van Evra, Boston Studies in the Philosophy of Science 32 (Dordrecht: Reidel, 1976), 585–601; Richard Burian, "More than a Marriage of Convenience: On the Inextricability of History and Philosophy of Science," *Philosophy of Science* 44 (1977): 1–42. Marx Wartofsky found the entire debate "so patently an unreasonable one that one has to step back to see it in its full absurdity" ("The Relation between Philosophy of Science and History of Science," in *Models: Representation and the Scientific Understanding,* ed. Marx W. Wartofsky and Robert S. Cohen, Boston Studies in the Philosophy of Science 48 [Dordrecht: Reidel, 1979], 119–39, citing 119). For him, history of science and philosophy of science cannot be legitimately separated, and to even think that they might be mutually incompatible, as came to be intimated in the course of the debate, was what made the whole discussion absurd in his view.

57. See Laudan, "Thoughts on HPS: 20 Years Later," 9–13.

58. Nicholas Jardine, "Philosophy of Science and the Art of Historical Interpretation," in *Theory Change, Ancient Axiomatics, and Galileo's Methodology: Proceedings of the 1978 Pisa Conference on the History and Philosophy of Science,* ed. Jaakko Hintikka, David Gruender, and Evandro Agazzi (Dordrecht: Reidel, 1978), 1:341.

59. McMullin, "Taxonomy," 16.

60. There is some question whether Reichenbach should be held accountable for this use of the distinction. See Nickles, "Scientific Discovery and the Future of Philosophy of Science," 10–12.

61. McMullin, "Taxonomy," 24–25.

62. Ibid., 42–43.

63. " . . . the modifications of concepts which lie at the root of scientific change cannot be accounted for along the deductivist lines traditionally favored by philosophers of science" (ibid., 62).

64. Giere, "History and Philosophy of Science," 285.

65. Ibid., 284.

66. Ibid., 286–93. Ironically enough, Kuhn offered some support for Giere's stance.

While Kuhn believed that "history of science can help to bridge the quite specific gap between philosophers of science and science itself," he added: "Actual experience in the practice of a science would probably be a more effective bridge than the study of its history" ("Relation of the History to the Philosophy of Science," 13). Indeed, the professional philosopher of science today is very likely to have pursued intense training in science.

67. Giere, "History and Philosophy of Science," 286.

68. Ibid., 293. Giere cautioned that temporal development should not be identified with history: "To argue that any consideration of temporal development brings in history would commit one to arguing that dynamics is a historical science" (289).

69. McMullin, "Marriage," 586. These categories map exactly with those Kuhn had indicated, but McMullin did not mention him.

70. Alan Musgrave, "Logical versus Historical Theories of Confirmation," *British Journal for the Philosophy of Science* 25 (1974): 1–23.

71. McMullin, "Marriage," 588.

72. Ernan McMullin, "The Fertility of Theory and the Unit of Appraisal in Science," in *Essays in Memory of Imre Lakatos,* ed. Cohen, Feyerabend, and Wartofsky, 395–432, citing 403.

73. Ernan McMullin, "Logicality versus Rationality: A Comment on Toulmin's Theory of Science," in *Philosophical Foundations of Science,* ed. Raymond Seeger and Robert Cohen, Boston Studies in the Philosophy of Science 11 (Dordrecht: Reidel, 1974), 415–30, citing 415.

74. Ibid., 426.

75. Ibid., 425.

76. Passmore, "The Relevance of History to the Philosophy of Science," 97.

77. Ernan McMullin, "The Ambiguity of 'Historicism,'" in *Current Research in Philosophy of Science,* ed. Peter Asquith and Henry Kyburg (East Lansing, Mich.: Philosophy of Science Association, 1979), 55–83, citing 79.

78. McMullin, "Marriage," 594. See Stephen Toulmin, *Human Understanding* (Princeton, N.J.: Princeton University Press, 1972); and McMullin, "Logicality and Rationality," 427.

79. McMullin, "Marriage," 597. "To attribute fertility to a theory is to say not just that it gives rise to new predictions but that it is capable of imaginative modification and extension in the light of new evidence" (McMullin, "Ambiguity of 'Historicism,'" 60).

80. Ibid., 598.

81. Burian, "More than a Marriage," 3, refers to Steven Brush, "Should History of Science Be Rated X?" *Science* 183 (1974), 1164–72.

82. Burian, "More than a Marriage," 5.

83. Ibid.

84. Ibid., 31.

85. Ibid., 30; he is notably indifferent to Kuhn.

86. "Proper evaluation of a theory's rationality is contextual; it requires knowledge of the relevant background of beliefs, problems, theories, experiments and instruments affecting the application of and support of the theory" (ibid., 34).

87. Ibid., 38.

88. McMullin, "Taxonomy," 18 n.

89. Ibid., 54.

90. Giere, "History and Philosophy of Science," 295.

91. Burian, "More than a Marriage," 28.

92. Wilhelm Windelband, *Geschichte und Naturwissenschaft* (Strassburg: Heitz, 1904).

93. This is not to underestimate the technical burden that the term "explanation" carries with it. See Philip Kitcher, "Explanatory Unification," and Michael Friedman, "Explanation

and Scientific Understanding," in *Theories of Explanation,* ed. Joseph Pitt (Oxford: Oxford University Press, 1988), 167–87 and 188–98, respectively.

94. Kuhn, "Relation between the History and the Philosophy of Science," 5.

95. Thomas S. Kuhn, "Objectivity, Value Judgment, and Theory Choice," in *The Essential Tension,* 324. Ernan McMullin cites this passage in "Rationality and Paradigm Choice in Science," in *World Changes,* ed. Horwich, 55–78, citing 57.

96. I. Bernard Cohen, "History and the Philosopher of Science," 312. Historians of science expressed considerable pique over the decade of the 1970s about the cavalier manner in which philosophers of science mangled history. See Brush, "Should History of Science Be Rated X?" 1164–72; and esp. L. Pearce Williams, "Should Philosophers Be Allowed to Write History?" *British Journal for the Philosophy of Science* 26 (1975): 241–53.

97. Cohen, "History and the Philosopher of Science," 345.

98. Ibid., 346.

99. "The critical historian must use some standard by which to judge whether or not he has been successful—successful, that is, at reconstructing the contemporary problem situation accurately. That is, he must decide how he will tell if his efforts to overcome the parochial and unfavorable modern bias have been sufficient. . . . A ready standard in this endeavor is the attainment of clarity and consistency." Thus, "the need for an independent standard transforms the aim of 'seeing a theory in its own time' into 'seeing a theory as clear and coherent in its own time' " (Lynn Lindholm, "Is Realistic History of Science Possible? A Hidden Inadequacy in the New History of Science," in *Scientific Philosophy Today,* ed. Joseph Agassi and Robert S. Cohen, Boston Studies in the Philosophy of Science 67 [Dordrecht: Reidel, 1981], 159–86, citing 181). That is, "historians do not know if they *can* declare a view unclear or incoherent in its own historical context" (167). "Whenever a seemingly unclear or inconsistent view presents itself, the critical historian is led almost irresistibly to the view that the weakness is one in his reconstruction" (182). Thus, "Cohen equates seeing an argument as it appeared in its own time with making such an argument coherent and valid" (165). "The misleading element in this picture is its smoothness, its 'unproblematic' finished quality. . . . [T]he new histories tend to camouflage the less intellectually respectable character of the traditions they disclose" (164).

100. See the apt comments in David Hull, "In Defense of Presentism," *History and Theory* 18 (1979): 1–15. And see also Nicholas Jardine, "Philosophy of Science and the Art of Historical Interpretation."

101. Larry Laudan, *Progress and Its Problems* (Berkeley: University of California Press, 1977), 159.

102. Ibid., 179.

103. Ibid., 180.

104. Ibid., 180–81. That is, he offers a critique of the "unit idea" approach of Arthur Lovejoy's "history of ideas" as an inadequate model for intellectual history.

105. Laudan, *Progress and Its Problems,* 182, 184.

106. Ibid., 165.

107. Barry Barnes, "Essay Review: The Vicissitudes of Belief," *Social Studies of Science* 9 (1979): 247–63, citing 252–53.

108. Lakatos, "History of Science and Its Rational Reconstruction," *In Memory of Rudolf Carnap, PSA 1970,* ed. R. C. Buck and Robert Cohen, Boston Studies in the Philosophy of Science 8 (Dordrecht: Reidel, 1971), 91. Famous as this squib from Kant by Lakatos is, it is not even original in its borrowing. Norwood Russell Hanson used it in just the same sense almost a decade before: "my . . . aphoristic mold: that history of science without philosophy of science is blind . . . that philosophy of science without history of science is empty" ("The Irrelevance of History of Science to Philosophy of Science," 279).

109. Kuhn, "Relation between the History and the Philosophy of Science," 10.

110. Hanson, "The Irrelevance of History of Science to Philosophy of Science," 274–87.

111. "[H]istory of science . . . is always written from the point of view of a particular philosophy of science" (Passmore, "The Relevance of History to the Philosophy of Science," 98).

112. James R. Brown, "History and the Norms of Science," *PSA 1980*, vol. 1, 236–48, citing 237.

113. Tomas Kulka, "Some Problems Concerning Rational Reconstruction: Comments on Elkana and Lakatos," *British Journal for the Philosophy of Science* 28 (1977): 325–44, citing 325. See, in addition, Marc Blaug, "Kuhn versus Lakatos, or Paradigms versus Research Programmes in the History of Economics," in *Paradigms and Revolutions*, ed. Gutting, 137–59.

114. Lakatos wrote: "The *first* world is the material world, the *second* is the world of consciousness, the *third* is the world of propositions, truth, standards: the world of objective knowledge" ("Falsification and the Methodology of Scientific Research Programmes," 180 n).

115. Kulka, "Some Problems Concerning Rational Reconstruction," 335.

116. McMullin, "Fertility of Theory," 401.

117. Lakatos, "Falsification and the Methodology of Scientific Research Programmes," 179 n.

118. "Imre Lakatos has even gone so far as to state that history of science should be written as it *should* have taken place, given a particular philosophy of science, rather than as it actually did take place" (David Hull, "In Defense of Presentism," 1).

119. Ian Hacking: "Imre Lakatos's Philosophy of Science," *British Journal for the Philosophy of Science* 39 (1979): 381–410. This is an excellent résumé of Lakatos's ideas in the form of a review of the posthumous publication of Lakatos's collected works.

120. Ibid., 384–85.

121. "Lakatos wants to find, in the rational dialectic of criticism within scientific theory, that invariance which demarcates good from bad science" (Wartofsky, "The Relation between Philosophy and History of Science," 133).

122. Alan Musgrave, "Method or Madness? Can the Methodology of Research Programmes Be Rescued from Epistemological Anarchism?" in *Essays in Memory of Imre Lakatos*, ed. Cohen, Feyerabend, and Wartofsky, 457–91, citing 473.

123. Hacking, "Imre Lakatos's Philosophy of Science," 387.

124. Ibid., 389.

125. John Jamieson Smart, "Science, History and Methodology," *British Journal for the Philosophy of Science* 23 (1972): 266–74, citing 269.

126. Ibid., 272.

127. On Joseph Agassi's "bootstrap" theory, see his monograph *Towards an Historiography of Science*, History and Theory, Beiheft 2 (Gravenhage: Mouton, 1963).

128. McMullin, "Fertility of Theory," 400.

129. Ibid., 401.

130. Ibid.

131. Ewa Chmielecka, "The Context of Discovery and the Context of Justification: A Reappraisal," 68.

132. "Historians are quite correct in thinking that the radical reconstructions which have been proposed first by Agassi and now by Lakatos, are very different indeed from the modest reconstruction techniques which they are accustomed to using" (Noretta Koertge, "Rational Reconstruction," in *Essays in Memory of Imre Lakatos*, ed. Cohen, Feyerabend, and Wartofsky, 359–69, citing 361).

133. Kulka, "Some Problems Concerning Rational Reconstruction," 338.

134. Lakatos, "History of Science and Its Rational Reconstruction," *PSA 1970*, 107.

135. Thomas S. Kuhn, "Notes on Lakatos," *PSA 1970*, 137–46. Wartofsky put it more pungently: "With Kuhn, we get history without dialectic; with Lakatos we get dialectic without history" ("The Relation between Philosophy and History of Science," 133).

136. Koertge, "Rational Reconstruction," 368.

137. Smart, "Science, History and Methodology," 268.

138. Hacking, "Imre Lakatos's Philosophy of Science," 394.

139. Stephen Toulmin, "History, Praxis and the 'Third World': Ambiguities in Lakatos' Theory of Methodology," in *Essays in Memory of Imre Lakatos,* ed. Cohen, Feyerabend, and Wartofsky, 655–75, citing 655.

140. Ibid.

141. Ibid., 662.

142. For Toulmin's extended criticisms of Kuhn, see esp. "Does the Distinction between Normal and Revolutionary Science Hold Water?" in *Criticism and the Growth of Knowledge,* 39–47.

143. In their different ways, Toulmin and Lakatos both represent efforts to develop a moderate historicism which finds its ultimate realization in the ideas of evolutionary or naturalized epistemology.

144. Toulmin, "History, Praxis and the 'Third World,'" 665.

145. Ibid.

146. Stephen Toulmin, *Human Understanding,* vol. 1, *The Collective Use and Evolution of Concepts* (Princeton, N.J.: Princeton University Press, 1972). That there was no second volume is one indication of its failure.

147. David Bloor, "Rearguard Rationalism [review of *Human Understanding,* by Toulmin]," *Isis* 65 (1974): 249–53, citing 252.

148. Toulmin, *Human Understanding,* viii.

149. Stephen Toulmin, "Rationality and Scientific Discovery," in *PSA 1972,* ed. Kenneth Schaffner and Robert Cohen (Dordrecht: Reidel, 1974), 387–406.

150. Struan Jacobs, "Stephen Toulmin's Theory of Conceptual Evolution," in *Issues in Evolutionary Epistemology,* 510–23, citing 513.

151. Ibid.

152. Ian Charles Jarvie, "Toulmin and the Rationality of Science," in *Essays in Memory of Imre Lakatos,* ed. Cohen, Feyerabend, and Wartofsky, 211–334, citing 314.

153. "Unless Toulmin can give us a convincing account of how rival disciplinary ideals can be rationally compared and evaluated, his theory simply retreats to the relativism which he so desperately seeks to avoid . . . he lands in the same relativism of which he accuses Kuhn" (Larry Briskman, "Toulmin's Evolutionary Epistemology," review of *Human Understanding,* by Toulmin, *Philosophical Quarterly* 24 [1974]: 160–69, citing 166). See Bloor, "Rearguard Rationalism," 251 n; Jacobs, "Stephen Toulmin's Theory of Conceptual Evolution," 522.

154. Larry Laudan, *Progress and Its Problems.* And see Laudan, "Historical Methodologies: An Overview and Manifesto," in *Current Research in Philosophy of Science,* ed. Peter Asquith and Henry Kyburg (East Lansing, Mich.: Philosophy of Science Association, 1979), 40–54.

155. Ibid., 95.

156. Ibid., 76.

157. Ibid., 75.

158. Laudan, "Dissecting the Holistic Picture of Scientific Change," in *Scientific Knowledge: Basic Issues in the Philosophy of Science,* ed. Janet Kourany (Belmont, Calif.: Wads-

worth Publishing, 1987), 279. (Originally published in Laudan, *Science and Values* [Berkeley: University of California Press, 1984], 67–102.)

159. "Is it true that the major historical shifts in the methodological rules of science and in the cognitive values of scientists have invariably been contemporaneous with one another and with shifts in substantive theories and ontologies?" (ibid., 283). His answer is clearly, no.

160. Laudan, *Progress and Its Problems*, 77.

161. Ibid., 124.

162. Ernan McMullin, "Discussion Review: Laudan's *Progress and Its Problems*," *Philosophy of Science* 46 (1979): 623–44, citing 629.

163. Ibid., 642.

164. Laudan, *Progress and Its Problems*, 130.

165. Ibid., 160–61.

166. Ibid., 160.

167. All modern philosophy of history sets out from a far more subtle grasp of the problem here than Laudan even imagines.

168. "But do we really have *pre-analytic* intuitions . . . ? And if we do, should such intuitions stand as our most basic criteria of judgment, as Laudan advocates . . . ?" (Barnes, "Essay Review: The Vicissitudes of Belief," 249).

169. For an extended meditation on these lines, see Alasdair MacIntyre, *Whose Justice? Which Rationality?* (Notre Dame, Ind.: University of Notre Dame Press, 1988).

170. McMullin, "Laudan's *Progress and Its Problems*," 636.

171. Ian Charles Jarvie, "Laudan's Problematic Progress and the Social Sciences," *Philosophy of the Social Sciences* 9 (1979): 484–97, citing 485.

172. Ibid., 488 and passim.

173. Ibid., 491.

174. Larry Laudan et al., "Scientific Change: Philosophical Models and Historical Research," *Synthese* 69 (1986), 141–223; and Arthur Donovan, Larry Laudan, and Rachel Laudan, eds. *Scrutinizing Science: Empirical Studies of Scientific Change* (1988; reprint, Baltimore: Johns Hopkins University Press, 1992).

175. Fuller, "Is History and Philosophy of Science Withering on the Vine?" 152–53.

176. Gerald Doppelt, "Review Discussion: Laudan's Pragmatic Alternative to Positivist and Historicist Theories of Science," *Inquiry* 24 (1981): 253–71, citing 263.

177. Ibid., 259.

178. Ibid., 258.

179. Ibid., 256.

180. Ibid., 265.

181. Ibid., 269.

182. Ibid., 270.

183. James Brown, "History and the Norms of Science," *PSA 1980*, vol. 1, 236–48, citing 239.

184. Ibid., 242.

185. Anthony Murphy and R. E. Hendrick, "Lakatos, Laudan and the Hermeneutic Circle," *Studies in History and Philosophy of Science* 15 (1984): 119–30, esp. 128.

186. See W. W. Bartley, "Theories of Rationality," in *Evolutionary Epistemology, Rationality, and the Sociology of Knowledge*, ed. Radnitzky and Bartley, 205–16.

187. Murphy and Hendrick, "Lakatos, Laudan and the Hermeneutic Circle," 128.

188. Fuller, "Is History and Philosophy of Science Withering on the Vine?" 149.

189. Fuller, *Philosophy of Science and Its Discontents*, 3.

190. Ibid., 2.

191. Laudan, "Thoughts on HPS: 20 Years Later," 9–13, citing 12. Laudan has contin-

ued this lament in "The History of Science and the Philosophy of Science," in *Companion to the History of Modern Science*, ed. Robert C. Olby, G. N. Cantor, J. R. R. Christie, and M. J. S. Hodge (New York: Routledge, 1990), 47–59.

192. At the 1992 conference of the Philosophy of Science Association, Michael Ruse introduced the session entitled "What Has the History of Science to Say to the Philosophy of Science?" with explicit reference to the cited passage from Larry Laudan (*PSA 1992*, vol. 2, 467).

193. Robert Richards, "Arguments in a Sartorial Mode, or the Asymmetries of History and Philosophy of Science," *PSA 1992*, vol. 2, 482–89, citing 487.

194. Ibid.

195. Rachel Laudan, "The 'New' History of Science: Implications for Philosophy of Science," *PSA 1992*, vol. 2, 476–81, citing 478–79. "To sum up this growing consensus: ideas, theories, and beliefs are out as objects of study; practices, particularly laboratory practices, are in. The role of tools, instruments and technologies has become central, as has the analysis of experiments. Knowledge is made, constituted, or constructed; discoveries, justifications, refutation, and verifications have vanished from the scene. Scientists' writing is rhetoric, not explication or representation or explanation. And the beliefs that win out and become accepted by the scientific community do so because their proponents have more effectively marshalled political power than their opponents" (479). These matters will preoccupy us for the balance of this study.

196. Marga Vicedo, "Is the History of Science Relevant to the Philosophy of Science?" *PSA 1992*, vol. 2., 490–96, citing 492.

197. Laudan, "Thoughts on HPS," 11. For the VPI project, see Laudan et al., "Scientific Change: Philosophical Models and Historical Research," *Synthese* 69 (1986): 141–223; and Arthur Donovan, Larry Laudan, and Rachel Laudan, eds., *Scrutinizing Science*. For criticism, see esp. Thomas Nickles, "Remarks on the Use of History As Evidence," *Synthese* 69 (1986): 253–66; Alan Richardson, "Philosophy of Science and Its Rational Reconstructions: Remarks on the VPI Program for Testing Philosophies of Science," *PSA 1992*, vol. 1, 36–46; Colin Howson, "The Poverty of Historicism," *Studies in History and Philosophy of Science* 21 (1990): 173–79.

198. Fuller, "Is History and Philosophy of Science Withering on the Vine?" 152.

199. Nickles, "Scientific Discovery and the Future of Philosophy of Science," 17.

200. Ibid., 16.

201. I submit that this lesson holds tremendous relevance for historians.

202. Philosophy of science has increasingly opted to work within the local domain of a scientific specialty. Indeed, Steve Fuller proclaimed in 1993, "the future of the philosophy of science lies either in some other branch of science studies (especially history and sociology) or in the conceptual foundations of the special sciences" (*Philosophy of Science and Its Discontents*, xii).

203. Mary Hesse, *Revolutions and Reconstructions in the Philosophy of Science* (Bloomington: Indiana University Press, 1980), 7.

204. Donald MacKenzie and Barry Barnes, "Scientific Judgment: The Biometry-Mendelism Controversy," in *Natural Order: Historical Studies in Scientific Culture*, ed. Barry Barnes and Steven Shapin (Beverly Hills, Calif.: Sage, 1979), 191–210, citing 191–92.

205. Nickles, "Scientific Discovery and the Future of Philosophy of Science," 30.

206. Ibid., 31.

207. Ibid., 7.

208. Ibid., 16.

209. Ibid., 1. Ronald Giere, "Kuhn's Legacy for North American Philosophy of Science," *Social Studies of Science* 27 (1997): 496–98.

210. Dudley Shapere, "The Character of Scientific Change" and "What Can the Theory

of Knowledge Learn from the History of Knowledge?" in *Reason and the Search for Knowledge*, Boston Studies in the Philosophy of Science 78 (Dordrecht: Reidel, 1984), 205-60 and 182-202, respectively.

211. Nickles, "Scientific Discovery and the Future of Philosophy of Science," 45.

212. Shapere, "What Can the Theory of Knowledge Learn," 193.

213. Nickles, "Scientific Discovery and the Future of Philosophy of Science," 7.

214. Thomas Nickles, "Good Science As Bad History," in *The Social Dimensions of Science*, ed. Ernan McMullin (Notre Dame, Ind.: University of Notre Dame Press, 1992), 91.

215. In bringing these descriptors together, I am not unaware of the shades of difference among them, but I want to stress their commonality.

216. Axtell captures this well: "it is not mistaken to see skepticism and foundationalism as flip-side images of this received view in empiricist philosophy" ("Normative Epistemology and the Bootstrap Theory," *Philosophical Forum* 23 [1992]: 329-43, citing 334). Elsewhere he elaborates: "The logicist's overstatement of the objectivity of metascientific discourse and the subjectivity of normative ethical discourse has exacerbated a dualistic approach to meta-level discourse that is antithetical to pragmatism" for "pragmatism has always advocated opening epistemology onto a plane of psychological, social and historical conditioning that cuts *across* these neat divides" ("Logicism, Pragmatism, and Metascience: Towards a Pan-critical Pragmatic Theory of Meta-level Discourse," *PSA 1990*, vol. 1, 39-49, citing 46).

217. "[D]oubt is not criticism. Moreover, the logical possibility of an alternative is not itself an alternative. . . . But while doubt is cheap, criticism is not. The *non*justificationist, having renounced proof in favour of improvement, acknowledges that alternatives and doubt are always possible but insists that generalized doubt and the possibility of alternatives be translated into concrete criticism and actual alternatives before they need be taken seriously" (Briskman, "Historicist Relativism and Bootstrap Rationality," 521).

218. Nickles, "Good Science As Bad History," 88.

219. Ibid., 116, 89. Axtell refers to "Dewey's emphasis on the generation of criteria for problem-solving through the 'method of intelligences' and the study of the mutual conditioning of means and ends" ("Normative Epistemology and the Bootstrap Theory," 336-37). "Dewey's pragmatic-normativism eschews both justificationism and relativism of value judgments, including axiological judgments" ("Logicism, Pragmatism, and Metascience," 46).

220. Nickles, "Good Science As Bad History," 116.

221. Ibid., 117.

222. "The pragmatic conception of belief and attitude is *dialectical*" (Axtell, "Logicism, Pragmatism, and Metascience," 47). "What we need . . . is a dialectical history of science. But having said this, it should be clear that the norms of such a science must be derived from the historical context itself and cannot be imposed upon it by philosophical fiat, or by some supra-historical *a priori* conception of rationality. . . . It requires, I think, a characterization of the *historical enterprise* called science, in a new way, in order to see in it the sources of its *intrinsic* normativeness" (Wartofsky, "The Relation between Philosophy and History of Science," 134).

223. Nickles, "Good Science As Bad History," 117.

224. Thomas Nickles, "Integrating the Science Studies Disciplines," in *The Cognitive Turn*, ed. Steve Fuller (Dordrecht: Kluwer, 1989), 225-56, citing 248.

225. Nickles, "Good Science As Bad History," 117.

226. Shapere, "What Can the Theory of Knowledge Learn," 185.

227. Ibid., 200.

228. Ibid.

229. Hilary Putnam, "The Craving for Objectivity," *New Literary History* 15 (1984): 229-39, citing 239.

230. Briskman, "Historicist Relativism and Bootstrap Rationality," 521.

231. Philip Kitcher, "The Naturalists Return," *Philosophical Review* 101 (1992): 53–114, citing 63. Paller observes, similarly, "The failure to provide adequate rules . . . has encouraged many philosophers to conclude that there are no unchanging, essential, and transcontextual rules which are discoverable a priori" ("Naturalized Philosophy of Science," 258).

232. Kitcher, "The Naturalists Return," 76.

233. Ibid., 76 n.

234. Ibid., 72 n.

235. Ibid., 72, 75.

236. "The 'naturalized turn' commits one to the rejection of the claim that the proper philosophical method is a priori conceptual analysis. . . . Then, since philosophers do not have privileged access to normative truths, any answer to the normative question is going to have to come through the only other available methodology, this sort of a posteriori methodology, specifically beginning with an historical analysis of particular cases" (Paller, "Naturalized Philosophy of Science," 259).

237. Kitcher, "The Naturalists Return," 113.

238. "What is given up is an unworkable supra-historical a priori conception of rationality and what is sought is, in Wartofsky's words, 'the sources of its *intrinsic* normativeness' " (Kegley, "History and Philosophy of Science," 253; citing Wartofsky, "The Relation between Philosophy of Science and History of Science," 134).

239. Passmore, "The Relevance of History to the Philosophy of Science," 92.

240. This is the conclusion that Richard Rorty has drawn from Quine, Kuhn, et al., and it signifies for him, as Kitcher phrases it, "the death of philosophy . . . [whereupon] succession passes variously to history, sociology or literary theory" ("The Naturalists Return," 113).

241. Kitcher, "The Naturalists Return," 56.

242. Ibid., 93.

243. Ibid., 100. Such radical historicism, as Guy Axtell aptly puts it, opts "*not to seek a 'way out'* at all, but rather to accept and cope with whatever perceived consequences there are to the skeptic's victory. . . . Richard Rorty and his followers manifest such a response" ("Normative Epistemology and the Bootstrap Theory," 330).

244. Kitcher, "The Naturalists Return," 100. Similarly, Harold Brown argues "there is no reason why we must equate an historicist theory of knowledge with . . . extreme historicism" ("For a Modest Historicism," 541).

245. Kitcher, "The Naturalists Return," 95.

246. On the emergence of this field, see Hilary Kornblith, ed., *Naturalizing Epistemology* (Cambridge, Mass.: MIT Press, 1985); Gerard Radnitzky and W. W. Bartley, eds., *Evolutionary Epistemology, Rationality, and the Sociology of Knowledge*; Kai Hahlweg and Clifford A. Hooker, eds., *Issues in Evolutionary Epistemology* (Albany: SUNY Press, 1989); Matthew Nitecki and Doris Nitecki, eds., *History and Evolution* (Albany: SUNY Press, 1992).

247. Philip Kitcher, "The Naturalists Return," 91 n.

248. Burian, "More than a Marriage," 40.

249. In particular I am thinking of philosophy of biology, exemplified by the work of David Hull and Michael Ruse. See Hull, *Science As a Process* (Chicago: University of Chicago Press, 1988); Ruse, *Taking Darwin Seriously*.

250. Ronald Giere, "Philosophy of Science Naturalized," *Philosophy of Science* 52 (1985): 331–56.

251. In the terms of Karl-Otto Apel: "Any attempt at ultimate grounding leads to a trilemma: either there must come about an infinite regress, or a logical circle, or else the

grounding-procedure has to be broken off by axiomatizing or, better, dogmatizing certain premises" (Apel, "Types of Rationality Today: The Continuum of Reason between Science and Ethics," in *La Rationalité Aujourd'hui* [Toronto: McGill University Press, 1980], 310).

252. Axtell, "Normative Epistemology and the Bootstrap Theory" (334), refers to Feigl, Reichenbach, and Ayer as employers of the "vicious circle" and "infinite regress" arguments and makes the important point that such arguments "helped them to insulate theoretical science, including its norms, from the metaphysical, the noncognitive, and the merely pragmatic. It allowed them to define scientific disagreement as disagreement resolvable by logic and evidence alone" (334).

253. Giere, "Philosophy of Science Naturalized," 333.

254. Harold Brown, "Normative and Naturalized Epistemology," *Inquiry* 31 (1988): 53–78, citing 55.

255. Quine, in the pioneer statement of this whole way of thinking, got to the heart of what giving up "first philosophy" meant: "scruples against circularity have little point once we have stopped dreaming of deducing science from observations" ("Epistemology Naturalized," in *Ontological Relativity and Other Essays* [New York: Columbia University Press, 1969], 75–76).

256. Giere, "Philosophy of Science Naturalized," 333–34.

257. Harold Brown, "Normative and Naturalized Epistemology," 61.

258. Ibid., 69.

259. Ibid., 61.

260. Giere, "Philosophy of Science Naturalized," 334. The claim here is that it proves impossible "to take a middle road between the old logical empiricism and the new historical relativism" (J. Kegley, "History and Philosophy of Science," 244). I reject that position unequivocally.

261. On this important notion of "de-transcendentalizing" and Rorty, see Nickles, "Scientific Discovery and the Future of Philosophy of Science," 48.

262. See Heinrich Scholz, ed., *Die Hauptschriften zum Pantheismus Streit zwischen Jacobi und Mendelssohn* (Berlin: Reuter und Reichard, 1916).

263. G. H. Merrill, "Moderate Historicism and the Empiricist Sense of 'Good Science,'" *PSA 1980*, vol. 1, 223–35.

264. Ibid., 225. "In the end it seems clear that philosophy is to be a handmaiden to history and that the only point of distinction between such a position of moderate historicism and radical historicism lies in the vague claim of the former that philosophy is to serve *some* (unspecified) evaluative role" (233).

265. Ibid., 224–25.

266. Ibid., 229–30.

267. On this question of skepticism, Kitcher puts the naturalist position well: "On naturalism's own grounds, there are bound to be unanswerable forms of skepticism. . . . [A] central naturalist thesis is that some parts of our current scientific beliefs must be assumed in criticizing or endorsing others" ("The Naturalists Return," 90–91).

268. G. H. Merrill, "Moderate Historicism," 227.

269. Here, Toulmin's idea of a "'common law' model" which works "by the collection and restatement of precedents" seems to offer a promising alternative to the "statutory law" of "general, timeless recipes" which grounded the orientation of the earlier philosophy of science and still monopolizes the very idea of law in their estimation. (Toulmin, "From Form to Function," 154). See James R. Brown, "Explaining the Success of Science," *Ratio* 27 (1985): 49–66.

270. Nicholas Rescher, "Extraterrestrial Science," *Philosophia Naturalis* 21 (1984), 400–424, citing 413.

271. Giere, "Philosophy of Science Naturalized," 342–43. See Jürgen Habermas, "Gegen einen positivistisch halbierten Rationalismus," *Kölner Zeitschriften für Soziologie und Sozialpsychologie* 16 (1964): 635–59.

272. Giere's dismissal of "emergentism" ("Philosophy of Science Naturalized," 343) is characteristically doctrinaire. While it would be too sanguine to take emergentism as established, we should be wary these days of considering it absurd. Evolutionary epistemology opens new prospects here that must not be foreclosed dogmatically.

273. Giere, "Philosophy of Science Naturalized," 354.

274. Ibid.

275. Adam Grobler, "Between Rationalism and Relativism: On Larry Laudan's Model of Scientific Rationality," *British Journal for the Philosophy of Science* 41 (1990): 493–507, citing 496.

276. Harold Brown, "Normative and Naturalized Epistemology," 65.

277. See, e.g., Alexander Rosenberg, "Normative Naturalism and the Role of Philosophy," *Philosophy of Science* 57 (1990): 34–43; Jarrett Leplin, "Renormalizing Epistemology," *Philosophy of Science* 57 (1990): 20–33; Harvey Siegel, "Laudan's Normative Naturalism," *Studies in History and Philosophy of Science* 21 (1990): 295–313.

278. Larry Laudan, "Progress or Rationality? The Prospects for Normative Naturalism," *American Philosophical Quarterly* 24 (1987): 19–31; Laudan, "Relativism, Naturalism and Reticulation," *Synthese* 71 (1987): 221–34; Laudan, "Normative Naturalism," *Philosophy of Science* 57 (1990): 44–59; Doppelt, "Laudan's Pragmatic Alternative to Positivist and Historicist Theories of Science"; Doppelt, "Relativism and Recent Pragmatic Conceptions of Scientific Rationality," in *Scientific Explanation and Understanding: Essays on Reasoning and Rationality in Science,* ed. Nicholas Rescher (Lanham, Md.: University Press of America, 1983), 107–42; Doppelt, "Relativism and the Reticulation Model of Scientific Rationality," *Synthese* 69 (1981): 225–52; Doppelt, "The Naturalist Conception of Methodological Standards in Science: A Critique," *Philosophy of Science* 57 (1990): 1–19.

279. Passmore, e.g., acknowledges that the hermeneutic circle is not necessarily a vicious one ("The Relevance of History to the Philosophy of Science," 98).

280. James Brown, "History and the Norms of Science," 243.

281. Richard Palmer, *Hermeneutics* (Evanston, Ill.: Northwestern University Press, 1969); David Hoy, *The Critical Circle: Literature, History, and Philosophical Hermeneutics* (Berkeley: University of California Press, 1978); Paul Ricoeur, *Hermeneutics and the Human Sciences* (Cambridge: Cambridge University Press, 1981).

282. Nickles, "Remarks on the Use of History As Evidence."

283. David Hull, "Testing Philosophical Claims about Science," *PSA 1992,* vol. 2, 468–75, citing 473.

284. Ibid., 471. He puts it pithily in another context: "reality has a way of forcing itself on us independent of our beliefs" ("In Defense of Presentism," 12). Harvey Brown makes the same point: "it is not the case that there are no checks on the range of possible interpretations . . . the range of possible ways of understanding it is limited by the properties of that object" ("For a Modest Historicism," 552).

285. Nickles, "Scientific Discovery and the Future of Philosophy of Science," 35.

286. Ibid., 37.

287. Ibid.

288. Ibid., 38.

289. Ibid., 22.

290. Ibid., 18.

291. Ibid., 39.

292. This is Kuhn's point about the exemplarity of paradigms, and reasoning by analogy.

293. Nickles, "Scientific Discovery and the Future of Philosophy of Science," 32.

294. Nickles, "Good Science As Bad History," 107.

295. Ibid.

296. Ibid., 113.

297. It is on the basis of "that logical error known as the Naturalist Fallacy," he writes, "that the 'historical' school appear to have erected their strange philosophy. The unfashionable, outcast, misnamed 'positivists,' knew better, about this as about much else besides" (Howson, "The Poverty of Historicism," 179).

298. Briskman, "Historicist Relativism and Bootstrap Rationality," 510.

Chapter Five

1. "At the philosophical center of the strong programme is Kuhn's *The Structure of Scientific Revolutions*" (Paul Roth, "Voodoo Epistemology: The Strong Programme in the Sociology of Science," in *Meaning and Method in the Social Sciences* [Ithaca, N.Y.: Cornell University Press, 1987], 174). "The central dogma in the Kuhnian mythology is that Kuhn's paradigm is a significant, indeed a *radical*, alternative to Merton's" (Sal Restivo, "The Myth of the Kuhnian Revolution," *Sociological Theory 1983*, ed. Randall Collins [San Francisco: Jossey-Bass, 1983], 293–305, citing 293).

2. Thomas S. Kuhn, preface to *Essential Tension* (Chicago: University of Chicago Press, 1977), xxi; Kuhn, *The Trouble with the Historical Philosophy of Science* (Robert and Maurine Rothschild Distinguished Lecture, 19 November 1991, Department of History of Science, Harvard University, 1992).

3. Wes W. Sharrock and R. J. Anderson, "The Wittgenstein Connection," *Human Studies* 7 (1984): 375–86, citing 375 (see too the epigraph to this chapter). Ernan McMullin, as well, recognizes the explicitly disciplinary character of the controversy: "Sociologists of science now lay claim to the whole areas which they take philosophers and 'internal' historians of science to have usurped, while philosophers denounce as a blatant take-over attempt the intrusion of sociologists into what had been for long an exclusively philosophic domain" ("The Rational and the Social in the History of Science," in *Scientific Rationality: The Sociological Turn*, ed. James Robert Brown [Dordrecht: Reidel, 1984], 127–63, citing 127–28).

4. Jürgen Habermas, "Science and Technology As Ideology," reprinted in *The Sociology of Science*, ed. Barry Barnes (Harmondsworth, U.K.: Penguin, 1972).

5. Thomas S. Kuhn, "The Relation between History and the History of Science" (1971), in *The Essential Tension*, 127–61, citing 160–61.

6. Alvin W. Gouldner, *The Coming Crisis of Western Sociology* (New York: Basic Books, 1970).

7. Karl Mannheim, *Ideology and Utopia* (New York: Harcourt, Brace & World, 1955); see Fritz Ringer, "The Origins of Mannheim's Sociology of Knowledge," in *The Social Dimensions of Science*, ed. Ernan McMullin (Notre Dame, Ind.: University of Notre Dame Press, 1992), 47–67.

8. David Frisby, *The Alienated Mind: Sociology of Knowledge in Germany, 1918–1933* (1983; reprint, New York: Routledge, 1992); Volker Meja and Nico Stehr, eds., *Knowledge and Politics: The Sociology of Knowledge Dispute* (New York: Routledge, 1990); Brian Longhurst, *Karl Mannheim and the Contemporary Sociology of Knowledge* (London: Macmillan, 1989); A. P. Simond, *Karl Mannheim's Sociology of Knowledge* (Oxford: Clarendon, 1978); Nico Stehr and Volker Meja, eds., *Society and Knowledge: Contemporary Perspectives in the Sociology of Knowledge* (New Brunswick, N.J.: Transaction, 1984).

9. Above all see Robert Merton, "The Sociology of Knowledge," *Isis* 27 (1937): 493–503; Merton, "Karl Mannheim and the Sociology of Knowledge" (1941), reprinted in *Social Theory and Social Structure* (New York: Free Press, 1957), 489–508; and Merton, "Paradigm for the Sociology of Knowledge" (1945), reprinted in *The Sociology of Science: Theoretical and Empirical Investigations* (Chicago: University of Chicago Press, 1973), 7–40. See also C. Wright Mills, "Language, Logic, and Culture," *American Sociological Review* 4 (1939): 670–80; Mills, "Methodological Consequences of the Sociology of Knowledge," *American Journal of Sociology* 14 (1940): 316–30.

10. See Merton, "The Social Institution of Science," in *On Social Structure and Science* (Chicago: University of Chicago Press, 1996), 267–336; Merton, "The Sociology of Science: An Episodic Memoir," in *The Sociology of Science in Europe*, ed. Robert Merton and Jerry Gaston (Carbondale: Southern Illinois University Press, 1977), 3–141.

11. Peter Berger and Thomas Luckmann's *The Social Construction of Reality: A Treatise in the Sociology of Knowledge* (New York: Doubleday, 1966).

12. In addition, a reflexive turn paralleling the one Mannheim himself took emerged among American sociologists of knowledge: they became obsessed with intellectuals like themselves and the warrant for their own judgments of society. See, e.g., Philip Rieff, ed., *On Intellectuals: Theoretical Studies; Case Studies* (Garden City, N.Y.: Doubleday/Anchor, 1970).

13. For a collection of primary texts in the Frankfurt School tradition, see, e.g., Paul Connerton, ed., *Critical Sociology: Selected Readings* (Harmondsworth, U.K.: Penguin, 1976).

14. For the opening blast, see Jürgen Habermas, "Gegen einen positivistisch halbierten Rationalismus," *Kölner Zeitschrift für Soziologie und Sozialpsychologie* 16 (1964): 635–59. For the whole dispute, see Theodor Adorno, ed., *The Positivist Dispute in German Sociology* (New York: Harper & Row, 1976).

15. "The first phase of the rationality debate was inspired not so much by historicism or the sociology of knowledge as by postwar cultural anthropology and Wittgensteinian philosophy of language" (James Bohman and Terence Kelly, "Intelligibility, Rationality, and Comparison: The Rationality Debates Revisited," *Philosophy and Social Criticism* 22 [1996]: 81–100, citing 82).

16. Peter Winch, *The Idea of a Social Science and Its Relation to Philosophy* (London: Routledge & Kegan Paul, 1958). The key essays in the "Rationality Dispute" are collected in *Rationality*, ed. Bryan Wilson (New York: Harper & Row, 1970). See also Robin Horton and Ruth Finnegan, eds., *Modes of Thought: Essays on Thinking in Western and Non-Western Societies* (London: Faber & Faber, 1973). The direct link of these debates with the emergence of the sociology of scientific knowledge is manifest in a third volume in this line of argumentation: Martin Hollis and Steven Lukes, eds., *Rationality and Relativism* (Cambridge, Mass.: MIT Press, 1982).

17. Winch, *Idea of a Social Science*, 100.

18. The fusion of Winch and Kuhn was a *confusion,* but that was not immediately apparent. For a careful effort to disentangle "scientific conventionalism" from "philosophical conventionalism," Duhem from Quine, and above all Kuhn from Wittgenstein/Winch, see Keekok Lee, "Kuhn—a Re-appraisal," *Explorations in Knowledge* 1 (1984): 33–88, esp. 68–76. For Lee, when "Winch's development of the Wittgensteinian thesis in *The Idea of a Social Science* is . . . regarded as the complement of Kuhn's *The Structure of Scientific Revolutions*" (76), Kuhn gets tumbled into a radical relativism beyond anything he intended. But to what degree was this his own fault? "Kuhn seems to be occupying such an impossible position when he allows two epistemologically incompatible currents of scientific and philosophical conventionalism to sit side by side in his thinking. As a result he is much confused" (74).

19. Max Black, *Models and Metaphors: Studies in Language and Philosophy* (Ithaca, N.Y.: Cornell University Press, 1962), esp. chap. 3, "Metaphor," 25–47, and chap. 13, "Models and Archetypes," 219–43; Donald Schon, *Displacement of Concepts* (London: Tavistock, 1963).

20. Willard van Orman Quine and J. S. Ullian, *The Web of Belief,* 2d ed. (New York: Random House, 1978).

21. Mary Douglas, *Implicit Meanings: Essays in Anthropology* (London: Routledge & Kegan Paul, 1975); Douglas, *Natural Symbols: Explorations in Cosmology* (Harmondsworth, U.K.: Penguin, 1973); Douglas, ed., *Rules and Meanings: The Anthropology of Everyday Knowledge* (Harmondsworth, U.K.: Penguin, 1973).

22. Douglas, *Purity and Danger* (London: Routledge & Kegan Paul, 1966).

23. See the review by Barry Barnes and Steven Shapin of her *Implicit Meanings: Essays in Anthropology* (1975): "Where Is the Edge of Objectivity?" *British Journal for the History of Science* 10 (1977): 61–66.

24. Mary Hesse, *The Structure of Scientific Inference* (Berkeley: University of California Press, 1974). See also her crucial earlier contributions: *Models and Analogies in Science* (London: Sheed & Ward, 1963); "The Explanatory Function of Metaphor," reprinted in *Revolutions and Reconstructions in the Philosophy of Science* (Bloomington: Indiana University Press, 1980) (originally published in *Logic, Methodology and Philosophy of Science,* ed. Yehoshua Bar-Hillel [1965]); "Positivism and the Logic of Scientific Theories," in *The Legacy of Logical Positivism for the Philosophy of Science,* ed. Peter Achinstein and Stephen F. Barker (Baltimore: Johns Hopkins University Press, 1969), 85–114; "Is There an Independent Observation Language?" reprinted in *Revolutions and Reconstructions,* 63–110 (originally published in *The Nature and Function of Scientific Theories,* ed. Robert G. Colodny [1970]); "An Inductive Logic of Theories," in *Analyses of Theories and Methods of Physics and Psychology,* ed. Michael Radner and Stephen Winokur, Minnesota Studies in the Philosophy of Science 4 (Minneapolis: University of Minnesota Press, 1970), 164–80.

25. Michael Polanyi, *Personal Knowledge* (Chicago: University of Chicago Press, 1958); Norwood Russell Hanson, *Patterns of Discovery* (Cambridge: Cambridge University Press, 1958).

26. Marx Wartofsky, "The Relation between Philosophy of Science and History of Science," in Wartofsky, *Models: Representation and the Scientific Understanding* (Dordrecht: Reidel, 1979), 131.

27. Ronald Giere, "Kuhn's Legacy for North American Philosophy of Science," *Social Studies of Science* 27 (1997): 497. For a recent discussion of Kuhn's fusion of these fields, see Paul Hoyningen-Huehne, "The Interrelations between the Philosophy, History, and Sociology of Science in Thomas Kuhn's Theory of Scientific Development," *British Journal for the Philosophy of Science* 43 (1992): 487–501. This issue is explored extensively in the same author's monograph *Reconstructing Scientific Revolutions: Thomas S. Kuhn's Philosophy of Science* (Chicago: University of Chicago Press, 1993).

28. Derek Phillips, "Paradigms and Incommensurability," *Theory and Society* 2 (1975): 37–61, citing 37.

29. Phillips, "Epistemology and the Sociology of Knowledge: The Contributions of Mannheim, Mills, and Merton," *Theory and Society* 1 (1974): 59–88.

30. Ian Charles Jarvie, "Laudan's Problematic Progress and the Social Sciences," *Philosophy of the Social Sciences* 9 (1979): 484–97, esp. 485: "in reading Laudan I came to appreciate Kuhn more, even if I still don't agree."

31. Ibid., 487.

32. Ibid.

33. One interesting acknowledgment of misappropriation comes from Harry Collins:

"As for myself, my first paper (1974) grew out of an attempt to apply Kuhnian ideas to the study of scientific communication. I had read Kuhn (mistakenly) as an application of Wittgenstein's (1953) notion of 'form-of-life' to science" (Harry M. Collins, "The Sociology of Scientific Knowledge: Studies in Contemporary Science," *Annual Review of Sociology* 9 [1983]: 265–85, citing 270). Collins generalizes: "Though Kuhn certainly provided the intellectual mood for some European developments (perhaps unwillingly . . .), his ideas were not developed in sufficient detail to give rise to an empirical research program" (273 n).

34. Thomas S. Kuhn, "Postscript, 1969," *Structure*, 176.

35. Thomas S. Kuhn, "Reflections on My Critics," in *Criticism and the Growth of Knowledge*, 237.

36. Ibid., 238.

37. Ibid., 235.

38. Kuhn, "Postscript, 1969," *Structure*, 176.

39. Keith Jones, "Is Kuhn a Sociologist?" *British Journal for the Philosophy of Science* 37 (1986): 443–52.

40. Ibid., 442.

41. Gerald Doppelt, "Kuhn's Epistemological Relativism: An Interpretation and Defense," *Inquiry* 21 (1978): 78.

42. "One of the curious features of Kuhn's sociology of science is that it does not treat sociological features as problematic" (S. Restivo, "The Myth of the Kuhnian Revolution," 294).

43. Urry, "Kuhn and the Sociology of Knowledge," *British Journal of Sociology* 24 (1973): 467.

44. Harvey, "Use and Abuse of Kuhnian Paradigms in Sociology," *Sociology* 16 (1982): 85.

45. Eckberg and Hill, "Paradigm Concept and Sociology," in *Paradigms and Revolutions*, ed. Gutting, 253.

46. Herminio Martins, "The Kuhnian 'Revolution' and Its Implications for Sociology," in *Imagination and Precision in the Social Sciences: Essays in Memory of Peter Nettl*, ed. Thomas J. Nossiter, Albert H. Hanson, and Stein Rokkan (London: Faber & Faber, 1972), 13–58, citing 19.

47. Ibid., 51.

48. "Kuhn is not only a Mertonian, but he is a Mertonian *sans* sociology. One of the few students of science who seems to have recognized this is Mary Hesse [in *Revolutions and Reconstructions*, 32]" (Randall Collins, "Development, Diversity, and Conflict in Sociology of Science," *Sociological Quarterly* 24 [1983]: 185–200, citing 190).

49. For Merton's school, see Bernard Barber, *Science and the Social Order* (New York: Free Press, 1952); Barber, "Resistance by Scientists to Scientific Discovery," *Science* 134 (1961): 596–602; Joseph Ben-David, *The Scientist's Role in Society: A Comparative Study* (Englewood Cliffs, N.J.: Prentice-Hall, 1971); Warren Hagstrom, *The Scientific Community* (New York: Basic Books, 1965); Diane Crane, *Invisible Colleges* (Chicago: University of Chicago Press, 1972); Derek J. de Solla Price, *Little Science, Big Science* (New York: Columbia University Press, 1963).

50. Merton, "The Ethos of Science," in *On Social Structure and Science*, 267.

51. "Kuhn's influence on the sociology of science has proved to be so profound that he has all but attained the rank of Merton. It seems he is even replacing him" (Peter Weingart, "On a Sociological Theory of Scientific Change," in *Social Processes of Scientific Development*, ed. Richard Whitley [London: Routledge & Kegan Paul, 1974], 45–68, citing 45).

52. M. D. King, "Reason, Tradition, and the Progressiveness of Science" (1971), reprinted in *Paradigms and Revolutions*, ed. Gutting, 97–116, citing 97.

53. Ibid., 98.

54. Ibid., 99.

55. Ibid., 102.

56. Ibid., 104.

57. Ibid., 109.

58. Ibid., 113.

59. Ibid., 114.

60. Ibid., 114-15.

61. "British sociologists of science were the chief creators of the anti-Mertonian Kuhn" (Restivo, "The Myth of the Kuhnian Revolution," 294).

62. Michael Mulkay, "Some Aspects of Cultural Growth in the Natural Sciences," *Social Research* 36 (1969): 22-52. See also Mulkay, *Science and the Sociology of Knowledge* (London: Allen & Unwin, 1979).

63. Mulkay, "Some Aspects of Cultural Growth in the Natural Sciences," 36; the Kuhn essay is reprinted in *The Essential Tension*.

64. Mulkay, "Some Aspects of Cultural Growth in the Natural Sciences," 39. See also Michael Mulkay, "Three Models of Scientific Development," *Sociological Review* 23 (1975): 509-26, and the exchange it evoked: John Parker, "Comment on 'Three Models of Scientific Development' by M. J. Mulkay," *Sociological Review* 23 (1975): 527-33; John Law and Barry Barnes, "Research Note: Areas of Ignorance in Normal Science: A Note on Mulkay's 'Three Models of Scientific Development,'" *Sociological Review* 24 (1976): 115-24; Michael Mulkay, "The Model of Branching," *Sociological Review* 24 (1976): 125-33.

65. "Kuhn's influence was greatest, initially, among British sociologists of science. Barry Barnes was, and continues to be, one of the strongest advocates of Kuhn's work" (Collins, "Development, Diversity, and Conflict in the Sociology of Science," 190).

66. Barry Barnes, "Paradigms—Scientific and Social," *Man,* n.s., 4 (1969): 94-102, citing 94.

67. Barnes, "Paradigms," 97.

68. Ibid., 95. Barnes has been, I would argue, the most effective interpreter to conceive Thomas Kuhn as a *sociological theorist* of indispensable importance. Here, and only here, I find some point of agreement with Trevor Pinch, "Kuhn—the Conservative and Radical Interpretations," 465-82, esp. 473. Barnes deals most extensively with Kuhn in his monograph *T. S. Kuhn and Social Science* (New York: Columbia University Press, 1982). There he concludes: "In my judgment the general significance of Kuhn's work lies neither in its specific historical narrative of the development of science, nor in the concepts invented for that narrative, but simply in its explicit discussions of general problems concerning cognition, semantics and culture" (120). His obituary notice on Kuhn sustains this position: "Immediate tributes to Kuhn are sure to give prominence to his 'large' ideas: incommensurability, revolutions, the claim that users of different paradigms inhabit different worlds. But it is worth querying whether in the long term other aspects of his work will not rate more highly: his stress on the importance of small recalcitrant scientific anomalies; his treatment of exemplars; his penetrating account of 'normal science'" (*Social Studies of Science* 27 [1997]: 488-90, citing 489).

69. Robin Horton, "African Traditional Thought and Western Science," in *Rationality,* ed. Wilson, 131-70.

70. Barry Barnes, "The Comparison of Belief Systems: Anomaly versus Falsehood," in *Modes of Thought,* ed. Horton and Finnegan, 182-98, citing 187.

71. Barry Barnes and R. G. A. Dolby, "The Scientific Ethos: A Deviant Viewpoint," *Archives européennes de Sociologie* 11 (1970): 3-25, citing 8.

72. Ibid., 9.

73. R. G. A. Dolby, "The Sociology of Knowledge in Natural Science," *Science Studies* 1 (1971): 3–21, citing 6.

74. Ibid., 9.

75. Ibid., 10. Collins commented in 1983 that "Dolby's work seems to have been influenced . . . by (a misreading of?) Kuhn" (Collins, "Sociology of Scientific Knowledge: Studies in Contemporary Science," 270).

76. David Bloor, "Two Paradigms for Scientific Knowledge?" *Science Studies* 1 (1971): 101–15, citing 101.

77. Ibid., 108.

78. Ibid.

79. Ibid., 110.

80. Ibid.

81. Barry Barnes, introduction to *Sociology of Science*, ed. Barnes, 10.

82. Ibid., 11.

83. Ibid., 13–14.

84. Barry Barnes, "On the Reception of Scientific Beliefs," in *Sociology of Science*, ed. Barnes, 283.

85. Ibid., 274; citing J. D. Y. Peel, "Understanding Alien Belief-Systems," *British Journal of Sociology* 20 (1969): 69–84, citing 71.

86. Barry Barnes, "Sociological Explanation and Natural Science: A Kuhnian Reappraisal," *Archives européennes de Sociologie* 13 (1972): 373–93, citing 373.

87. Ibid., 374.

88. Ibid., 375.

89. Ibid., 386.

90. Ibid., 388.

91. See Barnes's later essays exploring this matter: Barry Barnes, "Natural Rationality: A Neglected Concept in the Social Sciences," *Philosophy of the Social Sciences* 6 (1976): 115–26; Barnes and Donald MacKenzie, "On the Role of Interests in Social Change," in *On the Margins of Science: The Social Construction of Rejected Knowledge*, ed. Roy Wallis, Sociological Review Monograph 27 (Keele: University of Keele, 1979), 49–66; Barnes, "On the Causal Explanation of Scientific Judgments," *Social Science Information* 19 (1980): 685–95; Barnes, "On the Conventional Character of Knowledge and Cognition," *Philosophy of the Social Sciences* 11 (1981): 303–33; Barnes, "On the Extension of Concepts and the Growth of Knowledge," *Sociological Review* 30 (1982): 23–44; Barnes, "Social Life As Bootstrapped Induction," *Sociology* 17 (1983): 524–45.

92. Barry Barnes, *Scientific Knowledge and Sociological Theory* (London: Routledge & Kegan Paul, 1974), 154.

93. Barry Barnes, *Interests and the Growth of Knowledge* (London: Routledge & Kegan Paul, 1977), 10.

94. Barnes, *Scientific Knowledge and Sociological Theory*, vii. Here he was explicitly gesturing to the position of Harry Collins, of which I shall have more to say in the next chapter.

95. Barnes, "Sociological Explanation and Natural Science," 376.

96. John Law and David French, "Normative and Interpretive Sociologies of Science," *Sociological Review* 22 (1974): 581–95, citing 581.

97. Ibid., 586.

98. Ibid., 588.

99. Ibid., 589.

100. Ibid., 590.

101. Ibid., 591.

102. Richard Whitley, "Black Boxism and the Sociology of Science: A Discussion of the Major Developments in the Field," *Sociological Review Monographs* 18 (1972): 61–92, citing 76.

103. Ibid., 71.

104. Ibid., 77.

105. Ibid., 72.

106. Ibid., 85.

107. Nico Stehr, "The Ethos of Science Revisited: Social and Cognitive Norms," *Sociological Inquiry* 48 (1978): 172–96. The span of Stehr's bibliography is an effective indication of the proliferation of sociology of scientific knowledge from the early 1970s, and especially its diverse European elaborations.

108. Ibid., 178.

109. Ibid., 185.

110. "Just ten years ago, the sociology of science was dominated by and virtually synonymous with 'the Mertonian paradigm.' Today, it is part of an interdisciplinary field called 'social studies of science'" (Collins, "Development, Diversity, and Conflict in the Sociology of Science," 185).

111. Kuhn, preface to *The Essential Tension*, xxi.

112. Ibid.

113. Steve Fuller, *Thomas Kuhn: A Philosophical History for Our Times* (Chicago: University of Chicago Press, 2000), 3.

114. Ibid., xii.

115. Ibid., 31.

116. Ibid., 382.

117. "Bloor has consistently made traditional philosophy and its 'rational' epistemology a particular target. The enterprise of the Strong Programme is conceived specifically as supplanting all traditional epistemology, with the sociology becoming 'heir to the subject that used to be called philosophy'" (Peter Slezak, "Bloor's Bluff," *International Studies in the Philosophy of Science* 5 [1991]: 241–56, citing 242; Slezak, in turn, is citing David Bloor, *Wittgenstein: A Social Theory of Knowledge* [New York: Columbia University Press, 1983], 184).

118. Larry Laudan, "The Pseudo-Science of Science?" in *Scientific Rationality: The Sociological Turn*, ed. Brown, 41–74, citing 43.

119. "There is a fair amount of animus in Bloor's account of philosophy," write Sharrock and Anderson ("The Wittgenstein Connection," 376).

120. David Bloor, "Are Philosophers Averse to Science?" in *Meaning and Control: Essays in Social Aspects of Science and Technology*, ed. David O. Edge and James N. Wolfe (London: Tavistock, 1973), 1–17, with comment and reply, 18–30. (This paper was delivered in 1970.)

121. David Bloor, *Knowledge and Social Imagery*, 2d ed. (Chicago: University of Chicago Press, 1991), 52.

122. "The argument is not between philosophy and science at all, but is an argument within philosophy" (Sharrock and Anderson, "The Wittgenstein Connection," 377).

123. Bloor, "Are Philosophers Averse to Science?" 1.

124. Ibid., 10.

125. Hence the title of Bloor's major work, *Knowledge and Social Imagery*. See also David Bloor, "Durkheim and Mauss Revisited: Classification and the Sociology of Knowledge," *Studies in History and Philosophy of Science* 13 (1982): 267–97.

126. Bloor, "Two Paradigms of Scientific Knowledge?" 102. "The 'new philosophy of science' embraced by the strong programme originates in a reading of Kuhn," Paul Roth aptly explains ("Voodoo Epistemology," 173).

127. Ibid., 104.

128. "Recent attacks on Kuhn, for example by Lakatos, have not succeeded. . . . On the contrary, they have simply recast Kuhnian insights into a more rationalistic, and inferior, idiom" (Bloor, "Durkheim and Mauss Revisited," 281 n).

129. David Bloor, "Rearguard Rationalism," *Isis* 65 (1974): 249–53, citing 251 n.

130. Ibid., 249.

131. David Bloor, "Wittgenstein and Mannheim on the Sociology of Mathematics," *Studies in History and Philosophy of Science* 4 (1973): 173–91, citing 191 n.

132. John Law, "Is Epistemology Redundant? A Sociological View," *Philosophy of the Social Sciences* 5 (1975): 317–37.

133. Ibid., 327.

134. Bloor, "Wittgenstein and Mannheim," 174.

135. Ibid., 173–74.

136. Robert Nola asks one very pertinent question: "Is what is caused the *act* of x's believing that p, or the very *content* of the proposition that p which x believes?" ("Ordinary Human Inference As Refutation of the Strong Programme," *Social Studies of Science* 21 [1991]: 107–29, citing 108).

137. See Max Weber, *The Methodology of the Social Sciences* (New York: Free Press, 1949).

138. Such a "sociology of sociology" was already a reality. For an example, consider Gouldner, *The Coming Crisis of Western Sociology.*

139. Bloor, *Knowledge and Social Imagery,* 42.

140. Ibid., 16.

141. This criticism is made by a series of commentators, from Laudan to Friedman.

142. Some critics proclaim that Bloor was creating a straw man in making this critique, but that is not so. Lakatos and Laudan *explicitly* formulate such a view, and they are—to cite a phrase from Bloor—merely the "rearguard of rationalism," for it was the posture implicit in the entire Received View. (For Bloor's phrase, see his review of Stephen Toulmin's *Human Understanding:* "Rearguard Rationalism.")

143. Bloor, "Wittgenstein and Mannheim," 180 n.

144. David Kaiser, "A Mannheim for All Seasons: Bloor, Merton, and the Roots of the Sociology of Scientific Knowledge," *Science in Context* 11 (1998): 51–87, citing 53.

145. Ibid., 71.

146. Bloor, "Wittgenstein and Mannheim," 180 n.

147. David Bloor, "Popper's Mystification of Objective Knowledge," *Science Studies* 4 (1974): 65–76, citing 76. In a later essay, Bloor reiterated this view: "Objective knowledge *is* knowledge without a knowing (individual) subject. It is the knowledge sustained by a social collective or a coordinated and interacting group." That is, "all objective knowledge is social" ("Ordinary Human Inference As Material for the Sociology of Knowledge," *Social Studies of Science* 21 [1991]: 129–39, citing 138).

148. Barry Barnes tends to be more forthright in these matters: "In most substantial scientific controversies, what counts as a scientific problem and what as a solution to a scientific problem, far from being self-evident features of the situation, are part of what is being contested. . . . A sociological analysis of the controversy can do no more than make a contribution to our overall understanding of it" ("Problems of Intelligibility and Paradigm Instances," in *Scientific Rationality: The Sociological Turn,* ed. Brown, 117–19). In his debate with Laudan, Bloor recognizes that reasons can be causes, but he is not interested in that. Similarly, in his debate with Slezak and Nola, he at once acknowledges other sorts of causes and proceeds to concentrate exclusively upon social ones.

149. Bloor, "Popper's Mystification of Objective Knowledge," 76, 72.

150. Bloor, *Knowledge and Social Imagery,* 75.

151. As with his readings of Mannheim and of Wittgenstein, Bloor's reading of Durkheim has not gone unchallenged. See Warren Schmaus, *Durkheim's Philosophy of Science and the Sociology of Knowledge: Creating an Intellectual Niche* (Chicago: University of Chicago Press, 1994). Also of relevance is Schmaus, "Lévy-Bruhl, Durkheim, and the Positivist Roots of the Sociology of Knowledge," *Journal of the History of the Behavioral Sciences* 32 (1996): 424–40.

152. Bloor, "Durkheim and Mauss Revisited," 267.

153. Bloor, *Knowledge and Social Imagery*, 46–51.

154. Ibid., 94.

155. David Bloor, "The Strength of the Strong Programme," in *Scientific Rationality: The Sociological Turn*, ed. Brown, 88.

156. David Bloor, "Epistemology or Psychology?" *Studies in History and Philosophy of Science* 5 (1975): 382–95, citing 393.

157. Ibid., 383.

158. Bloor, "Durkheim and Mauss Revisited," 297 n.

159. David Bloor, "Reply to J. W. Smith," *Studies in History and Philosophy of Science* 15 (1984): 245–49, citing 245.

160. Bloor, "The Strength of the Strong Programme," 89.

161. "I am grateful to Bloor for supplementing my vague and mysterious remarks about the nature of the coherence conditions, by showing how they can be understood in a broadly Durkheimian picture of social interests" (Mary Hesse, "Comments on the Papers of David Bloor and Steven Lukes," *Studies in History and Philosophy of Science* 13 [1982]: 325–31, citing 325).

162. Mary Hesse, "The Strong Thesis in the Sociology of Science," chap. 2 in *Revolution and Reconstruction*, 29–60. But Hesse hedged her endorsement so much that she feared she had made it "so weak as to be indistinguishable from something any rationalist or realist could accept" (31). Laudan ("Pseudo-Science of Science?" 73 n) and others take that to be exactly the case, and there are grounds for their judgment.

163. Mary Douglas, review of *Knowledge and Social Imagery*, by Bloor, *Sociological Review* 26 (1978): 154–57. Douglas was important to the other members of the Strong Program as well, as evidenced in the mention of her "completely general approach" to the "relationship between social and natural order" in the Durkheimian tradition by Barry Barnes and Steven Shapin ("Body Order," in *Natural Order: Historical Studies of Scientific Culture*, ed. Barry Barnes and Steven Shapin [Beverly Hills, Calif.: Sage, 1979], 15).

164. Barnes and Shapin, "Where Is the Edge of Objectivity?"

165. Ibid., 61.

166. Ibid., 62.

167. Ibid., 63.

168. Ibid.

169. Ibid.

170. Ibid., 65.

171. David Bloor, "Polyhedra and the Abominations of Leviticus," *British Journal for the History of Science* 11 (1978): 245–72, citing 245.

172. Ibid., 248, 250.

173. Ibid., 251.

174. Bloor, "Epistemology or Psychology?" 394.

175. Bloor, "Durkheim and Mauss Revisited," 297 n.

176. Edward Manier, "Levels of Reflexivity: Unnoted Differences within the 'Strong Programme' in the Sociology of Knowledge," *PSA 1980*, vol. 1., 197–207, citing 199.

177. Ibid., 201.

178. For an aggressive attack upon the causal vacuity of the Strong Program from without, see Roth, "Voodoo Epistemology."

179. Slezak is most aggressive in belaboring Bloor's behaviorism, but Roth notes the connection between Bloor's version and Quine's.

180. Bloor, *Knowledge and Social Imagery,* 144.

181. Laudan, "Pseudo-Science of Science?" 53.

182. See Peter Galison and David Stump, eds., *The Disunity of Science: Boundaries, Contexts, and Power* (Stanford, Calif.: Stanford University Press, 1996).

183. Habermas, "Gegen einen positivistisch halbierten Rationalismus" (note 14 above).

184. David Kaiser stresses this point in "A Mannheim for all Seasons," 53, 75.

185. David Bloor, "Remember the Strong Program?" *Science, Technology, and Human Values* 22 (1997): 373–85, citing 373.

186. Slezak, "Bloor's Bluff," 246. See Noam Chomsky, "A Review of B. F. Skinner's *Verbal Behavior,*" in *The Structure of Language,* ed. Jerry A. Fodor and Jerrold J. Katz (Englewood Cliffs, N.J.: Prentice Hall, 1964), 547–78.

187. Slezak, "Bloor's Bluff," 246.

188. Robert Nola, "Ordinary Human Inference," 112–13.

189. Bloor, *Knowledge and Social Imagery,* 73.

190. Ibid., 74.

191. Bloor, "Wittgenstein and Mannheim," 176.

192. Ibid., 181.

193. Ibid., 182. The centrality of Wittgenstein to Bloor and the whole course of "science studies" is attested by his own later monograph *Wittgenstein: A Social Theory of Knowledge.* But Bloor's reception of Wittgenstein is hotly contested. See not only Sharrock and Anderson, "The Wittgenstein Connection"; but Michael Friedman, "On the Sociology of Scientific Knowledge and Its Philosophical Agenda," *Studies in History and Philosophy of Science* 29 (1998): 239–71; and above all, Michael Lynch, "Extending Wittgenstein: The Pivotal Move from Epistemology to the Sociology of Science," in *Science As Practice and Culture,* ed. Andrew Pickering (Chicago: University of Chicago Press, 1992), 215–65. See the ensuing exchange in *Science As Practice and Culture:* David Bloor, "Left and Right Wittgensteinians," 266–82; and Michael Lynch, "From the 'Will to Theory' to the Discursive Collage: A Reply to Bloor's 'Left and Right Wittgensteinians,'" 283–300.

194. Bloor, "Wittgenstein and Mannheim," 187.

195. David Bloor, "Rationalism, Supernaturalism, and the Sociology of Knowledge," in *Scientific Knowledge Socialized,* ed. Imre Hronszky, Marta Feher, and Balazs Dajka (Dordrecht: Kluwer, 1988), 59–74, citing 70.

196. Bloor, "Wittgenstein and Mannheim," 188.

197. Ibid., 190.

198. Bloor, *Knowledge and Social Imagery,* 156–57. Significantly, Bloor took Lakatos's history of mathematics as decisive *support* for his project. See Bloor, "Polyhedra and the Abomination of Leviticus," 250.

199. Michael Friedman ("On the Sociology of Scientific Knowledge," 239–71) questions whether Bloor's invocation of Wittgenstein is cogent, especially concerning the "dignity" of logical necessity. Gad Freudenthal has challenged Bloor's specific sociology of mathematics: "How Strong Is Dr. Bloor's 'Strong Programme'?" *Studies in History and Philosophy of Science* 10 (1979): 67–83. Angus Gellatly comes to Bloor's defense on the matter: "Logical Necessity and the Strong Programme for the Sociology of Knowledge," *Studies in History and Philosophy of Science* 11 (1980): 325–39.

200. "Why, in particular, should the enterprise of empirically and naturalistically describing how beliefs become locally credible as a matter of fact compete or stand in conflict

with the enterprise of articulating the non-empirical and prescriptive structure in virtue of which beliefs ought to be accepted as a matter of norm?" (Friedman, "On the Sociology of Scientific Knowledge," 244).

201. This is clearly Barnes's view. It is the basis for the provocative essay he and Bloor contributed to the volume *Rationality and Relativism*.

202. Friedman, "On the Sociology of Scientific Knowledge," 241.

203. Ibid., 249–50.

204. Ibid., 251.

205. Ibid.

206. Ibid.

207. Ibid., 258–59. Logic and normativity are "features of our socio-linguistic practice that lie so deep and are so pervasive, as it were, that there is, as a matter of fact, no point of view outside them" (263).

208. The phrase is from Wittgenstein, *Remarks on the Foundations of Mathematics,* 1:121; cited in Friedman, "On the Sociology of Scientific Knowledge," 259.

209. Friedman, "On the Sociology of Scientific Knowledge," 261.

210. Ibid., 262–63.

211. Steven Shapin, "Homo Phrenologicus: Anthropological Perspectives on an Histor-ical Problem," in *Natural Order,* ed. Barnes and Shapin, 42.

212. Slezak, "Bloor's Bluff," 249.

213. Barnes, *Interests and the Growth of Knowledge,* 43.

214. Ibid., 15.

215. Jürgen Habermas, *Knowledge and Human Interests* (1968; English translation, Boston: Beacon, 1971). Barnes was influenced in his reading of Habermas particularly by Mary Hesse's review of the work, "In Defense of Objectivity" (1972), reprinted in *Revolu-tions and Reconstructions, 167–86.*

216. Barnes, *Interests and the Growth of Knowledge,* 38. Of Habermas's view, Barnes writes: "Habermas does not realize that in describing 'hermeneutic' knowledge, he is merely pointing out certain universal features of all knowledge" (18). Here Barnes anticipates a line of thought that became central to Richard Rorty and universal pragmatism.

217. Ibid., 2, 6.

218. Donald MacKenzie and Barry Barnes, "Scientific Judgment: The Biometry-Mendel-ism Controversy," in *Natural Order,* ed. Barnes and Shapin, 203.

219. Ibid., 205.

220. Barnes, *Scientific Knowledge and Sociological Theory,* 115. This candor and skep-ticism may have distinguished Barnes from Bloor in the early days of the Strong Program. See Edward Manier, "Levels of Reflexivity." See also, Saeid Zibakalam, "Emergence of a Radi-cal Sociology of Scientific Knowledge: The Strong Programme in the Early Writings of Barry Barnes," *Dialectica* 47 (1993): 3–25.

221. Barnes, *Interests and the Growth of Knowledge,* 35.

222. Bloor, "The Strength of the Strong Programme," 79.

223. Ibid., 89; David Bloor, "Afterword: Attacks on the Strong Programme," in *Knowl-edge and Social Imagery,* 2d ed., 171.

224. Bloor, "Ordinary Human Inference As Material for the Sociology of Knowledge," 133.

225. Barry Barnes and Steven Shapin, introduction to *Natural Order,* ed. Barnes and Shapin, 9–13, citing 11. Steven Shapin made the case for the empirical fecundity of the new approach in "History of Science and Its Historical Reconstructions," *History of Science* 20 (1982): 157–211; and in "Social Uses of Science," in *The Ferment of Knowledge,* ed. George S. Rousseau and Roy Porter (Cambridge: Cambridge University Press, 1980), 93–139.

226. Shapin, "History of Science and Its Sociological Reconstructions," 157.

227. Knorr-Cetina, "The Constructivist Program in the Sociology of Knowledge: Retreats or Advances?" *Social Studies of Science* 12 (1982): 322.

228. Roth, "Voodoo Epistemology," 210ff.

229. Steven Yearley, "The Relationship between Epistemological and Sociological Cognitive Interests: Some Ambiguities Underlying the Use of Interest Theory in the Study of Scientific Knowledge," *Studies in History and Philosophy of Science* 13 (1982): 353–88, citing 375, 387. See also Steven Yearley, "Settling Accounts: Action, Accounts, and Sociological Explanation," *British Journal of Sociology* 39 (1988): 578–99. But see Donald MacKenzie, "Reply to Steven Yearley," *Studies in History and Philosophy of Science* 15 (1984): 251–59.

230. Nola, "Ordinary Human Inference," 123.

231. Peter Slezak, "Scientific Discovery by Computer As Empirical Refutation of the Strong Programme," *Social Studies of Science* 19 (1989): 563–600, citing 584–85. Of course, Slezak uses a very formal concept of causality to appraise SSK: "Genuine causal generalization must support counterfactual conditionals, so that, had the cause not obtained, the effect would not have followed" (584). Is that the appropriate—*universal*—standard for explanation? By whose authority? As Augustine Brannigan replied, "the claims advanced by Slezak would take issue with any sociology of ideas which attempted to ground the form or the content of science in some element of the social or cultural structure" ("Artificial Intelligence and the Attributional Model of Scientific Discovery," *Social Studies of Science* 19 [1989]: 601–13, citing 603).

232. The Strong Program "is *not* a sociological theory, in any customary sense of that term. It specifies no detailed causes, or functional mechanisms and no laws. It is, rather, a *meta-sociological* manifesto" (Laudan, "The Pseudo-Science of Science," 42). By "meta-sociological" I take it that Laudan means *theoretical* in my sense of the ensemble of epistemology, methodology, and rhetorical reflexivity.

233. Coherence conditions, according to Bloor, "derive from social processes and are properties of the social collectivity, rather than being in the nature of the human mind" ("Reply to J. W. Smith," 248). Peter Slezak ("Bloor's Bluff," 241–56) mounts a strong attack on the sociology of scientific knowledge from the vantage of psychological theories of causation embodied in cognitive science, and others have insisted upon either biological (innate) causes or upon rational (immanent, or in Bloor's terminology "teleological") causes.

234. "Taking their philosophical cues from Kuhn's account of scientific revolutions (in particular, the claim that paradigm changes are more akin to conversions than to reasoned judgments), Wittgenstein's insistence on the thoroughly social and conventional determinations of language use, and Quine's and Duhem's claims that theories can always be revised to accommodate seemingly adverse evidence, advocates of the strong programme conclude that neither reason nor fact (both of which they view as conventionally determined anyway) serve to explain the choice of one scientific theory over another" (Roth, "Voodoo Epistemology," 154).

235. "The strong programme parallels Quine's thought on at least four basic epistemological theses. First, it endorses Quine's holism (which, following Hesse, is sometimes referred to as the Duhem-Quine thesis). . . . Second, it accepts the underdetermination of theories by data. Third, indeterminacy of translation is also part of the general philosophical position. . . . Fourth, . . . more than a passing resemblance to Quine's call to naturalize epistemology" (ibid., 163).

236. "Laudan's own work on Pierre Duhem will have told him that in practice crucial experiments can always be challenged" (Bloor, "Strength of the Strong Programme," 78). As we have seen, and Laudan painstakingly established, this is not an adequate conception of what Duhem meant.

237. For a similar argument, see John M. Nicholas, "Scientific and Other Interests," in *Scientific Rationality: The Sociological Turn*, ed. Brown, 265-94, esp. 276: "There is a widespread conviction that the procedures of logic and rationality guarantee no unequivocal answers to inductive and explanatory enquiries. One of the sources of the conviction is the so-called 'Quine-Duhem thesis,' which is generally understood to entail that an agent faced with a seeming difficulty or contradiction between data and theory has a free choice as to whether he deems the theory falsified or retains it by parking the blame elsewhere within the system of knowledge with which the theory is connected. The significance of this claim is badly exaggerated."

238. Slezak, "Scientific Discovery by Computer," 587.

239. Jan Golinski, "The Theory of Practice and the Practice of Theory: Sociological Approaches in the History of Science," *Isis* 81 (1990): 492-505, citing 503. Golinski notes correctly that "Sociologists tend to invoke more radical versions of the thesis" (503 n).

240. See esp. Barnes, *T. S. Kuhn and Social Science*.

241. "The general argument can be made on the basis of the work of Mary Hesse (1974). Her 'network' model of the verbal component of scientific culture illustrates the conventional and endlessly negotiable character of our classifications and knowledge" (MacKenzie and Barnes, "Scientific Judgment: The Biometry-Mendelism Controversy," in *Natural Order*, ed. Barnes and Shapin, 207 n). It may be that in some abstract, philosophical discourse of possibility these matters are "endlessly negotiable." It is patent that in actual scientific practice they are not. See, e.g., Peter Galison, *How Experiments End* (Chicago: University of Chicago Press, 1987).

242. Roth, "Voodoo Epistemology," 165.

243. Slezak makes a similar argument in "Bloor's Bluff," 253-54. Paul Roth delves extensively into this question in "Voodoo Epistemology," 210ff.

244. "Does the underdetermination of theories provide an *a priori* warrant for the claim that all theory choices have a 'social component,' i.e., are the result of social circumstances and conventions?" For that, "one would have to show that the instrumentalities scientists use for circumventing the problem of underdetermination are always social in character" (Laudan, "Pseudo-Science of Science?" 68).

245. For a balanced assessment, see Michael Friedman, "On the Sociology of Scientific Knowledge."

246. Slezak, "Bloor's Bluff," 254.

247. "The underdetermination of judgements by sensory evidence does not, *on its own*, warrant invoking *social* factors to explain the outcome. . . . Quine has argued that, notwithstanding the underdetermination of theory by observational data, we invoke other criteria such as simplicity, probability, explanatory coherence, comprehensiveness and so on, in making choices among theories. The idea that underdetermination itself, *ipso facto*, warrants to resort to *sociological* factors is a common fallacy among sociologists of science." Accordingly, "*independent* grounds beyond underdetermination itself must be provided for preferring sociological . . . explanation" (Peter Slezak, "Artificial Experts," *Social Studies of Science* 21 [1991]: 175-201, citing 190).

248. Roth, "Voodoo Epistemology," 173 n.

249. Paul Roth and Robert Barrett, "Deconstructing Quarks," *Social Studies of Science* 20 (1990): 579-632, citing 625.

250. Bloor, "Wittgenstein and Mannheim," 173.

251. Roth, "Voodoo Epistemology," 184.

252. Ibid., 186.

253. Ibid., 188.

254. Ibid., 204.

255. Ibid., 214.

256. Ibid., 213.

257. Ibid., 214.

258. Ibid., 216.

259. Ibid., 213.

260. Paul Thagard, "Welcome to the Cognitive Revolution," *Social Studies of Science* 19 (1989): 653–57, citing 655.

261. Peter Slezak, "How Strong Is the 'Strong Programme'?" *Social Studies of Science* 21 (1991): 154–56, citing 155.

262. For a similar interpretation, see Philip Kitcher, "A Plea for Science Studies," in *A House Built on Sand*, ed. Noretta Koertge (Oxford: Oxford University Press, 1998), 32–56.

Chapter Six

1. Bruno Latour and Steven Woolgar, *Laboratory Life: The Social Construction of Scientific Facts* (Beverly Hills, Calif.: Sage, 1979).

2. Karin Knorr-Cetina, *The Manufacture of Knowledge* (Oxford: Pergamon, 1981); Michael Lynch, *Art and Artifact in Laboratory Science* (London: Routledge & Kegan Paul, 1985).

3. Harry M. Collins, *Changing Order: Replication and Induction in Scientific Practice* (Chicago: University of Chicago Press, 1985).

4. For Pinch, see *Confronting Nature: The Sociology of Solar-Neutrino Detection* (Dordrecht: Kluwer, 1986). For Andrew Pickering, see *Constructing Quarks* (Chicago: University of Chicago Press, 1984).

5. Andrew Pickering, "From Science As Knowledge to Science As Practice," in *Science As Practice and Culture*, ed. Pickering (Chicago: University of Chicago Press, 1992), 2.

6. Harry M. Collins, "Knowledge, Norms and Rules in the Sociology of Science," *Social Studies of Science* 12 (1982): 299–309, citing 302.

7. The notion of "tacit knowledge" was developed most extensively by Michael Polanyi (see, e.g., Polanyi, *The Tacit Dimension* [Garden City, N.Y.: Doubleday, 1966]). For sharp skepticism of this tendency as "practice mysticism," see Steve Fuller, "Of Conceptual Intersections (Response to Slezak)," *Social Studies of Science* 19 (1990): 625–38, esp. 635. For a more temperate but still critical appraisal, see Stephen Turner, *The Social Theory of Practices: Tradition, Tacit Knowledge, and Presuppositions* (Chicago: University of Chicago Press, 1994).

8. Pickering, "From Science As Knowledge to Science As Practice," 1. This phrase seems to be widely invoked in contemporary theoretical discourse, e.g., by Quine and Rorty.

9. Latour and Woolgar, *Laboratory Life*. Sergio Sismondo calls it "the best-read of the laboratory ethnographies and a paradigm of constructivism" ("Some Social Constructions," *Social Studies of Science* 23 [1993]: 515–53, citing 532). Ian Hacking wrote about it some ten years later because it had so deeply stamped the field ("The Participant Irrealist at Large in the Laboratory," *British Journal for the Philosophy of Science* 39 [1988]: 277–94, citing 277). Jan Golinski, too, comments on "the lasting influence" of the work ("The Theory of Practice and the Practice of Theory: Sociological Approaches in the History of Science," *Isis* 81 [1990]: 492–505, citing 496).

10. Hacking, "The Participant Irrealist," 277.

11. Bruno Latour, "Postmodern? No, Simply Amodern: Steps Towards an Anthropology of Science," *Studies in History and Philosophy of Science* 21 (1990): 145–71, citing 145–46. Latour observed that Woolgar then and later commented upon the "naivety" of this approach.

12. Bruno Latour, "Insiders and Outsiders in the Sociology of Science: Or, How Can We Foster Agnosticism?" *Knowledge and Society: Studies in the Sociology of Culture Past and Present* 3 (1981): 199–216, citing 203.

13. "No one denies that the sociologist of religion can be both an agnostic and a good sociologist, but a sociologist of science is not permitted to be an agnostic" (Latour, "Insiders and Outsiders," 201).

14. Latour and Woolgar, *Laboratory Life*, 30.

15. Latour, "Insiders and Outsiders," 201–3.

16. Ibid., 211.

17. Latour and Woolgar, *Laboratory Life*, 189.

18. Pierre Bourdieu, "The Specificity of the Scientific Field and the Social Conditions of the Progress of Reason," *Social Science Information* 14 (1975): 19–47. For a discerning critique of this "economic model" see Karin Knorr-Cetina, "Scientific Communities or Transepistemic Arenas of Research? A Critique of Quasi-Economic Models of Science," *Social Studies of Science* 12 (1982): 101–30.

19. "Credibility accrues from credible information. . . . It can in turn be converted into money, positions, recognitions, etc., and through these resources into further information" (Karin Knorr-Cetina, "New Developments in Science Studies: The Ethnographic Challenge," *Canadian Journal of Sociology* 8 [1983]: 153–77, citing 164).

20. Sismondo terms this an "enthusiasm for linguistic and near-linguistic activities," and he sees this as resulting in "a far more radical constructivism" than had hitherto appeared in science studies ("Some Social Constructions," 533).

21. Latour and Woolgar, *Laboratory Life*, 52. As Knorr-Cetina puts it, this notion of "literary inscription . . . characterize[s] scientific work as mainly concerned with the creation and transformation of written 'traces,' such as measurement data and scientific support" ("New Developments in Science Studies," 154–55 n).

22. Ibid., 51.

23. Hacking, "The Participant Irrealist," 278 n.

24. James Scott, "Exploring Socio-Technical Analysis: Monsieur Latour Is Not Joking!" *Social Studies of Science* 22 (1992): 59–80, citing 60.

25. Latour and Woolgar, *Laboratory Life*, 76–79; see Paul Tibbetts, "The Sociology of Scientific Knowledge: The Constructivist Thesis and Relativism," *Philosophy of the Social Sciences* 16 (1986): 39–57, esp. 40–41, for a good discussion of this.

26. That is, "eliminating the linguistic modalities which qualify any given statement" (Latour and Woolgar, *Laboratory Life*, 237).

27. Latour and Woolgar, *Laboratory Life*, 81.

28. Steven Shapin has observed how pervasive the military metaphors are in Latour's work: "The military (Machiavellian, Hobbesian, Nietzschean) metaphor is, of course, Latour's trade mark, and it is basic to his understanding of scientific and technological activity. Technoscience is war conducted by much the same means. Its object is domination and its methods involve the mobilization of allies, their multiplication and drilling, their strategic and forceful juxtaposition to the enemy" ("Following Scientists Around," *Social Studies of Science* 18 [1988]: 533–50, citing 534).

29. Latour and Woolgar, *Laboratory Life*, 64.

30. Ibid., 180 n.

31. Hacking, "The Participant Irrealist," 282. See Bas C. van Fraassen, *The Scientific Image* (Oxford: Clarendon, 1980); and the volume *Images of Science: Essays on Realism and Empiricism*, with a reply from Bas C. van Fraassen, ed. Paul M. Churchland and Clifford A. Hooker (Chicago: University of Chicago Press, 1985), which explores his thought; for Goodman, see *Ways of Worldmaking* (Indianapolis: Hackett, 1978).

32. Hacking, "The Participant Irrealist," 282.

33. Bruno Latour, *Science in Action* (Cambridge, Mass.: Harvard University Press, 1987), 13–15.

34. Bruno Latour, "Is It Possible to Reconstruct the Research Process? Sociology of a Brain Peptide," in *The Social Process of Scientific Investigation*, ed. Karin Knorr, Roger Krohn, and Richard Whitley, Sociology of the Sciences Yearbook 4 (1980) (Dordrecht: Reidel, 1981), 53–73, citing 56.

35. Ibid., 58.

36. Ibid., 61.

37. Ibid., 62.

38. Ibid.

39. Ibid., 66.

40. "'Fiction' is not taken as synonymous of 'empty,' 'false' or 'fraudulent.' It is the word that describes the construction of paths and plausible stories" (ibid., 72 n. 21). This is, in fact, a far more constructive approach than that of Medawar, "Is the Scientific Paper Fraudulent?" *Saturday Review* (August 1, 1964), 43–44. See Thomas Nickles, "Good Science As Bad History: From Order of Knowing to Order of Being," in *The Social Dimensions of Science*, ed. Ernan McMullin (Notre Dame, Ind.: University of Notre Dame Press, 1992), 85–129, esp. 93, for a good discussion of Medawar.

41. Latour, "Insiders and Outsiders," 211.

42. Latour, "Is It Possible to Reconstruct the Research Process?" 69.

43. Knorr-Cetina, "New Developments in Science Studies," 159.

44. Ibid., 158.

45. Latour, *Science in Action,* 141.

46. Knorr-Cetina, "New Developments in Science Studies," 167.

47. Latour, *Science in Action,* 202. "We never see 'science' and 'society' separately. . . . 'Scientific' work is also 'social' work . . . solutions to the problem of knowledge are solutions to the problem of social order. 'Society' is constructed and stabilized at the same time that 'facts' and 'machines' are constructed" (Shapin, "Following Scientists Around," 539). Michel Callon and Bruno Latour, "Don't Throw the Baby Out with the Bath School! A Reply to Collins and Yearley," in *Science As Practice and Culture*, ed. Pickering, 343–68.

48. Latour, *Science in Action,* 60–62. "You can always get allies to march in the direction they already want to go, but that is unlikely to do you much good. So what you must do is to *translate* their interests into your preferred course of action. . . . Having enrolled your allies, you must now make sure that they keep in line, particularly that they don't transform your creation so that it is no longer recognizable as yours" (Shapin, "Following Scientists Around," 537).

49. Sismondo, "Some Social Constructions," 526.

50. She objected: "interests are not generally obvious to agents themselves . . . interests, like other phenomena, appear to be negotiated and accomplished in social action rather than to simply 'exist' . . . 'objectively' attributed and 'subjectively' perceived interests do not always coincide . . . a question as to who may or may not legitimately identify somebody's interests, and on what grounds" (Knorr-Cetina, "Scientific Communities or Transepistemic Arenas of Research?" 129 n. 32).

51. Karin Knorr-Cetina, "Ethnographic Study of Scientific Work: Towards a Constructivist Interpretation," in *Science Observed: Perspectives on the Social Study of Science*, ed. Karin Knorr-Cetina and Michael Mulkay (London: Sage, 1983), 116.

52. Ibid., 116–17.

53. Knorr-Cetina, "Scientific Communities or Transepistemic Arenas for Research?" 116.

54. Knorr-Cetina, "Ethnographic Study of Scientific Work," in *Science Observed,* ed. Knorr-Cetina and Mulkay, 119.

55. Karin Knorr-Cetina, "Epistemic Cultures: Forms of Reason in Science," *History of Political Economy* 23 (1991): 105–22, citing 107.

56. Karin Knorr, "Tinkering toward Success: Prelude to a Theory of Scientific Practice," *Theory and Society* 8 (1979): 347–75, citing 352.

57. Ibid., 361, 368.

58. Knorr-Cetina, "New Developments in Science Studies," 161.

59. Karin Knorr-Cetina, "The Scientist As an Analogical Reasoner: A Critique of the Metaphor Theory of Innovation," in *The Social Process of Scientific Investigation,* ed. Knorr, Krohn, and Whitley 25–52, citing 41.

60. See Ian Hacking, *Representing and Intervening* (Cambridge: Cambridge University Press, 1983), 220–30.

61. Karin Knorr-Cetina, "The Couch, the Cathedral, and the Laboratory: On the Relationship between Experiment and Laboratory in Science," in *Science As Practice and Culture,* ed. Pickering, 113–38, citing 127.

62. Francis Bacon, *The New Organon* (Cambridge: Cambridge University Press, 2000), 20–21.

63. Knorr-Cetina, "Ethnographic Study of Scientific Work," 120.

64. Knorr, "Tinkering toward Success," 367.

65. Knorr-Cetina, "Ethnographic Study of Scientific Work," 121.

66. Knorr-Cetina, "The Couch, the Cathedral, and the Laboratory," 115.

67. Ibid., 123.

68. Karin Knorr-Cetina, "The Fabrication of Facts: Toward a Microsociology of Scientific Knowledge," in *Society and Knowledge: Contemporary Perspectives in the Sociology of Knowledge,* ed. Nico Stehr and Volker Meja (New Brunswick, N.J.: Transaction, 1984), 223–44, citing 226.

69. Ibid., 227.

70. Ibid.

71. Ibid., 229.

72. Knorr-Cetina, "The Scientist As an Analogical Reasoner," 43.

73. Ibid., 38.

74. Knorr-Cetina, "Fabrication of Facts," 231.

75. Knorr-Cetina, "Scientific Communities or Transepistemic Arenas for Research?" 123.

76. Ibid., 121.

77. Ibid., 119.

78. Sismondo, "Some Social Constructions," 528.

79. Karin Knorr-Cetina, "Strong Constructivism—from a Sociologist's Point of View: A Personal Addendum to Sismondo's Paper," *Social Studies of Science* 23 (1993): 555–63, citing 561.

80. Karin Knorr-Cetina, "The Constructivist Programme in the Sociology of Science: Retreats or Advances?" *Social Studies of Science* 12 (1982): 320–24, citing 321.

81. Knorr-Cetina, "Strong Constructivism," 558–60.

82. Bruno Latour, "Clothing the Naked Truth," in *Dismantling Truth: Reality in the Postmodern World,* ed. Hilary Lawson and Lisa Apignanesi (New York: St. Martin's, 1989), 101–26.

83. Harry M. Collins, "Stages in the Empirical Programme of Relativism," *Social Studies of Science* 11 (1981): 3–10; Harry M. Collins, "An Empirical Relativist Programme in the Sociology of Scientific Knowledge," in *Science Observed,* ed. Knorr-Cetina and Mulkay, 85–114.

84. Collins, "Stages in the Empirical Programme of Relativism," 4.

85. Harry M. Collins, "The Place of the 'Core-Set' in Modern Science: Social Contingency with Methodological Propriety in Science," *History of Science* 19 (1981): 6–19, citing 7.

86. Harry M. Collins, "Special Relativism—the Natural Attitude," *Social Studies of Science* 12 (1982): 139–43, citing 141.

87. "The core-set of scientists are those who are actively involved in experimentation or observation, or making contributions to the theory of the phenomenon, or of the experiment, such that they have an effect on the outcome of the controversy" (Collins, "The Place of the 'Core-Set' in Modern Science," 8). This will always be an empirical approximation, Collins observes, and he prefers "set" to "group" because "members of a core-set will not necessarily interact frequently with one another, for some members may be enemies" (8).

88. Ibid., 12.

89. Harry M. Collins, "The Sociology of Scientific Knowledge: Studies of Contemporary Science," *Annual Review of Sociology* 9 (1983): 265–85, citing 281.

90. Collins, "The Place of the 'Core-Set' in Modern Science," 14.

91. Ibid., 15.

92. Harry M. Collins, "Son of Seven Sexes: The Social Destruction of a Physical Phenomenon," *Social Studies of Science* 11 (1981): 33–62, citing 54. See Laudan's denunciation of this assertion as "wildly implausible" ("A Note on Collins's Blend of Relativism and Empiricism," *Social Studies of Science* 12 [1982]: 131).

93. Collins, "Stages in the Empirical Programme of Relativism," 3.

94. Collins, "An Empirical Relativist Programme in the Sociology of Scientific Knowledge," 91.

95. Harry M. Collins, "What Is TRASP?: The Radical Programme As a Methodological Imperative," *Philosophy of the Social Sciences* 11 (1981): 215–24, citing 221.

96. Ibid., 220.

97. Collins, "Sociology of Scientific Knowledge," 280.

98. Mary Hesse, "Changing Concepts and Stable Order," *Social Studies of Science* 16 (1986): 714–26, citing 720.

99. Collins, "Special Relativism—the Natural Attitude," 140.

100. As Collins and Yearley note, "The importance of the philosophical arguments about relativism in the 1970s was, in retrospect, not that they showed that relativism was true but that it was tenable and therefore could be used as a methodology for the study of science" (Harry M. Collins and Steven Yearley, "Epistemological Chicken," in *Science As Practice and Culture*, ed. Pickering, 303).

101. Collins, "Knowledge, Norms and Rules in the Sociology of Science," 304.

102. Harry M. Collins and Graham Cox, "Recovering Relativity: Did Prophecy Fail?" *Social Studies of Science* 6 (1976): 423–44, citing 439.

103. Yves Gingras and Sam Schweber, "Constraints on Construction," *Social Studies of Science* 16 (1986): 372–83, citing 376.

104. Collins, "An Empiricist Relativist Programme," 95.

105. Collins, "Stages in the Empirical Programme of Relativism," 4.

106. Callon and Latour, "Don't Throw the Baby Out with the Bath School!" 355.

107. Andrew Pickering, "Forms of Life: Science, Contingency and Harry Collins," *British Journal for the History of Science,* 20 (1987): 213–21, citing 214.

108. Ibid., 216.

109. Ibid., 218 n.

110. Ibid., 220.

111. Ibid., 219.

112. Ibid., 220.

113. Andrew Pickering, "Against Putting the Phenomena First: The Discovery of the Weak Neutral Current," *Studies in History and Philosophy of Science* 15 (1984): 85–117; Pickering, "Knowledge, Practice, and Mere Construction (Response to Roth and Barrett)," *Social Studies of Science* 20 (1990): 682–729, citing 684.

114. Andrew Pickering, "The Role of Interests in High-Energy Physics," in *The Social Process of Scientific Investigation,* ed. Knorr, Krohn, and Whitley, 107–38, citing 114.

115. Pickering was an affiliate of the Edinburgh Science Studies Unit.

116. Pickering, "Against Putting the Phenomena First," 86.

117. Ibid., 87.

118. Pickering, "The Role of Interests in High-Energy Physics," 113 n.

119. Ibid., 114.

120. Ibid., 113.

121. Pickering, *Constructing Quarks,* 5–6.

122. Ibid., 404.

123. Ibid., 7.

124. Ibid., 8.

125. Ibid., 11.

126. Ibid., 405.

127. Ibid.

128. Ibid., 14.

129. Ibid., 19 n. 13.

130. See, e.g., Richard Burian, review of *Constructing Quarks,* by Pickering, *Synthese* 82 (1990): 163–74.

131. Gingras and Schweber, "Constraints on Construction," 374–75.

132. Ibid.

133. Ibid., 377.

134. Ibid., 379.

135. Ibid., 376.

136. Ibid.

137. Ibid., 381.

138. Ibid.

139. "The underdetermination of science by all possible evidence, even in that strong form that Quine has argued for, does not support the claim that Pickering seems to base on it—namely, the claim that experimental results do not constrain the scientist to make particular choices between pairs of theories (or sets of theories)" (David Henderson, "On the Sociology of Science and the Continuing Importance of Epistemologically Couched Accounts," *Social Studies of Science* 20 [1990]: 113–48, citing 117).

140. Gingras and Schweber, "Constraints on Construction," 379.

141. Ibid., 380.

142. Paul Roth and Robert Barrett, "Deconstructing Quarks," *Social Studies of Science* 20 (1990): 581.

143. Ibid., 590.

144. Ibid., 616.

145. Ibid., 595.

146. Ibid.

147. Ibid., 582.

148. Paul Roth and Robert Barrett, "Reply: Aspects of Sociological Explanation," *Social Studies of Science* 20 (1990): 729–45, citing 741.

149. Ibid., 744.

150. Ibid., 745.

151. Allan Megill has noted this in his "Recounting the Past: 'Description,' Explanation, and Narrative in Historiography," *American Historical Review* 94 (1989): 627–53.

152. This general crisis had been predicted earlier by Alvin W. Gouldner, *The Coming Crisis of Western Sociology* (New York: Basic Books, 1970), and it was diagnosed in the mid-1980s, especially by Anthony Giddens in *Central Problems in Social Theory* (Berkeley: University of California Press, 1979). More recently, Jeffrey Alexander has surveyed the condition of the discipline in *Fin de Siècle Social Theory: Relativism, Reduction, and the Problem of Reason* (London: Verso, 1995).

153. On ethnomethodology see Harold Garfinkel, *Studies in Ethnomethodology* (Englewood Cliffs, N.J.: Prentice Hall, 1967); Harold Garfinkel and Harvey Sacks, "On Formal Structures of Practical Behavior," in *Theoretical Sociology: Perspectives and Development,* ed. John C. McKinney and Edward A. Tiryakian (New York: Appleton-Century-Crofts, 1970), 337–66; Paul Attewell, "Ethnomethodology since Garfinkel," *Theory and Society* 1 (1974): 179–210; Roy Turner, ed., *Ethnomethodology* (Harmondsworth, U.K.: Penguin, 1974); John Heritage, *Garfinkel and Ethnomethodology* (Cambridge, U.K.: Polity, 1984); and Paul Atkinson, "Ethnomethodology: A Critical Review," *Annual Review of Sociology* (1988): 441–65.

154. Hilbert, "Ethnomethodology and the Micro-Macro Order," *American Sociological Review* 55 (1990): 794–808, citing 795.

155. Ibid., 796.

156. Michael Lynch, "From the 'Will to Theory' to the Discursive Collage," in *Science As Practice and Culture,* ed. Pickering, 290.

157. Steve Bruce and Roy Wallis, "Rescuing Motives," *British Journal of Sociology* 34 (1983): 61–71, citing 64.

158. Michael Lynch, "Extending Wittgenstein: The Pivotal Move from Epistemology to the Sociology of Science," in *Science As Practice and Culture,* ed. Pickering, 215–65, citing 239–40.

159. Steven Woolgar, "Some Remarks about Positionism," in *Science As Practice and Culture,* ed. Pickering, 331.

160. Chris Doran, "Jumping Frames: Reflexivity and Recursion in the Sociology of Science," *Social Studies of Science* 19 (1989): 515–31.

161. Steven Shapin, "Homo Phrenologicus: Anthropological Perspectives on an Historical Problem," in *Natural Order: Historical Studies in Scientific Culture,* ed. Barry Barnes and Steven Shapin (Beverly Hills, Calif.: Sage, 1979), 42; cited by Steven Woolgar, "Interests and Explanation in the Social Study of Science," *Social Studies of Science* 11 (1981): 365–94, on 366.

162. Woolgar, "Interests and Explanation," 380.

163. David Bloor, "Left and Right Wittgensteinians," in *Science As Practice and Culture,* ed. Pickering, 266–82, citing 278.

164. Barry Barnes, "Ethnomethodology As Science," review of *Garfinkel and Ethnomethodology,* by John Heritage, *Social Studies of Science* 15 (1985): 751–62, citing 756.

165. Ibid., 752.

166. Bloor, "Left and Right Wittgensteinians," 267.

167. Knorr-Cetina, "Ethnographic Study of Scientific Work," 134.

168. Lynch, "From the 'Will to Theory' to the Discursive Collage," in *Science As Practice and Culture,* ed. Pickering, 298.

169. John Law, "Power/Knowledge and the Dissolution of the Sociology of Knowledge," introduction to *Power, Action and Belief: A New Sociology of Knowledge?* ed. Law, Sociological Review Monograph 32 (London: Routledge & Kegan Paul, 1986), 2.

170. Ibid., 11–12.

171. Ibid., 18.

172. Ibid., 13. See Barry Hindess, "Power, Interests and the Outcomes of Struggles," *Sociology* 16 (1982): 498–511; and Hindess, "'Interests' in Political Analysis," in *Power, Action and Belief*, ed. Law, 112–31.

173. Bruno Latour, *The Pasteurization of France* (Cambridge, Mass.: Harvard University Press, 1988), 256; originally published as *Les microbes: guerre et paix* (Paris: Métailié, 1984); cited in Simon Schaffer, "The Eighteenth Brumaire of Bruno Latour," *Social Studies of Science* 22 (1991): 174–92, on 185.

174. Michel Callon, "Some Elements of a Sociology of Translation: Domestication of the Scallops and the Fishermen of St. Brieuc Bay," in *Power, Action and Belief*, ed. Law, 196–233, citing 199.

175. Latour, "Postmodern? No, Simply Amodern," 155.

176. Bruno Latour, "Give Me a Laboratory and I Will Raise the World," in *Science Observed*, ed. Knorr-Cetina and Mulkay, 141–71, citing 160.

177. Law, "Power/Knowledge and the Dissolution of the Sociology of Knowledge," 18.

178. Knorr-Cetina, "The Constructivist Programme in the Sociology of Science," 322.

179. Latour and Woolgar, *Laboratory Life: The Construction of Scientific Facts*, 2d ed. (Princeton, N.J.: Princeton University Press, 1986), 285.

180. Ibid., 273.

181. Ibid., 280.

182. Paul Thagard, "Welcome to the Cognitive Revolution," *Social Studies of Science* 19 (1989): 656; Peter Slezak, "The Social Construction of Social Constructionism," *Inquiry* 37 (1994): 139–57, citing 141.

183. Barnes and Shapin, eds., *Natural Order*.

184. Steven Shapin, "History of Science and Its Historical Reconstructions," *History of Science* 20 (1982): 157–211.

185. Steven Shapin, "Social Uses of Science," in *The Ferment of Knowledge*, ed. George S. Rousseau and Roy Porter (Cambridge: Cambridge University Press, 1980), 93–139.

186. Steven Shapin and Simon Schaffer, *Leviathan and the Air Pump: Hobbes, Boyle, and the Experimental Life* (Princeton, N.J.: Princeton University Press, 1985).

187. Cassandra Pinnick, "What's Wrong with the Strong Programme's Case Study of the 'Hobbes-Boyle' Dispute?" in *A House Built on Sand: Exposing Postmodernist Myths about Science*, ed. Noretta Koertge (Oxford: Oxford University Press, 1998), 227–39, citing 189.

188. Shapin and Schaffer, *Leviathan and the Air Pump*, 15.

189. Ibid., 16.

190. Pinnick, "What's Wrong with the Strong Programme's Case Study?" 235.

191. Ibid., 237, 236.

192. Paul Gross and Norman Levitt, *Higher Superstition: The Academic Left and Its Quarrels with Science*, 2d ed. (Baltimore: Johns Hopkins University Press, 1998), 65.

193. Ibid., 66.

194. Ibid.

195. Ibid., 67.

196. Ibid.

197. Ibid., 264–66 n. 35.

198. Ibid., 292.

199. Ibid.

200. Shapin and Schaffer, *Leviathan and the Air Pump*, 344.

201. Ibid., 21.

202. Ibid., 4.

203. Ibid., 5–6.

204. Margaret Jacob, "Reflections on Bruno Latour's Version of the Seventeenth Century," in *A House Built on Sand*, ed. Koertge, 240–54, citing 250.
205. Shapin and Schaffer, *Leviathan and the Air Pump*, 7.
206. Ibid., 8.
207. Ibid.
208. Ibid.
209. Ibid., 9.
210. Ibid., 112.
211. Ibid., 3.
212. Ibid., 204–5.
213. Ibid., 24.
214. Ibid., 25.
215. My recourse to Kantian terminology is not casual. Precisely what is transpiring is a transformation of the *a priori* into the *a posteriori*.
216. Shapin and Schaffer, *Leviathan and the Air Pump*, 25.
217. Ibid., 39.
218. Ibid., 73.
219. Ibid., 332.
220. Ibid., 290.
221. Ibid., 283.
222. Ibid., 298.
223. Ibid., 341.
224. Ibid., 339.
225. Ibid., 338.
226. Ibid., 342.
227. "The English Revolution of midcentury contributed decisively to the formation of both representative institutions and modern science" (Jacob, "Reflections on Bruno Latour's Version of the Seventeenth Century," 241).
228. Michel Callon and Bruno Latour, "Unscrewing the Big Leviathan: How Actors Macro-Structure Reality and How Sociologists Help Them to Do So," in *Advances in Social Theory and Methodology: Toward an Integration of Micro- and Macro-Sociologies*, ed. Karin Knorr-Cetina and Aaron V. Cicourel (Boston: Routledge & Kegan Paul, 1981), 277–303; and Latour, *The Pasteurization of France*. Shapin and Schaffer refer expressly to these in *Leviathan and the Air Pump*, 340 n. 10.
229. Latour, "Postmodern? No, Simply Amodern"; and Latour, *We Have Never Been Modern* (Cambridge, Mass.: Harvard University Press, 1993).
230. Callon and Latour, "Unscrewing the Big Leviathan," 278.
231. Steve Fuller, "Talking Metaphysical Turkey about Epistemological Chicken, and the Poop on Pidgins," in *The Disunity of Science*, ed. Peter Galison and David Stump (Stanford, Calif.: Stanford University Press, 1996), 170–86, citing 177.
232. Callon and Latour, "Unscrewing the Big Leviathan," 279.
233. Ibid., 292.
234. Latour, "Postmodern? No, Simply Amodern," 147.
235. Ibid., 147.
236. Ibid., 151.
237. Ibid., 152.
238. Ibid.
239. Ibid., 158.
240. Ibid.
241. Ibid., 159.

242. Ibid., 169.

243. Margaret Jacob, "Science Studies after Social Construction," in *The Disunity of Science*, 95–120, citing 102.

244. Ibid.

245. Ibid.

246. Ibid., 103.

247. Jacob, "Reflections on Bruno Latour's Version of the Seventeenth Century," 242.

248. Dick Pels, "Have We Never Been Modern? Towards a Demontage of Latour's Modern Constitution," *History of the Human Sciences* 8 (1995): 129–41.

249. Christopher Norris, "Why Strong Sociologists Abhor a Vacuum: Shapin and Schaffer on the Boyle/Hobbes Controversy," *Philosophy and Social Criticism* 23 (1997): 9–40, citing 9.

250. Ibid., 14.

251. Shapin and Schaffer, *Leviathan and the Air Pump*, 13.

252. Norris, "Why Strong Sociologists Abhor a Vacuum," 27.

253. Ibid., 32.

254. Ibid., 38. Sokal and Bricmont make a similar argument: "The evidence of the Earth's rotation is vastly stronger than anything Kuhn could put forward in support of his historical theories. This does not mean, of course, that physicists are more clever than historians, or that they use better methods, but simply that they deal with less complex problems, involving a small number of variables which, moreover, are easier to measure and control" (*Fashionable Nonsense: Postmodern Intellectuals' Abuse of Science* [New York: Picador, 1998], 77).

255. Philip Kitcher, *The Advancement of Science* (Oxford: Oxford University Press, 1993), 295.

256. Ibid., 297.

257. Philip Kitcher, "A Plea for Science Studies," in *A House Built on Sand*, ed. Koertge, 54 n. 17.

258. Shapin and Schaffer, *Leviathan and the Air Pump*, 344.

259. Thomas S. Kuhn, *The Trouble with the Historical Philosophy of Science* (Robert and Maurine Rothschild Distinguished Lecture, 19 November 1991, Department of History of Science, Harvard University, 1992), 3.

260. Ibid., 9.

261. Ibid., 7.

262. Skuli Sigurdsson, "The Nature of Scientific Knowledge: An Interview with Thomas Kuhn," *Harvard Science Review* (winter, 1990): 18–25, citing 24.

263. Steve Fuller, *Thomas Kuhn: A Philosophical History for Our Times* (Chicago: University of Chicago Press, 2000), 388.

264. "Autobiographical Interview," in *The Road since Structure*, by Kuhn, ed. James Conant and John Haugland (Chicago: University of Chicago Press, 2000), 315.

265. Ibid., 317.

266. Paul Gross has, in fact, written a rather favorable obituary notice on Kuhn: "Bête Noire of the Science Worshipers," *History of the Human Sciences* 10 (1997): 125–28. For Gross, there are new enemies to be engaged, e.g., Bruno Latour.

267. Trevor Pinch, "Kuhn—the Conservative and Radical Interpretations," 478. Pinch and Harry Collins have taken up the polemic with Gross and Levitt from the side of radical constructionists, in their little diatribe *The Golem: What Everyone Should Know about Science* (Cambridge: Cambridge University Press, 1993).

268. For Fuller's fulminations, see "Being There with Thomas Kuhn: A Parable for Postmodern Times," *History and Theory* 31 (1992): 241–75; and his ubiquitous obituary notices

on Kuhn's "malignant" legacy: "Confessions of a Recovering Kuhnian," *Social Studies of Science* 27 (1997): 492–94; "Kuhn: A Personal Judgment," *History of the Human Sciences* 10 (1997): 129–31; "Kuhn As Trojan Horse," *Radical Philosophy* 82 (March/April 1997): 5–7.

269. Fuller, *Thomas Kuhn*, xv, xvi.

270. Ibid., 37.

271. Ibid., 4.

272. Ibid., 287.

273. Ibid., 293–95.

274. Let this be taken, as no doubt it will, as a swipe from a "guild" historian who cannot escape what Fuller charmingly characterizes as the vantage of "Prig history" (ibid., 21–26). I find Fuller's views on historical method utterly uncongenial, but this is not the locus to pursue that matter.

275. Ibid., 5.

276. Ibid., 380–81.

277. Rachel Laudan, "The 'New' History of Science: Implications for Philosophy of Science," *PSA 1992*, vol. 2, 479; Latour, according to Richard Rorty, had "taken the baton from Kuhn and run the next leg of the relay," carrying the essential message of liberation that " 'science does not have a privileged relation to the way things really are' " (Gross, "Bête Noire of the Science Worshipers," 126).

Chapter Seven

1. Paul Gross, "Bête Noire of the Science Worshipers," *History of the Human Sciences* 10 (1997): 125–28.

2. Rachel Laudan, "The 'New' History of Science: Implications for Philosophy of Science," *PSA 1992*, vol. 2, 479.

3. I got the anecdote from my friend Albert van Helden, and then I found it reported with all imaginable *Schadenfreude* in Paul Gross and Norman Levitt, *The Higher Superstition*, 2d ed. (Baltimore: Johns Hopkins University Press, 1998), xiii and 290. For Latour's take-off of Einstein, see Bruno Latour, "A Relativistic Account of Einstein's Relativity," *Social Studies of Science* 18 (1988): 3–44; John A Schumacher, "The Observer's Frame of Reference in Natural and Social Science: A Response to Latour," *Social Studies of Science* 18 (1988): 523–31; John Huth, "Latour's Relativity," in *A House Built on Sand: Exposing Postmodernist Myths about Science*, ed. Noretta Koertge (Oxford: Oxford University Press, 1998), 181–92. Huth observes that " 'A Relativistic Account' has, in fact, become a poster child for what can go wrong in science studies," "a classic example of how a misinterpreted theory ends up as grist for the social constructivist program" (182). Nonetheless, he notes, citation indices suggest that less than 1 percent of some 2,170 citations of Latour's work refer to this piece and "a colleague . . . working in the field of science studies referred to the article as 'silly' " (182). I used the term *farce* to describe the piece and this note is the only place I will consider it, for I cannot take it seriously as a representation of Einstein's science or, more pertinently, of important work in science studies. I see it as a self-indulgent literary lark with *one popular text* Einstein wrote. Texts are things the heirs of the French sixties (poststructuralism) are confident they can derange with impunity. The science of Einstein is another matter altogether. How tongue-in-cheek Latour may have been in writing this piece cannot be proven, though my sense is that his tongue is *usually* planted firmly in his cheek. If the editors of *Social Studies of Science* took the piece seriously as scientific exegesis, so much the worse for them, but my suspicion is they laughed all the way to the printers. The outrage among physicists, however, seems clear and its harsh institutional expression certainly puts such

sociological impertinence in its place! This is what gives pungency to the phrase "*bête noire* of the science worshipers."

4. David Bloor, "Reply to Bruno Latour," *Studies in History and Philosophy of Science* 30 (1999): 131–36, citing 132. See below, pp. 203–5.

5. The importance of the ludic construction of Latour reception can be seen in the title of Olga Amsterdamska's utterly humorless review, "Surely You Are Joking, Monsieur Latour!" *Science, Technology, and Human Values* 15 (1990): 495–504.

6. This Hegelian likeness is not merely my idiosyncratic response. Dick Pels notes that Latour is "a true heir to Hegel in our otherwise unHegelian times" ("Have We Never Been Modern?" *History of the Human Sciences* 8 [1995]: 132).

7. Michel Callon and Bruno Latour, "Don't Throw the Baby Out with the Bath School! A Reply to Collins and Yearley," in *Science As Practice and Culture*, ed. Andrew Pickering (Chicago: University of Chicago Press, 1992): 343–68, citing 344.

8. Stephen Cole calls Latour "the demigod of the constructivist movement" (Cole, "Voodoo Sociology: Recent Developments in the Sociology of Science," in *The Flight from Science and Reason*, ed. Paul Gross, Norman Levitt, and Martin Lewis [Baltimore: Johns Hopkins University Press, 1997], 274–87, citing 282). That, along with Richard Rorty's characterization of him as the "*bête noire* of the science-worshipers," makes Latour the most important figure in science studies at the close of the twentieth century.

9. Latour, *Pasteurization of France*, 162–63.

10. Ibid., 7.

11. Ibid., 191.

12. Ibid., 215.

13. Ibid., 5.

14. Ibid., 215.

15. Bruno Latour, "Postmodern? No, Simply Amodern: Steps Towards an Anthropology of Science," *Studies in History and Philosophy of Science* 21 (1990): 145–71"

16. "In the old days the struggle against magic was called the 'Enlightenment,' but this image has backfired. The Enlightenment has since become the age of (ir)radiation . . . the warhead of the missile that will blind us with light (Perhaps it is too late. Perhaps the missiles have already been launched . . .)" (Latour, *Pasteurization of France*, 213).

17. Ibid., 5.

18. Ibid., 234.

19. Ibid., 6. "Epistemologies . . . have always been war machines defending science against its enemies" (6). "All the failings of epistemology—its scorn of history, its rejection of empirical analysis, its pharisaic fear of impurity—are . . . qualities that are sought for in a frontier guard" (216).

20. Ibid., 5. See Michel Serres with Bruno Latour, *Conversations on Science, Culture, and Time* (Ann Arbor: University of Michigan Press, 1995); and Bruno Latour, "The Enlightenment without the Critique: An Introduction to Michel Serres," in *Contemporary French Philosophy*, ed. A. Phillips Griffiths (Cambridge: Cambridge University Press, 1987): 83–98.

21. Paul Valéry, "The Crisis of the Mind," in *Paul Valéry: An Anthology*, ed. James Lawler (Princeton, N.J.: Bollingen, 1956), 94–107.

22. Paul Fussell, *The Great War and Modern Memory* (London: Oxford University Press, 1975), 24.

23. Latour, *Pasteurization of France*, 236.

24. Ibid., 6.

25. Ibid., 153.

26. Ibid., 7.

27. Ibid., 8–9. See his early essay, "Insiders and Outsiders."

28. Latour, *Pasteurization of France*, 6. "The sociology of science is so congenitally weak" (229). Latour's contempt for sociology permeates the text: "The meaning of the 'social' continually shrinks—it has now been reduced to the level of 'social' problems. It is what is left when everything else has been divided up among the powerful; whatever is neither economic, technical, legal, nor anything else is left to it. . . . If *sociology* were (as its name suggests) the science of *associations* rather than the science of the social to which it was reduced in the nineteenth century, then perhaps we would be happy to call ourselves 'sociologists' " (205).

29. Bruno Latour, "Give Me a Laboratory and I Will Raise the World," in *Science Observed*, ed. Karin Knorr-Cetina and Michael Mulkay (London: Sage, 1983), 141–70, citing 157. "In our modern societies most of the really fresh power comes from sciences—no matter which—and not from the classical political process" (168).

30. Latour, *Pasteurization of France*, 9.

31. Ibid., 10.

32. Ibid., 192. Henceforth, propositions from "Irreductions" will be enumerated parenthetically in the text, without specific page references.

33. Ibid., 6.

34. "We have to give evidence that 'science' and 'society' are both explained more adequately by an analysis of the relations among forces and that they become mutually inexplicable and opaque when made to stand apart" (ibid., 7).

35. "Latour's insistence that winning is the only thing that matters in science has some very pernicious implications," Amsterdamska observes ("Surely You Are Joking," 501). "Latour's network theory of science and technology treats research as a kind of war whose only objective is domination" (499). He "sees only attempts to dominate, strategies for winning battles, means of attack, trials of strength, and other forms of violence" (496). More concisely, Steven Shapin observes that "Latour's trademark" entails that "technoscience is war conducted by much the same means" ("Following Scientists Around," *Social Studies of Science* 18 [1988]: 534).

36. Latour, *Pasteurization of France*, 7.

37. Ibid., 155.

38. Ibid., 154.

39. Ibid., 166. Nietzsche's metaphysic is central here: "an undiscriminated field of wills, point forces, and resistances," as Nick Lee and Steve Brown put it in "Otherness and the Actor Network," *American Behavioral Scientist* 37 (1994): 772–90, citing 774. Latour tries to circumscribe his Nietzscheanism: only "certain forces" seek to dominate; "I have said 'certain' rather than 'all' as in Nietzsche's bellicose myth" because many forces are "too happy and proud to take command of others. . . . I speak only of those weaknesses that want to increase their strength" (Latour, *Pasteurization of France*, 167).

40. The explicit homage to Hobbes is in the classic essay which Latour wrote with Michel Callon: "Unscrewing the Big Leviathan: How Actors Macro-Structure Reality and How Sociologists Help Them to Do So," in *Advances in Social Theory and Methodology*, ed. Karin Knorr-Cetina and Aaron V. Cicourel (Boston: Routledge & Kegan Paul, 1981).

41. Latour, no doubt, could reply: "There has never been such a thing as deduction" (2.1.2).

42. "We have to make a ninety-degree turn from the SSK yardstick and define a second dimension" (Callon and Latour, "Don't Throw the Baby Out with the Bath School," 348).

43. "In a world already structured by macro-actors, nothing could be poorer and more abstract than individual social interaction" (Callon and Latour, "Unscrewing the Big Leviathan," 300).

44. Latour works through *Leviathan and the Air Pump* first in his essay "Postmodern? No, Simply Amodern!" and then in *We Have Never Been Modern*, (Cambridge, Mass.: Harvard University Press, 1993), 5–27.

45. Latour, *We Have Never Been Modern*, 37–39.

46. Latour, *Pasteurization of France*, 203.

47. Ibid., 201.

48. Ibid., 208.

49. Ibid.

50. Ibid., 210.

51. Latour, *We Have Never Been Modern*, 6.

52. Ibid., 14.

53. Latour, "Postmodern? No, Simply Amodern!" 165.

54. Latour, *We Have Never Been Modern*, 7.

55. Ibid., 32. "The essential point of this modern Constitution is that it renders the work of mediation that assembles hybrids invisible, unthinkable, unrepresentable" (34).

56. Ibid., 56.

57. Ibid.

58. Ibid., 112.

59. Ibid., 12.

60. Ibid., 30. For a discerning critique of these ideas, see Dick Pels, "Have We Never Been Modern?"

61. Latour, *We Have Never Been Modern*, 7.

62. "If we thought that termites were better philosophers than Leibniz, we could compare a network to a termite's nest" ("Irreductions," 1.4.3).

63. "Bruno Latour has been following scientists around for years. Now he wants us to follow him following them around" (Shapin, "Following Scientists Around," 533).

64. Bruno Latour, *Science in Action* (Cambridge, Mass.: Harvard University Press, 1987), 220.

65. Of course, Latour repudiates Hegel explicitly: "by believing that he was abolishing Kant's separation between things-in-themselves and the subject, Hegel brought the separation even more fully to life. He raised it to the level of a contradiction, pushed it to the limit and beyond, then made it the driving force of history" (Latour, *We Have Never Been Modern*, 57).

66. Ibid., 78.

67. Ibid.

68. Of course, Latour repudiates dialectics: "Dialectics literally beat about the bush" (*We Have Never Been Modern*, 55). Still, "the straight lines of philosophy are of no use when it is the crooked labyrinth of machinery and machinations . . . we have to explore" (Bruno Latour, "On Technical Mediation—Philosophy, Sociology, Genealogy," *Common Knowledge* 3 [1994]: 30).

69. This is why Latour invokes the line from Tournier: "Beware of purity. It is the vitriol of the soul" (*Pasteurization of France*, 201).

70. Latour, *Science in Action*, 143.

71. Ibid., 223.

72. Ibid., 219.

73. Ibid., 254.

74. Ibid., 13.

75. Ibid., 29.

76. Ibid., 108.

77. Ibid., 42.

78. Ibid., 29.

79. Ibid., 38.

80. Ibid., 63.

81. Ibid., 67, 69.

82. Ibid., 90.

83. Ibid., 93.

84. Ibid., 94.

85. Ibid., 99.

86. Ibid., 100.

87. Ibid., 92.

88. Ibid., 97.

89. For scientists' exasperation, see Alan Sokal and Jean Bricmont, *Fashionable Non-sense*, 92–99, 124–33; and Huth, "Latour's Relativity."

90. Jan Golinski, "The Theory of Practice and the Practice of Theory: Sociological Approaches in the History of Science," *Isis* 81 [1990]: 492–505.

91. Alan Sokal, "What the *Social Text* Affair Does and Does Not Prove," in *A House Built on Sand*, ed. Koertge, 9–22, citing 17.

92. Latour, *Science in Action*, 142.

93. Callon and Latour, "Don't Throw the Baby Out with the Bath School," 348.

94. Latour, *Science in Action*, 166.

95. Ibid., 167.

96. Ibid., 171–72.

97. Ibid., 170.

98. Ibid., 166.

99. Ibid., 169.

100. Ibid., 166.

101. Shapin, "Following Scientists Around," 539.

102. See Callon and Latour, "Unscrewing the Big Leviathan."

103. Michel Callon and John Law, "On Interests and Their Transformation: Enrolment and Counter-Enrolment," *Social Studies of Science* 12 (1982): 615–25, citing 622.

104. Ibid., 621.

105. Michel Callon, "Struggles and Negotiations to Define What Is Problematic and What Is Not: The Socio-logic of Translation," in *The Social Process of Scientific Investigation*, ed. Karin Knorr, Roger Krohn, and Richard Whitley (Dordrecht: Reidel, 1981), 197–219, citing 197.

106. Ibid.

107. Ibid.

108. Ibid., 198.

109. Ibid., 213.

110. Ibid., 206.

111. Ibid., 198.

112. Ibid., 206.

113. Thomas S. Kuhn, *The Structure of Scientific Revolutions*, 2d ed. (Chicago: University of Chicago Press, 1970), 19–20.

114. "We do not try to undermine the solidity of the accepted parts of science" (Latour, *Science in Action*, 100).

115. Callon, "Struggles and Negotiations," 206.

116. Michel Callon, "Some Elements of a Sociology of Translation: Domestication of the Scallops and Fishermen of St. Brieuc Bay," in *Power, Action, and Belief: A New Sociology of Knowledge?* ed. John Law (London: Routledge & Kegan Paul, 1986), 196–233.

346 NOTES TO PAGES 197–199

117. Michel Callon and John Law, "On the Construction of Sociotechnical Networks: Content and Context Revisited," *Knowledge and Society: Studies in the Sociology of Culture Past and Present* 8 (1989): 57–83.

118. Ibid., 64.

119. Ibid., 57.

120. Ibid., 77.

121. John Law and Michel Callon, "The Life and Death of an Aircraft: A Network Analysis of Technical Change," in *Shaping Technology/Building Society,* ed. Wiebe Bijker and John Law (Cambridge, Mass.: MIT Press, 1992), 21–52, citing 49.

122. Ibid., 144.

123. Latour and Callon, "Unscrewing the Big Leviathan," 287.

124. Latour, *We Have Never Been Modern,* 3.

125. Michel Callon, "Techno-economic Networks and Irreversibility," in *A Sociology of Monsters: Essays on Power, Technology, and Domination,* ed. John Law (New York: Routledge, 1991), 132–61.

126. Ibid., 142.

127. Bruno Latour, "Technology Is Society Made Durable," in *A Sociology of Monsters,* ed. John Law, 103–31, citing 124.

128. Michel Callon and John Law, "Agency and the Hybrid *Collectif,*" *South Atlantic Quarterly* 94 (1995): 481–507.

129. Ibid., 481.

130. Ibid., 484.

131. Ibid., 485.

132. Ibid., 500.

133. Ibid., 487. "It's French philosophy, with its poststructuralism, which has most ruthlessly tried to decenter the subject," but "radical English language thinkers," especially "radical 'postmodern' feminists" have "picked [it] up and pressed [it] home" (499). See below, pp. 209–15.

134. Ibid., 496.

135. Shapin, "Following Scientists Around," 547.

136. Ibid., 542.

137. Bruno Latour, "Clothing the Naked Truth," in *Dismantling Truth: Reality in the Postmodern World,* ed. Hilary Lawson and Lisa Appignanesi (New York: St Martin's, 1989), 101–26, citing 120.

138. Latour, "Postmodern? No, Simply Amodern!" 170.

139. Latour, "On Technical Mediation," 32–33.

140. Bruno Latour, "Where Are the Missing Masses? The Sociology of a Few Mundane Artifacts," in *Shaping Technology/Building Society,* ed. Bijker and Law, 225–58; Bruno Latour [Jim Johnson, pseud.], "Mixing Humans and Nonhumans Together: The Sociology of a Door-Closer," *Social Problems* 35 (1988): 298–310.

141. Latour [Johnson, pseud.], "Mixing Humans and Nonhumans," 310.

142. Latour, "On Technical Mediation," 34.

143. Ibid., 46.

144. Latour, "Where Are the Missing Masses?" 231, 233.

145. Latour, "Technology Is Society Made Durable," esp. 106.

146. Latour, "On Technical Mediation," 46.

147. Ibid., 34.

148. Latour, "Where Are the Missing Masses?" 243.

149. Shirley C. Strum and Bruno Latour, "Redefining the Social Link: From Baboons to Humans," *Social Science Information* 26 (1987): 783–802. See also the discussion in Callon and Latour, "Unscrewing the Big Leviathan," 281–83.

150. Strum and Latour, "Redefining the Social Link," 788.

151. Latour, "On Technical Mediation," 50.

152. Callon and Latour, "Unscrewing the Big Leviathan," 284.

153. Latour, "On Technical Mediation," 52.

154. Latour and Callon, "Unscrewing the Big Leviathan," 285.

155. Ibid., 47.

156. Ibid., 53.

157. Ibid., 56. See Bruno Latour, "Pragmatogonies: A Mythical Account of How Humans and Nonhumans Swap Properties," *American Behavioral Scientist* 37 (1994): 791–808.

158. Latour, "On Technical Mediation," 59.

159. Ibid., 58.

160. Ibid., 61.

161. Callon and Latour, "Unscrewing the Big Leviathan," 292.

162. Ibid., 294.

163. Ibid.

164. Ibid., 297.

165. Ibid., 299.

166. Latour, "On Technical Mediation," 53.

167. Lee and Brown, "Otherness and the Actor Network," 775. They note and critique Latour's rhetoric of liberal democracy on behalf of the nonhuman.

168. Harry M. Collins and Steven Yearley, "Epistemological Chicken," in *Science As Practice and Culture*, ed. Pickering, 310.

169. Ibid., 313 n.

170. Simon Schaffer, "The Eighteenth Brumaire of Bruno Latour," *Studies in History and Philosophy of Science* 22 (1991): 174–92, citing 189.

171. Ibid., 187.

172. Ibid., 186.

173. Lee and Brown, "Otherness and the Actor Network," 781.

174. Ibid., 785.

175. Oscar Kenshur, "The Allure of the Hybrid: Bruno Latour and the Search for a New Grand Theory," in *The Flight from Science and Reason*, ed. Gross, Levitt, and Lewis, 288–97, citing 293.

176. Yves Gingras, "Following Scientists through Society? Yes, but at Arm's Length!" in *Scientific Practice: Theories and Stories of Doing Physics*, ed. Jed Z. Buchwald (Chicago: University of Chicago Press, 1995), 123–48, citing 125.

177. Ibid., 137.

178. Ibid., 133.

179. Ibid., 131.

180. Latour, "Pragmatogonies," 795; Latour, "On Technical Mediation," 54.

181. Latour, "Pragmatogonies," 795.

182. Ibid., 794.

183. Latour, "On Technical Mediation," 49.

184. Latour, "Technology Is Society Made Durable," 111.

185. Ibid., 120.

186. Bruno Latour, "One More Turn after the Social Turn," in *The Social Dimension of Science*, ed. Ernan McMullin (Notre Dame, Ind.: University of Notre Dame Press, 1992), 272–94, citing 282–83.

187. Bruno Latour, "The Force and Reason of Experiment," in *Experimental Inquiries: Historical, Philosophical and Social Studies of Experimentation in Science*, ed. Homer E. LeGrand (Dordrecht: Kluwer, 1990), 49–80, citing 65.

188. Ibid., 66. Latour invokes the work of Prigogine and Stengers for the argument that

"irreversible events are not defined by their conditions of possibility" (78). See Ilya Prigogine and Isabelle Stengers, *Entre le temps et l'eternite* (Paris: Fayard, 1988).

189. Collins and Yearley, "Epistemological Chicken"; Callon and Latour, "Don't Throw the Baby Out with the Bath School!"; and Collins and Yearley, "Journey into Space," in *Science As Practice and Culture,* ed. Pickering, 369–89.

190. Latour, "One More Turn after the Social Turn."

191. Isaiah Berlin, *The Hedgehog and the Fox: As Essay in the Old Criticism* (London: Weidenfeld & Nicolson, 1954).

192. Bruno Latour, "For David Bloor . . . and Beyond: A Reply to David Bloor's 'Anti-Latour,'" *Studies in History and Philosophy of Science* 30 (1999): 113–29, citing 115.

193. Latour, "One More Turn after the Social Turn," 279.

194. Latour, *We Have Never Been Modern,* 54.

195. Latour, "One More Turn after the Social Turn," 278.

196. Ibid., 279.

197. Ibid., 276.

198. Ibid., 272. And see Callon and Latour, "Don't Throw the Baby Out with the Bath School!" 344.

199. Callon and Latour, "Don't Throw the Baby Out with the Bath School!" 351.

200. Ibid., 348.

201. Latour, "One More Turn after the Social Turn," 282.

202. David Bloor, "Anti-Latour," *Studies in History and Philosophy of Science* 30 (1999): 81–112.

203. Ibid., 81.

204. Ibid., 82.

205. Ibid., 87.

206. Ibid., 82.

207. Ibid., 85.

208. Ibid., 95.

209. Ibid., 107.

210. Ibid., 97.

211. Ibid., 92.

212. Ibid., 98.

213. Ibid., 106.

214. Latour, "For David Bloor," 116.

215. Ibid., 120.

216. Ibid., 123.

217. Ibid., 125.

218. Ibid., 126.

219. By 1999, Latour is sensitive to the issue of the "science wars" (ibid., 127).

220. Ibid., 122.

221. Ibid., 134.

222. Bloor, "Reply to Bruno Latour," 131.

223. Ibid., 132.

224. Sandra Harding, *The Science Question in Feminism* (Ithaca, N.Y.: Cornell University Press, 1986), 198. Ruth Bleier, in her introduction to an important anthology, observes parenthetically: "It is noteworthy, incidentally, that the sociology of scientific knowledge . . . remains peculiarly oblivious, so far as I can tell, to the entire body of feminist scholarship in all disciplines, which contributes fundamentally, both theoretically and methodologically, to the sociology of knowledge including scientific knowledge; nor does it appear to recognize gender as a social category of possible significance in understanding the social production of

scientific knowledge" (introduction to *Feminist Approaches to Science*, ed. Ruth Bleier [New York: Pergamon, 1986], 5).

225. I take this periodization from Josephine Donovan, *Feminist Theory: The Intellectual Traditions of American Feminism* (New York: Continuum, 1998). Donovan discusses feminist theory of science in her final chapter, 187–208.

226. Sandra Harding, *Whose Science? Whose Knowledge? Thinking from Women's Lives* (Ithaca, N.Y.: Cornell University Press, 1991), 19.

227. Donna Haraway, "In the Beginning was the Word: The Genesis of Biological Theory" (1981), reprinted in *Simians, Cyborgs, and Women*, 80.

228. A crucial documentation of this problem was Margaret Rossiter, *Women Scientists in America* (Baltimore: Johns Hopkins University Press, 1982).

229. Haraway, "In the Beginning," 71.

230. Ibid., 79.

231. Ibid., 78.

232. Harding, *Science Question*, 9.

233. Ibid., 29.

234. Ibid., 9.

235. Ibid., 16.

236. Evelyn Fox Keller, "Making a Difference: Feminist Movement and Feminist Critique of Science," in *Feminism in Twentieth-Century Science, Technology, and Medicine*, ed. Angela Creager, Elizabeth Lunbeck, and Londa Schiebinger (Chicago: University of Chicago Press, 2001), 98–109, citing 98. Keller was one of the pioneers in this domain of "gender and science." Her essay, "Gender and Science" (in *Discovering Reality: Feminist Perspectives on Epistemology, Metaphysics, Methodology, and Philosophy of Science*, ed. Sandra Harding and Merrill Hintikka [Dordrect: Reidel, 1983], 187–205) was one of the first contributions to the field, and her book, *Reflections on Gender and Science* (New Haven, Conn.: Yale University Press, 1985), won acclaim for its articulation of a feminist vantage on doing science. With Helen Longino, Keller edited one of the most effective anthologies of feminist theory of science, *Feminism and Science* (Oxford: Oxford University Press, 1992). Later still she chronicled the field in "The Origin, History, and Politics of the Subject Called 'Gender and Science,'" in *Handbook of Science and Technology Studies*, ed. Sheila Jasanoff et al. (Thousand Oaks, Calif.: Sage, 1995), 80–94.

237. Sandra Harding, "Why Has the Sex/Gender Distinction Become Visible Only Now?" in *Discovering Reality*, ed. Harding and Hintikka, 311–24.

238. Scott Gilbert et al., "The Importance of Feminist Critique for Contemporary Cell Biology" (1988), reprinted in *Feminism and Science*, ed. Nancy Tuana (Bloomington: Indiana University Press, 1989), 172–87.

239. A "state of war" emerged between the social and the life sciences in this context (Donna Haraway, "Primatology Is Politics by Other Means," originally in *PSA 1984*, vol. 2; revised and included in *Feminist Approaches to Science*, 77–118, citing 89). Feminists attacked the prevailing presumptions of "sociobiology."

240. Quite simply, sex was *not* to gender as nature to culture. Donna Haraway raised this objection in "Primatology Is Politics," 85; and more extensively in "'Gender' for a Marxist Dictionary: The Sexual Politics of a Word," reprinted in *Simians, Cyborgs, and Women*, 127–48. Evelyn Fox Keller explored it deftly in "The Gender/Science System: Or, Is Sex to Gender As Nature Is to Science?" *Hypatia* 2 (1987): 37–49.

241. Harding, *Whose Science?* 31.

242. Ibid., 127.

243. Ibid., 106.

244. Ibid., 117.

245. Ibid., 68. See also Ruth Hubbard, "Some Thoughts on the Masculinity of the Natural Sciences," in *Feminist Thought and the Structure of Knowledge,* ed. Mary McCanney Gergen (New York: New York University Press, 1988), 1–15.

246. Harding, *Science Question,* 63.

247. Ibid., 82.

248. Ibid., 18.

249. Decisive for the interpretation of the first issue was Carolyn Merchant, *The Death of Nature: Women, Ecology, and the Scientific Revolution* (1980; reprint, San Francisco: Harper-Collins, 1990). On the question of gendered division of labor, important work was done by Hilary Rose, "Hand, Brain, and Heart: A Feminist Epistemology for the Natural Sciences" (1983), reprinted in *Sex and Scientific Inquiry,* ed. Sandra Harding and Jean O'Barr (Chicago: University of Chicago Press, 1987), 265–82; and Nancy Hartsock, "The Feminist Standpoint: Developing the Ground for a Specifically Feminist Historical Materialism," in *Discovering Reality,* ed. Harding and Hintikka, 283–310. On the question of psychological identity, the most important source was "object relations theory" in psychoanalysis—Nancy Chodorow, *The Reproduction of Mothering* (Berkeley: University of California Press, 1978); Dorothy Dinnerstein, *The Mermaid and the Minotaur* (New York: Harper & Row, 1976)— and the application of this to the problem of masculinity in science by Evelyn Fox Keller, "Gender and Science" (1978), reprinted in *Discovering Reality,* 187–205. The "man of reason" is, of course, Genevieve Lloyd's construction: *The 'Man of Reason'* (Minneapolis: University of Minnesota Press, 1984).

250. Harding, *Science Question,* 121, referring to Keller, "Gender and Science."

251. Harding, *Science Question,* 90.

252. Ibid., 105.

253. Ibid., 103.

254. Ibid., 67.

255. "The intellectual danger resides in viewing science as pure social product; science then dissolves into ideology and objectivity loses all intrinsic meaning. In the resulting cultural relativism, any emancipatory function of modern science is negated, and the arbitration of truth recedes into the political domain" (Evelyn Fox Keller, "Feminism and Science" [1982], reprinted in *Sex and Scientific Inquiry,* 233–46, citing 237). Keller feared cultural relativism would lead feminism out of "the realpolitik modern culture" (237).

256. Harding, *Science Question,* 24–27, 136–62.

257. Harding, *Whose Science?* 97.

258. Ibid., 126.

259. Ibid., 59. See Georg Wilhelm Friedrich Hegel, *Phenomenology of Spirit* (Oxford: Oxford University Press, 1977), 111–19; Georg Lukács, *History and Class Consciousness* (Cambridge, Mass.: MIT Press, 1971).

260. Harding, *Whose Science?* 158.

261. Ibid., 127.

262. Ibid., 71.

263. Harding, *Science Question,* 26.

264. "Consciousness raising" by definition entails not spontaneity but incitement.

265. Postmodernism turns largely on the ostensibly pernicious implications of the imbrication: "author/authorization/authority."

266. Hence the queries in Harding's second book title: *Whose Science? Whose Knowledge?*

267. Harding, *Whose Science?* 37.

268. Ibid.

269. Ibid., 11.

270. Ibid., 6.

271. Ibid., 40.

272. Ibid., 59.

273. Ibid., 81.

274. Harding gets into trouble trying to *prove* that the contents of mathematics are (without considerable mediation) socially constructed (*Science Question,* 48–52). Here she is following Bloor into folly (she specifically invokes his work, 49 n). Latour puts it precisely: "we do not have to waste time looking for 'social explanations' of these forms, if by social is meant features of society mirrored by mathematics in some distorted way" (*Science in Action,* 246). A contextualist theory of science need not *derive* every formal argument from social forces. It need only situate the *relative entrenchment* of ideas in historical terms. Some fields simply *are* more "internal"—not because they are "harder" or "purer," but because they *exclude* so much reality.

275. Harding, *Science Question,* 239.

276. Harding, *Whose Science?* 40.

277. Ibid., 94.

278. Ibid., 78.

279. Ibid., 94.

280. Ibid., 149.

281. Ibid., 75.

282. Ibid., 143.

283. Ibid., 149. Developing her idea of "strong objectivity" in response to Haraway's challenge represents the major intellectual move from Harding's first book (*The Science Question*) to her second (*Whose Science?*).

284. Ibid., 156.

285. Ibid., 152.

286. Ibid., 159.

287. Ibid., 187.

288. Ibid., 142.

289. Ibid.

290. "The essentialism—anti-essentialism debates define 80s feminism" (Naomi Schor, introduction to *The Essential Difference,* ed. Naomi Schor and Elizabeth Weed [Bloomington: Indiana University Press, 1994], vii). See also Diana Fuss, *Essentially Speaking: Feminism, Nature, and Difference* (New York: Routledge, 1989).

291. Linda Alcoff, "Cultural Feminism versus Post-Structuralism: The Identity Crisis in Feminist Theory," *Signs* 13 (1988): 405–36.

292. Haraway discusses this in "A Cyborg Manifesto: Science, Technology, and Socialist-Feminism in the Late Twentieth Century," reprinted in *Simians, Cyborgs, and Women: The Reinvention of Nature* (New York: Routledge, 1991), 155–61. Keller notes: "as we have learned, it makes dubious sense to ask about women's experience without first asking, and quickly, Which women?" ("Making a Difference: Feminist Movement and Feminist Critique of Science," 100).

293. Cherríe Moraga and Gloria Anzaldúa, eds., *This Bridge Called My Back: Writings by Radical Women of Color* (Watertown, Mass.: Persephone, 1981); Gloria Hull, Patricia Bell Scott, and Barbara Smith, eds., *All the Women Are White, All the Men Are Black, But Some of Us Are Brave* (Westbury, N.Y.: Feminist Press, 1982); bell hooks, *Feminist Theory: From Margin to Center* (Boston: South End, 1984); Chandra Mohanty, "Under Western Eyes: Feminist Scholarship and Colonial Discourse," *Boundary* 2 3 (1984): 333–58.

294. For a recent and complex assessment of these issues, see *Reclaiming Identity: Realist Theory and the Predicament of Postmodernism,* ed. Paula Moya and Michael Hames-Garcia (Berkeley: University of California Press, 2000).

295. This overly schematic sketch seeks to evoke the thought of Barthes, Derrida, Foucault, and their myriad epigoni.

296. Harding ambivalently offers this characterization of postmodern feminism in each of her two books. The leading feminist postmodernists, according to Harding, were Jane Flax and Donna Haraway. In *Science Question* Harding upholds the standpoint position against postmodern feminism. In *Whose Science?* she seeks to find an accommodation between standpoint theory and postmodernism.

297. Kenneth Gergen, "Feminist Critique of Science and the Challenge of Social Epistemology," in *Feminist Thought and the Structure of Knowledge,* ed. M. M. Gergen.

298. The decisive confrontation of feminism with poststructuralist theory is to be found in the essays collected in *Feminism/Postmodernism,* ed. Linda Nicholson (New York: Routledge, 1990). For unequivocal advocacy of postmodern feminism, see Susan Hekman, *Gender and Knowledge: Elements of a Postmodern Feminism* (Cambridge, U.K.: Polity, 1990).

299. Donna Haraway, "Primatology Is Politics," 80. See also "Animal Sociology and a Natural Economy of the Body Politic" and "The Past Is a Contested Zone," in *Simians, Cyborgs and Women,* by Haraway, 7–20 and 21–42, respectively; and her monograph *Primate Visions: Gender, Race, and Nature in the World of Modern Science* (New York: Routledge, 1989).

300. Haraway, "Primatology Is Politics," 79.

301. Ibid., 80, 87.

302. Ibid., 95.

303. Ibid., 94. The key text here is Sherry Ortner, "Is Female to Male As Nature Is to Culture?" in *Woman, Culture, and Society,* ed. Michelle Zimbalist Rosaldo and Louise Lamphere (Stanford, Calif.: Stanford University Press, 1974), 67–88.

304. Haraway, "Primatology Is Politics," 95.

305. Not everyone—and particularly not the primatologists—has found her characterization satisfactory. See Linda Fedigan, "The Paradox of Feminist Primatology: The Goddess's Discipline?" in *Feminism in Twentieth-Century Science, Technology, and Medicine,* 46–72, and Londa Schiebinger, *Has Feminism Changed Science?* (Cambridge, Mass.: Harvard University Press, 1999), 126–39.

306. Haraway, "Primatology Is Politics," 79. It follows that "as orientalism is deconstructed politically and semiotically, the identities of the occident destabilize, including those of feminists" (Donna Haraway, "Cyborg Manifesto," 156).

307. Haraway, "Primatology Is Politics," 91.

308. Ibid., 86.

309. Haraway, "Cyborg Manifesto," 149–81. See her introduction to *Simians, Cyborgs, and Women,* 1–4.

310. Haraway, "Cyborg Manifesto," 150.

311. Haraway, "Primatology Is Politics,"84.

312. Haraway, "Cyborg Manifesto," 153.

313. Ibid., 164, 161.

314. Ibid., 177.

315. Ibid., 165.

316. Ibid.

317. Haraway, "Primatology Is Politics," discussed above.

318. Haraway, "Cyborg Manifesto," 164.

319. Haraway, "Primatology Is Politics," 83.

320. Haraway, "Situated Knowledges: The Science Question in Feminism and the Privilege of the Partial Perspective" (1988), reprinted in *Simians, Cyborgs and Women*, 183–201, citing 248 n. 2.

321. Haraway, "Cyborg Manifesto," 154.

322. Haraway, "Primatology Is Politics," 86.

323. Haraway, "Cyborg Manifesto," 157.

324. Ibid., 150.

325. Ibid., 176.

326. Ibid., 181.

327. Haraway, "Primatology Is Politics," 85.

328. Ibid., 86. "Organicism is the analytical longing for a natural body, for purity outside the disruptions of the 'artificial.' It is the reversed, mirror image of other forms of longing for transcendence" (86).

329. Haraway, "Situated Knowledges," 183–201.

330. Ibid., 183. Haraway explains the context of the essay in a footnote (248 n. 1).

331. Ibid., 184–87.

332. Ibid., 188.

333. Ibid., 191.

334. Ibid., 190.

335. Ibid., 196.

336. Ibid., 193.

337. Ibid., 196–98.

338. Ibid., 194.

339. Ibid., 193.

340. Ibid., 191, 193. "The standpoints of the subjugated are not 'innocent' positions. On the contrary, they are preferred because in principle they are least likely to allow denial of the critical and interpretive core of all knowledge" (191).

341. Longino develops "contextual empiricism" in a series of works, most accessibly in *Science As Social Knowledge: Values and Objectivity in Scientific Inquiry* (Princeton, N.J.: Princeton University Press, 1990); and "Essential Tensions—Phase Two: Feminist, Philosophical, and Social Studies of Science," in *The Social Dimension of Science*, ed. McMullin, 198–216.

342. Harding discussed Longino's collaborative essay with Ruth Doell ("Body, Bias, and Behaviour: A Comparative Analysis of Reasoning in Two Areas of Biological Science" [1983], reprinted in *Feminism and Science*, ed. Keller and Longino, 73–90) in *Science Question*, 92–107.

343. In "Can There Be a Feminist Science?" (1987; reprinted in *Feminism and Science*, ed. Tuana, 45–57), Longino claimed that Harding had misunderstood the character of her feminist empiricism. In later writing Longino charged that "much of Sandra Harding's work" actually "reviled empiricism as a form of scientism" ("Essential Tensions," 202). She argued that what Harding criticized "is the claim that methods currently in use in the natural sciences are sufficient to eliminate masculinist or other bias in the sciences." But that was misguided on two counts: first, in its notion of "methods" (i.e., specific procedures, like "mathematical modeling . . . titration techniques, etc."), and second, more importantly, in alleging feminist empiricists could believe that bias in the sciences could be eliminated on the terms of these very biased sciences (215). Longino had never proposed either idea and found both inept. Harding herself backed away from claims about specific method: see Sandra Harding, "Is There a Feminist Method?" in *Feminism and Science*, 17–32. Rather, she there claimed that what feminism critiqued was "the fetishization of method itself" (18). Whereas earlier Harding had expressed a disingenuous puzzlement about what Longino might have

meant by claiming to be "doing science as a feminist," she turned about in this essay to identify feminism not with any particularly feminist science but rather as "fundamentally a moral
and political movement for the emancipation of women" (30), a position perfectly compatible with Longino's.

344. Longino, "Essential Tensions," 199.

345. Ibid.

346. Longino, *Science As Social Knowledge,* 212.

347. Ibid., 213.

348. Ibid., 188.

349. Ibid., 190; see also her earlier version of this: "Can There Be a Feminist Science?"

350. Longino, "Essential Tensions," 200. Longino adds, in a footnote, that "Kuhn, himself, rejects this interpretation," but that "he continues to be cited as the intellectual and
philosophical legitimation of the sociological approach" (215). This is welcome corroboration of my argument in chapter 5.

351. Ibid., 203.

352. Longino, *Science As Social Knowledge,* 4.

353. Ibid., 5–6.

354. Longino, "Can There Be a Feminist Science?" 47.

355. Longino, *Science As Social Knowledge,* 12.

356. Ibid., 219.

357. Ibid., 191.

358. Ibid., 219.

359. Ibid., 221.

360. Ibid.

361. Longino, "Essential Tensions," 208.

362. Ibid., 201.

363. "The objectivity of scientific inquiry is a consequence of that inquiry's being a social
and not an individual enterprise" (ibid., 205).

364. Ibid., 207.

365. Longino, *Science As Social Knowledge,* 68.

366. Ibid., 62.

367. Ibid., 13, 12.

368. Longino, "Essential Tensions," 205.

369. This is so, notwithstanding Paul Roth's attack on the causal conception in these
studies ("What Does the Sociology of Scientific Knowledge Explain? Or, When Epistemological Chickens Come Home to Roost," *History of the Human Sciences* 7 [1994]: 95–108).
Enough—not everything, but enough—has been done to secure *this* point.

370. Longino points to Kuhn's effort (in "Objectivity, Value Judgment, and Theory
Choice") to articulate five "cognitive values" that helped determine theory choice rationally
(Longino, *Science As Social Knowledge,* 77).

371. Longino takes this argument in a very interesting direction in "Cognitive and Non-
Cognitive Values in Science: Rethinking the Dichotomy," in *Feminism, Science, and the Philosophy of Science,* ed. Lynn Hankinson Nelson and Jack Nelson (Dordrecht: Kluwer, 1996),
39–58. See below, pp. 221–22.

372. Longino, *Science As Social Knowledge,* 185.

373. Ibid., 232.

374. Ibid., 191.

375. Ibid., 76.

376. Ibid., 77.

377. Ibid.

378. Ibid. Here, again, Longino is working along the lines Kuhn articulated in his essay, "Objectivity, Value Judgment, and Theory Choice" (in *The Essential Tension* [Chicago: University of Chicago Press, 1977]).

379. Longino, *Science As Social Knowledge*, 78.

380. Ibid., 79.

381. Helen Longino, "Subjects, Powers, and Knowledge: Description and Prescription in Feminist Philosophies of Science," in *Feminist Epistemologies,* ed. Linda Alcoff and Elizabeth Potter (New York: Routledge, 1993), 101–20, citing 112.

382. Longino, *Science As Social Knowledge*, 80.

383. Ibid., 191.

384. Ibid., 76.

385. Ibid., 100.

386. Ibid., 191.

387. Ibid., 192.

388. Ibid., 225.

389. Ibid., 216.

390. Ibid., 79.

391. Longino, "Essential Tensions," 207.

392. Longino, *Science As Social Knowledge,* 79. "Pragmatic-epistemic needs, that is, what we need to know the truth about, and metaphysical assumptions derived from such needs as well as from social experience and aspiration provide stabilizing frameworks for the selection and interpretation of data" (223–24).

393. Ibid., 80.

394. Longino, "Essential Tension," 207.

395. Longino, "Subjects, Powers, and Knowledge," 114.

396. Longino, "Cognitive and Non-Cognitive Values," 41–43.

397. Kuhn, "Objectivity, Value Judgment, and Theory Choice."

398. Longino, "Cognitive and Non-Cognitive Values," 44.

399. Ibid., 54.

400. Ibid., 51, 50.

401. Ibid., 55.

402. Ibid., 54.

403. Noretta Koertge, "Feminist Epistemology: Stalking an Un-Dead Horse," in *Flight from Science and Reason,* ed. Gross, Levitt, and Lewis, 413–19, citing 413.

404. Gross and Levitt, *Higher Superstition,* 110.

405. Ibid., 108.

406. Ibid., 112.

407. Ibid., 116, 112. For the stark suspicion of their ultimate intent, see Elisabeth Lloyd, "Science and Anti-Science: Objectivity and Its Real Enemies," in *Feminism, Science, and the Philosophy of Science,* 217–59, esp. 250–51.

408. Harding, *Science Question,* 43; cited in Gross and Levitt, *Higher Superstition,* 130.

409. Gross and Levitt, *Higher Superstition,* 136.

410. Ibid., 145.

411. Noretta Koertge, "Wrestling with the Social Constructor," in *The Flight from Science and Reason,* 266–73, citing 271, 270.

412. Ibid., 271, referring to Longino, *Science As Social Knowledge,* 191.

413. Susan Haack is very blunt here: "Unless one is befogged by the emotional appeal of the word 'democratic,' it is clear that the idea is ludicrous that the question, say, what theory of sub-atomic particles should be accepted, should be put to a vote. Only those with appropriate expertise are competent to judge the worth of evidence" (Haack, "Science As

Social?—Yes and No," in *Feminism, Science, and the Philosophy of Science,* 79–94, citing 86). That is about as thoughtless a reading of the issue of democratization in scholarship as I have ever seen in print.

414. Keller, "Making a Difference: Feminist Movement and Feminist Critique of Science"; and Londa Schiebinger, *Has Feminism Changed Science?*

415. Schiebinger, *Has Feminism Changed Science?* 3–8.

416. Ibid., 8.

417. Ibid., 11.

418. Ibid.

419. Ibid., 135.

420. Ibid., 1.

421. Andrew Pickering, "Against Putting the Phenomena First: The Discovery of the Weak Neutral Current," *Studies in History and Philosophy of Science* 15 (1984): 85–117; Pickering, "Against Correspondence: A Constructivist View of Experiment and the Real," in *PSA 1986,* ed. Arthur Fine and Peter K. Machamer, vol. 2., 196–208; Pickering, "Models in/ of Scientific Practice," *Philosophy and Social Action* 13 (1987): 69–77; Pickering, "Living in the Material World: On Realism and Experimental Practice," in *The Uses of Experiment: Studies in the Natural Sciences,* ed. David Gooding, Trevor Pinch, and Simon Schaffer (Cambridge: Cambridge University Press, 1989), 275–98; Pickering, "Knowledge, Practice and Mere Construction (Response to Roth and Barrett)," *Social Studies of Science* 20 (1990): 682–729; Pickering, "Openness and Closure: On the Goals of Scientific Practice," in *Experimental Inquiries,* ed. LeGrand; Pickering, "The Mangle of Practice: Agency and Emergence in the Sociology of Science," *American Journal of Sociology* 99 (1993): 559–89; Pickering, "Objectivity and the Mangle of Practice," in *Rethinking Objectivity,* ed. Allan Megill (Durham, N.C.: Duke University Press, 1994), 109–26; Pickering, "Beyond Constraint: The Temporality of Practice and the Historicity of Knowledge," in *Scientific Practice: Theories and Stories of Doing Physics,* ed. Buchwald; and Pickering, *The Mangle of Practice: Time, Agency, and Science* (Chicago: University of Chicago Press, 1995).

422. Pickering, "From Science As Knowledge," 5; "Openness and Closure," 237 n. 23.

423. Pickering, "From Science As Knowledge," 5.

424. Ibid., 6.

425. Pickering, "Knowledge, Practice, and Mere Construction," 693; and "Objectivity and the Mangle of Practice," 112.

426. Ian Hacking, "Philosophers of Experiment," *PSA 1988,* vol. 2, 147–56, citing 148.

427. Pickering, "Knowledge, Practice, and Mere Construction," 707.

428. Pickering discusses Alvarez in the crucial essay on "Openness and Closure," while Morpurgo's experiments are the topic of a myriad of his publications.

429. Pickering, "Beyond Constraint," 46.

430. Pickering, "Openness and Closure," 217.

431. Pickering, "Against Putting the Phenomena First," 87; "Forms of Life: Science, Contingency and Harry Collins," *British Journal for the History of Science* 20 (1987): 213–21, citing 220.

432. Pickering, "Openness and Closure," 235 n. 12.

433. Ibid., 216.

434. Ibid., 217. See also "Beyond Constraint," 48: "a fundamental aspect of modeling, namely, that it is an *open-ended* process having no determinate destination."

435. For this phrase, "it just happened," see Pickering, "Knowledge, Practice, and Mere Construction," 703, 712 n. 16; and "The Mangle of Practice: Agency and Emergence in the Sociology of Science," 576.

436. See Pickering, "Knowledge, Practice, and Mere Construction": "[C]oherence, I am

inclined to believe, is a constant *telos* of scientific practice" (694); "the search for coherence is *constitutive* of practice in its full temporality" (697).

437. Pickering, "Openness and Closure," 217.

438. On "association," "connection," and "closure," see "Openness and Closure," 215–17.

439. Pickering, "Living in a Material World," 276–77.

440. Pickering, "Openness and Closure," 215.

441. Pickering, "Objectivity and the Mangle of Practice," 115.

442. Pickering, "Beyond Constraint," 51.

443. Pickering, "Objectivity and the Mangle of Practice," 115.

444. Pickering, "From Science As Knowledge," 9.

445. Ibid., 703.

446. Pickering, "Beyond Constraint," 54.

447. Pickering, "Knowledge, Practice, and Mere Construction," 721 n. 26.

448. Andrew Pickering and Adam Stephanides, "Constructing Quaternions: On the Analysis of Conceptual Practice," in *Science As Practice and Culture*, ed. Pickering, 139–67, citing 163–64.

449. Pickering, "Openness and Closure," 213. Ian Hacking writes similarly of a self-validating stabilization of complex experimental elements in "The Self-Vindication of the Laboratory Sciences," in *Science As Practice and Culture*, ed. Pickering, 29–64.

450. Pickering, "Knowledge, Practice," 697.

451. Pickering, "Beyond Constraint," 43.

452. Ibid., 43 n.

453. Ian Hacking, introduction to *Scientific Practice: Theories and Stories of Doing Physics*, 5.

454. Brian Baigre, "Scientific Practice: The View from the Tabletop," in *Scientific Practice: Theories and Stories of Doing Physics*, 120–21.

455. Pickering, "Beyond Constraint," 45.

456. Ibid., 53.

457. Peter Galison, *How Experiments End* (Chicago: University of Chicago Press, 1987), 10.

458. Ibid., 11.

459. Pickering, "Beyond Constraint," 46.

460. Ibid.

461. Ibid., 50.

462. Ibid., 47.

463. Ibid., 51.

464. Pickering, "Forms of Life," 213–21.

465. Pickering, "Beyond Constraint," 54.

466. Galison, *How Experiments End*, 257.

467. Ibid., 246.

468. Ibid., 255.

469. See his essay, "History, Philosophy, and the Central Metaphor," *Science in Context* 2 (1988): 197–212.

470. Peter Galison, "Context and Constraint," in *Scientific Practice*, 13–41.

471. Ibid., 16.

472. Ibid., 40.

473. Ibid., 41.

474. Ibid., 14, 27.

475. Ibid., 27.

476. Ibid., 37.
477. Ibid., 15.
478. Ibid., 17.
479. Pickering, "Beyond Constraint," 55.

Chapter Eight

1. In the rhetoric of the currently most "radical" wing, the so-called "reflexivists," a flamboyant avant-gardist "progressivism" is brandished to cudgel all resistors, even as the entire practice is then disavowed as "ironic." This avant-gardism, I submit, is characteristic of postmodernism as a rhetorical pose. Always to propose to "go beyond" the current, to "make it new," is every whit as *post*modern as it was "high modern." For a signal instance of this, see the collaborative essay by Steven Woolgar and Malcolm Ashmore, "The Next Step," in *Knowledge and Reflexivity: New Frontiers in the Sociology of Knowledge,* ed. Steven Woolgar (London: Sage, 1988), 1–11, esp. 10: "I am concerned that readers will miss the irony of our progressive account. They may suspect that (deep down) we actually like the possibility that reflexivity is an advance on previous approaches." For a blatantly avant-garde posture (or should one say "positionism"?) see Steven Woolgar's "Some Remarks about Positionism: A Reply to Collins and Yearley," in *Science As Practice and Culture,* ed. Andrew Pickering (Chicago: University of Chicago Press, 1992), 327–42. I take this exchange between Collins and Yearley and Woolgar to be the climactic showdown in the internecine wars within the sociology of scientific knowledge.

2. Harry M. Collins and Steven Yearley, "Epistemological Chicken," in *Science As Practice and Culture,* ed. Andrew Pickering (Chicago: University of Chicago Press, 1992), 302.

3. Hence Hilary Rose's apt statement cited in the epigraph to this chapter ("Hyper-Reflexivity—a New Danger for the Counter-Movements," in *Counter-Movements in the Sciences,* ed. Helga Nowotny and Hilary Rose, Sociology of the Sciences Yearbook 3 [1979] [Dordrecht: Reidel, 1979], 282).

4. Laudan, "The Pseudo-Science of Science?" *Philosophy of the Social Sciences* 11 (1981); David Bloor, "The Strength of the Strong Programme," *Philosophy of the Social Sciences* 11 (1981). James R. Brown, ed., *Scientific Rationality: The Sociological Turn* (Dordrecht: Reidel, 1984).

5. David Bloor, "Durkheim and Mauss Revisited: Classification and the Sociology of Knowledge," *Studies in History and Philosophy of Science* 13 (1982): 267–97. See Gerd Buchdahl, "Editorial Response to David Bloor," *Studies in History and Philosophy of Science* 13 (1982): 299–304; Steven Lukes, "Comments on David Bloor," *Studies in History and Philosophy of Science* 13 (1982): 313–18; Joseph Wayne Smith, "Primitive Classification and the Sociology of Knowledge: A Response to Bloor," *Studies in History and Philosophy of Science* 15 (1984): 237–43. Bloor responded to each critic: "A Reply to Gerd Buchdahl," *Studies in History and Philosophy of Science* 13 (1982): 305–18; "Reply to Steven Lukes," *Studies in History and Philosophy of Science* 13 (1982): 319–24; "Reply to J. W. Smith," *Studies in History and Philosophy of Science* 15 (1984), 245–49. Mary Hesse defended Bloor in the context of this debate: "Comments on the Papers of David Bloor and Steven Lukes," *Studies in History and Philosophy of Science* 13 (1982): 325–31.

6. Martin Hollis and Steven Lukes, eds., *Rationality and Relativism* (Cambridge, Mass.: MIT Press, 1982).

7. And it provoked sharp philosophical counterattacks, not only from Laudan, but also from W. H. Newton-Smith, *The Rationality of Science* (Boston: Routledge & Kegan Paul, 1981), chap. 10; Paul Roth, *Meaning and Method in the Social Sciences* (Ithaca, N.Y.: Cornell University Press, 1987), chaps. 7–8 (entitled "Voodoo Epistemology I" and "Voodoo

Epistemology II"!); Timothy McCarthy, "Scientific Rationality and the 'Strong Program' in the Sociology of Knowledge," in *Construction and Constraint: The Shaping of Scientific Rationality,* ed. Ernan McMullin (Notre Dame, Ind.: University of Notre Dame Press, 1988), 75–96. The approach was defended by Mary Hesse in "The Strong Thesis of Sociology of Science."

8. "The whole rationale of the sociology of science is to challenge scientists' taken-for-granted assumptions" (Sara Delamont, "Three Blind Spots? A Comment on the Sociology of Science by a Puzzled Outsider," *Social Studies of Science* 17 [1987]: 163–70, citing 167).

9. "Cultural studies" represents a generally different approach from that of "science studies," though they converge at the extreme, postmodern end of the science studies spectrum. "Cultural studies" represents the adventure of literary theorists pronouncing themselves authorized to prescribe to all other disciplines by virtue of their mastery of the tropes of language. When science studies becomes indiscernible from this approach, it has really entered the *mise en abîme.*

10. Nigel Pleasants, "The Post-positivist Dispute in Social Studies of Science and Its Bearing on Social Theory," *Theory, Culture, and Society* 14 (1997): 143–56, citing 144–45.

11. Ibid., 149.

12. For versions of this narrative, see an early formulation in Harry M. Collins, "The Sociology of Scientific Knowledge: Studies of Contemporary Science," *Annual Review of Sociology* (1983): 265–85; its depiction, with resistance, by a discourse analyst unwilling to accept supersession: Jonathan Potter, "What Is Reflexive about Discourse Analysis? The Case of Reading Readings," in *Knowledge and Reflexivity: New Frontiers in the Sociology of Knowledge,* ed. Woolgar, 37–52; and Steven Shapin, "Here and Everywhere: Sociology of Scientific Knowledge," *Annual Review of Sociology* (1995): 289–321.

13. See Ian Hacking, *Representing and Intervening* (Cambridge: Cambridge University Press, 1983), 149–66 and esp. 262–75.

14. Paul Tibbetts and Patricia Johnson, "The Discourse and *Praxis* Models in Recent Reconstructions of Scientific Knowledge Generation," *Social Studies of Science* 15 (1985): 739–49, citing 742.

15. David Gooding, Trevor Pinch, and Simon Schaffer, preface to *The Uses of Experiment,* xiii.

16. Jan Golinski, "The Theory of Practice and the Practice of Theory: Sociological Approaches in the History of Science," *Isis* 81 (1990): 496–97.

17. Ibid., 499.

18. Tibbetts and Johnson, "Discourse and *Praxis* Models," 744, 741.

19. Ibid., 747.

20. Michael Lynch, "From the 'Will to Theory' to the Discursive Collage," in *Science As Practice and Culture,* ed. Pickering, 289.

21. Michael Mulkay, *Science and the Sociology of Knowledge* (London: Allen & Unwin, 1979).

22. Michael Mulkay, "Action and Belief or Scientific Discourse? A Possible Way of Ending Intellectual Vassalage in Social Studies of Science," *Philosophy of the Social Sciences* 11 (1981): 163–71, citing 169.

23. Ibid., 163.

24. Ibid., 170. It should not pass unobserved that this language is stunningly reminiscent of the most adamant logical empiricism!

25. Michael Mulkay and Nigel Gilbert, "What Is the Ultimate Question? Some Remarks in Defense of the Analysis of Scientific Discourse," *Social Studies of Science* 12 (1982): 309–19, citing 310.

26. Ibid., 311.

27. Michael Mulkay, Nigel Gilbert, and Steven Yearley, "Why an Analysis of Scientific Discourse Is Needed," in *Science Observed,* ed. Karin Knorr-Cetina and Michael Mulkay (London: Sage, 1983), 171–204, citing 177.

28. Ibid., 171.

29. Nigel Gilbert and Michael Mulkay, "Warranting Scientific Belief," *Social Studies of Science* 12 (1982): 383–408, citing 383. Is it self-parodic that this passive construction is quintessentially in the woeful rhetoric of the scientific research paper?

30. Nigel Gilbert and Michael Mulkay, "Experiments Are the Key: Participants' Histories and Historians' Histories of Science," *Isis* 75 (1984): 105–25, citing 124.

31. Ibid., 124.

32. Ibid., 125.

33. Steven Shapin, "Talking History: Reflections on Discourse Analysis," *Isis* 75 (1984): 125–28, citing 128.

34. Ibid., 126.

35. Ibid., 127.

36. Ibid., 129.

37. Thomas Gieryn, "Relativist/Constructivist Programmes in the Sociology of Science: Redundance and Retreat," *Social Studies of Science* 12 (1982): 279–97, citing 291.

38. Thomas Gieryn, "Not-Last Words: Worn Out Dichotomies in the Sociology of Science," *Social Studies of Science* 12 (1982): 329–35, citing 332.

39. Ibid., 333.

40. Ibid.

41. Ellsworth Fuhrman and Kay Oehler, "Discourse Analysis and Reflexivity," *Social Studies of Science* 16 (1986): 293–307, citing 294.

42. Jonathan Potter, "Discourse Analysis and the Turn of the Reflexive Screw: A Response to Fuhrman and Oehler," *Social Studies of Science* 17 (1987): 171–77, citing 171, 174.

43. Ellsworth Fuhrman and Kay Oehler, "Reflexivity Redux: Reply to Potter," *Social Studies of Science* 17 (1987): 177–81, citing 179.

44. Peter Halfpenny, "Talking of Talking, Writing of Writing: Some Reflections on Gilbert and Mulkay's Discourse Analysis," *Social Studies of Science* 18 (1988): 169–82, citing 171.

45. Ibid., 178.

46. Ibid., 177.

47. Fuhrman and Oehler, "Reflexivity Redux," 177.

48. Jonathan Potter and Andy McKinlay, "Discourse—Philosophy—Reflexivity: Comments on Halfpenny," *Social Studies of Science* 19 (1989): 137–45.

49. Peter Halfpenny, "Reply to Potter and McKinlay," *Social Studies of Science* 19 (1989): 145–52, citing 146, 148.

50. Collins and Yearley, "Epistemological Chicken," 304.

51. Ibid., 305.

52. Ibid., 303.

53. Ibid., 308.

54. Dick Pels, "The Politics of Symmetry," *Social Studies of Science* 26 (1996): 277–304, citing 278.

55. Ibid., 279.

56. Ibid.

57. Collins and Yearley, "Journey into Space," 378–79; citing Bruno Latour, "The Politics of Explanation: An Alternative," in *Knowledge and Reflexivity: New Frontiers in the Sociology of Knowledge,* ed. Woolgar, 170.

58. Bruno Latour, "One More Turn after the Social Turn," in *The Social Dimension of Science,* ed. McMullin, 272.

59. Ibid., 272–73.

60. Ibid., 288.

61. Steve Fuller, "Back to Descartes? The Very Idea!" review of *Science: The Very Idea,* by Steven Woolgar, *Social Studies of Science* 19 (1989): 357.

62. Woolgar and Ashmore, "The Next Step"; Woolgar, "Some Remarks about Positionism: A Reply to Collins and Yearley," 327–42; Woolgar, "On the Alleged Distinction between Discourse and Praxis," *Social Studies of Science* 16 (1986): 309–17; Woolgar, *Science: The Very Idea* (London: Tavistock, 1988).

63. Steven Woolgar, "Writing an Intellectual History of Scientific Development: The Use of Discovery Accounts," *Social Studies of Science* 6 (1976): 395–422, citing 395.

64. Ibid., 397.

65. Ibid., 398.

66. Ibid., 400.

67. Ibid., 396.

68. Ibid., 399.

69. Steven Woolgar, "Interests and Explanation in the Social Study of Science," *Social Studies of Science* 11 (1981): 365–94, citing 375.

70. Ibid., 382.

71. Ibid., 379.

72. Ibid., 383.

73. Ibid., 385.

74. Ibid. There is a familiar savor to these phrases; they smack of poststructuralism, and indeed, Woolgar, having worked on the pathbreaking laboratory ethnography, *Laboratory Life,* with the eminent French sociologist of science, Bruno Latour, turns out to be steeped in this approach to "discourse." See Woolgar, "On the Alleged Distinction between Discourse and *Praxis,*" for his explicit identification with "continental"—i.e., French poststructuralist—theory of discourse.

75. Donald MacKenzie, "Interests, Positivism, History," *Social Studies of Science* 11 (1981), 499.

76. Chris Doran, "Jumping Frames: Reflexivity and Recursion in the Sociology of Science," *Social Studies of Science* 19 (1989): 515–31.

77. Barry Barnes, "On the 'Hows' and 'Whys' of Cultural Change," *Social Studies of Science* 11 (1981): 497 n.

78. Ibid., 494.

79. Ibid.

80. See Fuller, "Back to Descartes? The Very Idea!" See Richard J. Bernstein's characterization of "Cartesian anxiety" as the impulse behind "foundationalism" in *Beyond Objectivism and Relativism* (Philadelphia: University of Pennsylvania Press, 1983). What needs to be discerned is that it is also the cudgel with which *anti*foundationalism besets any empirical inquiry: negative absolutism or global scepticism. An ironically perceptive identification of this commonality between foundationalism and antifoundationalism is Stanley Fish, "Consequences," in *Against Theory: Literary Studies and the New Pragmatism,* ed. W. J. T. Mitchell (Chicago: University of Chicago Press, 1985), 106–31.

81. This "folk-X" disparagement, like the "dope" jargon, has become endemic to the disputes in SSK as a carryover from ethnomethodology. What is ironic is the elitist character of the "folk" deprecation among such ostensibly "progressive" radicals.

82. Dennis Wrong, "The Oversocialized Conception of Man in Modern Society," *Amer-*

ican Sociological Review 26 (1961): 183–93; see Alvin W. Gouldner, *The Coming Crisis of Western Sociology* (New York: Basic Books, 1970).

83. Steven Woolgar, "Critique and Criticism: Two Readings of Ethnomethodology," *Social Studies of Science* 11 (1981): 506.

84. Ibid., 510.

85. Ibid., 508.

86. Ibid., 511–12.

87. Doran, "Jumping Frames," 516.

88. "Social" gets eliminated from the subtitle of the second edition of Woolgar's famous collaboration with Latour, *Laboratory Life,* to signify that "construction" embraces not only the natural but the social as well: all is "inscription" (*Il n'y a pas de hors-texte.*)

89. Woolgar, "Critique and Criticism," 507.

90. Ibid., 509.

91. Ibid.

92. Steven Woolgar, "Irony in the Social Study of Science," in *Science Observed,* ed. Knorr-Cetina and Mulkay, 239–66, citing 240.

93. Ibid.

94. Ibid., 241.

95. Steven Woolgar, "Time and Documents in Researcher Interaction: Some Ways of Making Out What Is Happening in Experimental Science," in *Representation in Scientific Practice,* ed. Michael Lynch and Steven Woolgar (Cambridge, Mass.: MIT Press, 1990), 123–52, citing 124.

96. Ibid.

97. Steven Woolgar, "Reflexivity Is the Ethnographer of the Text," in *Knowledge and Reflexivity,* ed. Woolgar, 14.

98. Ibid., 150 n. 1.

99. Ibid.

100. Michael Lynch and Steven Woolgar, "Introduction: Sociological Orientations to Representational Practice in Science," in *Representation in Scientific Practice,* ed. Lynch and Woolgar (Cambridge, Mass.: MIT Press, 1990), 1–18, citing 13.

101. Steven Woolgar and Dorothy Pawluch, "Ontological Gerrymandering: The Anatomy of Social Problems Explanations," *Social Problems* 32 (1985): 214–27, citing 216.

102. Ibid., 217.

103. Ibid.

104. Ibid., 224.

105. Woolgar, "Irony in the Social Study of Science," 248.

106. See Kenneth Burke, *Permanence and Change* (Indianapolis: Bobbs-Merrill, 1965), 69 ff.

107. Woolgar, "Irony in the Social Study of Science," 258. See Wayne Booth, *A Rhetoric of Irony* (Chicago: University of Chicago Press, 1974).

108. Woolgar, "Irony in the Social Study of Science," 260.

109. Ibid., 254.

110. Ibid., 260.

111. Steven Woolgar, "Laboratory Studies: A Comment on the State of the Art," *Social Studies of Science* 12 (1982): 481–98, citing 483, 485.

112. Ibid., 485.

113. Ibid., 492.

114. Ibid., 489. This last line is taken as the inspiration of the most experimental aspect of reflexivity, the pursuit of "new literary forms" for SSK.

115. Ibid., 493.

116. Steven Woolgar, "Discovery: Logic and Sequence in a Scientific Text," in *The Social Process of Scientific Investigation*, ed. Karin Knorr, Roger Krohn, and Richard Whitley (Dordrecht: Reidel, 1981), 242.

117. Woolgar and Ashmore, "The Next Step," 4.

118. Woolgar, "Reflexivity Is the Ethnographer of the Text," 28.

119. Ibid., 16.

120. Woolgar, "Irony in the Social Study of Science," 239.

121. See note 1 above, on this avant-gardism.

122. Woolgar, "Time and Documents," 150 n. 1.

123. Steven Woolgar, foreword to *The Reflexive Thesis: Wrighting the Sociology of Knowledge*, by Malcolm Ashmore (Chicago: University of Chicago Press, 1989), xviii.

124. Fuller, "Back to Descartes? The Very Idea!," 357.

125. Ibid.

126. Ibid., 358.

127. This is Nagel's famous phrase associated with foundationalism. It is also the (im)posture of poststructuralism.

128. Woolgar, "Critique and Criticism," 512.

129. Ibid. There are numerous instances of Woolgar's admission of the negativity or emptiness of the agenda of this drastic radicalism; nonetheless, heroically, he moves forward into the "new frontiers."

130. Harry M. Collins, "An Empirical Relativist Programme for the Sociology of Scientific Knowledge: Appendix: Special Relativism," in *Science Observed*, ed. Knorr-Cetina and Mulkay, 101. I am completely in agreement with Collins on this matter.

131. The criticism came from Tibbetts and Johnson, "Discourse and *Praxis* Models."

132. Woolgar, "On the Alleged Distinction between Discourse and *Praxis*," 311.

133. Ibid., 310.

134. Ibid., 312.

135. Are we dealing yet again with the insidious recourse to "definitive" accounting?

136. Woolgar, "On the Alleged Distinction between Discourse and *Praxis*," 311; citing Hayden White, "Foucault Decoded," in *Tropics of Discourse: Essays in Cultural Criticism* (Baltimore: Johns Hopkins University Press, 1978), 230. It is a matter of wonderment how glib such thinkers can be about what things "truly are"!

137. Ibid., 313.

138. Woolgar, foreword to *The Reflexive Thesis*, by Ashmore, xviii.

139. Lynch and Woolgar, "Introduction: Sociological Orientations to Representational Practice in Science," 2.

140. Woolgar, "On the Alleged Distinction between Discourse and *Praxis*," 312.

141. Pleasants, "Post-positivist Dispute," 150.

142. Woolgar, foreword to *The Reflexive Thesis*, by Ashmore, xix.

143. Malcolm Ashmore, *The Reflexive Thesis: Wrighting the Sociology of Knowledge* (Chicago: University of Chicago Press, 1989), 109.

144. Woolgar, "Reflexivity Is the Ethnographer of the Text," 30.

145. Ibid.

146. Ibid.

147. Latour, "The Politics of Explanation: An Alternative," in *Knowledge and Reflexivity*, ed. Woolgar, 168.

148. Ibid., 173.

149. Ibid., 166. Interestingly, Michel Callon, too, distances the project of "actor network theory" from Woolgar's "reflexivity" approach: Michel Callon and John Law, "On Interests and Their Transformation: Enrolment and Counter-Enrolment," *Social Studies of*

Science 12 (1982): 615–25, esp. 621: "Unlike Woolgar, we are not concerned with the general explanatory form of interest explanations. Our aim would not be to establish a general set of rhetorical rules . . . but to discover how it is that actors enrol one another, and why it is that some succeed whereas others do not."

150. Doran, "Jumping Frames," 517.

151. Fuller, "Back to Descartes? The Very Idea!" 358.

152. Halfpenny, "Reply to Potter and McKinlay," 149.

153. Halfpenny, "Talking of Talking, Writing of Writing," 179.

154. See Fuhrman and Oehler, "Discourse Analysis and Reflexivity," 301. Woolgar and Ashmore explore these problems in "The Next Step," 5ff.

155. Furhman and Oehler, "Discourse Analysis and Reflexivity," 303.

156. Peter Burke, *New Perspectives on Historical Writing* (University Park, Pa.: University of Pennsylvania Press, 1992).

157. Jorge Luis Borges, "Pierre Menard, Author of the *Quixote*," in *Labyrinths* (New York: New Directions, 1986), 36–44. Compare Michael Mulkay, "Don Quixote's Double: A Self-Exemplifying Text," in *Knowledge and Reflexivity*, ed. Woolgar, 81–100; Mulkay, *The Word and the World: Explorations in the Form of Sociological Analysis* (London: Allen & Unwin, 1985); Mulkay, "New Literary Forms: Exploring the Many Worlds of Textuality," pt. 4 of *Sociology of Science: A Sociological Pilgrimage* (Bloomington: Indiana University Press, 1990), 167–216.

158. The reviewers of Ashmore's *The Reflexive Thesis* seem to share my uneasiness. Mitchell Berbrier notes: "the book does not make for a comfortable read. It is neither clear nor linear. Ashmore seems to get nowhere slowly by going everywhere fast (he covers a lot of ground), and it is hard to criticize him for this as that seems to be his aim. At the same time it is easy to feel angry at him for writing such a frustratingly unconventional book, though that may also be his aim" (review of *The Reflexive Thesis*, by Malcolm Ashmore, *Isis* 83 [1992]: 178–79, citing 179). Trevor Pinch's review (*Contemporary Sociology* 19 [1990]: 882–83) is more generous, not finding the project tedious, because "Ashmore, like a new-wave Hofstader, carries the argument through the use of a clever amalgam of stylistic devices and explorations of unconventional textual forms" (882). Pinch was of a more critical disposition in his collaboration with himself: Pinch and Pinch, "Reservations about Reflexivity and New Literary Forms, or Why Let the Devil Have All the Good Tunes?" in *Knowledge and Reflexivity*, ed. Woolgar, 178–97.

159. Lynch and Woolgar, "Introduction: Sociological Orientations to Representational Practice," 13.

160. Collins and Yearley, "Epistemological Chicken," 306.

161. Paul Roth, "What Does the Sociology of Scientific Knowledge Explain?" *History of the Human Sciences* 7 [1994]: 99.

162. Ibid., 96.

163. Ibid., 97.

164. Ibid., 102.

165. Ibid., 106.

166. Paul Roth, "Will the Real Scientists Please Stand Up? Dead Ends and Live Issues in the Explanation of Scientific Knowledge," *Studies in History and Philosophy of Science* 27 (1996): 43–68, citing 51.

167. Ibid., 58 n.

168. Ibid., 59 n.

169. Ibid., 62–63.

170. Ibid., 63.

171. Ibid., 64.

172. Ibid., 66.

173. Ibid., 67.

174. Ibid., 67 n.

175. Halfpenny, "Reply to Potter and McKinlay," 150.

176. Michael Lynch, *Scientific Practice and Ordinary Action: Ethnomethodology and Social Studies of Science* (Cambridge: Cambridge University Press, 1993), 36.

177. See *The Sokal Hoax: The Sham That Shook the Academy,* ed. the editors of *Lingua Franca* (Lincoln: University of Nebraska Press, 2000); and Alan Sokal and Jean Bricmont, *Fashionable Nonsense: Postmodern Intellectuals' Abuse of Science* (New York: Picador, 1998).

178. I stand especially close to Sokal's position in his contribution to *A House Built on Sand: Exposing Postmodernist Myths about Science,* ed. Noretta Koertge (Oxford: Oxford University Press, 1998), where he avows his virtually complete agreement with Philip Kitcher's essay in that same volume. My agreement with Kitcher's position in that essay is complete.

179. Ian Hacking, *The Social Construction of What?* (Cambridge, Mass.: Harvard University Press, 1999), 62.

180. Ibid., 63.

181. Ibid., 67.

182. Ibid., 94. David Albert notes similarly that "many science studies people are misunderstood as attempting to *delegitimize* science whereas what they're really doing is *demystifying* science" ("*Lingua Franca* Roundtable, May 1997," in *The Sokal Hoax,* 257).

183. John Brenkman makes a somewhat analogous statement: "There's a certain kind of anti-intellectualism on the part of scientists vis-à-vis work in the humanities and the 'softer' social sciences, and clearly there is a very strong strand of anti-intellectualism as regards science on the part of a lot of people who do humanities and cultural studies" ("*Lingua Franca* Roundtable, May 1997," in *The Sokal Hoax,* 256).

184. Steven Weinberg used this metaphor in "Steven Weinberg Replies," in *The Sokal Hoax,* 166–71, citing 169.

185. Again, Weinberg is exemplary.

186. Andrew Pickering's "mangle of practice" is the most elaborate endeavor to explain that in terms of actual scientific practices.

187. Paul Roth and Robert Barrett, "Deconstructing Quarks," *Social Studies of Science* 20 (1990): 595.

188. Andrew Pickering, "Knowledge, Practice and Mere Construction (Response to Roth and Barrett)," *Social Studies of Science* 20 (1990): 682–729. See the excellent discussion in Hacking, *Social Construction of What?* 68–80.

189. Hacking, *Social Construction of What?* 31.

190. Bruno Latour, "Is There Science after the Cold War?" in *The Sokal Hoax,* 124–26.

191. George Levine, letter to the editors of *Lingua Franca,* reprinted in *The Sokal Hoax,* 63–64.

192. Hacking, *Social Construction of What?* 93.

193. See Paul Gross and Norman Levitt, *Higher Superstition: The Academic Left and Its Quarrels with Science,* 2d ed. (Baltimore: Johns Hopkins University Press, 1998). For more careful judgments here, see M. Norton Wise, "The Enemy Without and the Enemy Within," *Isis* 87 (1996): 323–27; and Nicholas Jardine and Marina Frasca-Spada, "Splendours and Miseries of the Science Wars," *Studies in History and Philosophy of Science* 28 (1997): 219–35.

194. Andrew Ross, "Reflections on the Sokal Affair," in *The Sokal Hoax,* 245–48. In a roundtable discussion in *Lingua Franca* concerning the Sokal affair, Elisabeth Lloyd con-

demned the "complete misrepresentation of the position of the feminist philosophers of science, but also of people like Bruno Latour" in Gross and Levitt (in *The Sokal Hoax*, 256–57). David Albert agreed that "all of the well-known and highly regarded practitioners of social constructivism aren't susceptible to the kinds of attacks that Gross and Levitt mount" (ibid., 258). Unfortunately, things are not nearly so straightforward, but see the energetic defense by Lloyd, "Science and Anti-Science: Objectivity and Its Real Enemies," in *Feminism, Science, and the Philosophy of Science*, ed. Lynn Hankinson Nelson and Jack Nelson (Dordrecht: Kluwer, 1996), 217–59.

195. "[T]he Science Wars [are] a second front opened up by conservatives cheered by the successes of their legions in the holy Culture Wars" (Andrew Ross, "Science Backlash on Technoskeptics," *The Nation* 262 [Oct. 2, 1995], 346). And see Ross, "Introduction," *Social Text* 46/47 (1996): 1–13.

196. Gross and Levitt, *Higher Superstition*, 256. They were not alone in this reaction. See Lewis Wolpert, *The Unnatural Nature of Science* (London: Faber & Faber, 1992); Steven Weinberg, *Dreams of a Final Theory* (New York: Pantheon, 1992).

197. Gross and Levitt, *Higher Superstition*, 2.

198. Thus, to grasp the postmodernist impulse, we have to situate it in an explicitly antiscientific movement.

199. Gross and Levitt, *Higher Superstition*, 49.

200. Ibid., 12.

201. Ibid., 10.

202. To be capable of discerning no difference, for example, between the position of Helen Longino and that of Sandra Harding (claiming Longino's work was "neither stronger nor more convincing than those of other feminist epistemologists" [ibid., 146]) is indeed an "illegitimate polemical shortcut." Philip Kitcher writes: "This strikes me as terribly wrong. Thinking of Keller, Longino, and Shapin and Schaffer as belonging to the same intellectual species as Harding is a bit like thinking of gibbons, chimpanzees, seals, and dolphins as being conspecific with opossums (they are all mammals, of course, but there the similarities end)" ("A Plea for Science Studies," in *A House Built on Sand*, ed. Koertge, 48–49).

203. See Barbara Epstein, "Postmodernism and the Left" (1997), reprinted in *The Sokal Hoax*, 214–29.

204. Gross and Levitt, *Higher Superstition*, 5.

205. Ibid., 71.

206. Ibid., 89.

207. Ibid., 74.

208. Robin Fox, "Anthropology and the 'Teddy Bear' Picnic," 51–52; cited in Gross and Levitt, *Higher Superstition*, 81.

209. Ibid., 78.

210. Ibid., 81.

211. See my "Are We Being Theoretical Yet?" *Journal of Modern History* 65 (1993): 783–814.

212. See my "Reading 'Experience': The Debate in Intellectual History among Scott, Toews, and LaCapra," in *Reclaiming Identity: Realist Theory and the Predicament of Postmodernism*, ed. Paula Moya and Michael Hames-Garcia (Berkeley, Los Angeles, London: University of California Press, 2000), 279–311.

213. Gross and Levitt, *Higher Superstition*, 144.

214. Ibid., 40.

215. Ibid., 8.

216. This has largely been accomplished in Noretta Koertge, ed., *A House Built on Sand*.

217. Gross and Levitt, *Higher Superstition*, 27.

218. Ibid., 22.

219. Ibid., 65.

220. Ibid., 11.

221. Ibid., 38.

222. Ibid., 55-56.

223. Ibid., 44.

224. Ibid., 24.

225. One should not regard the "scientific community" as a monolith: it is composed of highly segmented groups, with no assurance of mutual comprehension. See Galison and Stump, eds., *The Disunity of Science* (Stanford, Calif.: Stanford University Press, 1996).

226. Arthur Fine, "Natural Ontological Attitude," in *The Shaky Game* (Chicago: University of Chicago Press, 1996).

227. Great scientific achievements have been made by instrumentalists. Indeed, there is a category error in the very idea that one's ontology defines one's scientific integrity.

228. Gross and Levitt, *Higher Superstition,* 59.

229. Ibid., 90.

230. Ibid., 47.

231. Ibid., 139.

232. Ibid., 72.

233. Ibid., 69.

234. Ibid., 235.

235. Norman Levitt, "Mathematics as the Stepchild of Contemporary Culture," in *The Flight from Science and Reason,* 39-53, citing 49.

236. Latour, "Postmodern? No, Simply Amodern!"

237. Susan Haack, "Towards a Sober Sociology of Science," in *Flight from Science and Reason,* 259-65, citing 259. While I am all for such a sober sociology of science, I cannot find much guidance here; compare Kitcher's "A Plea for Science Studies" for a dramatically superior formulation.

238. Haack, "Towards a Sober Sociology of Science," 262.

239. Susan Haack, "Science As Social?—Yes and No," in *Feminism, Science, and the Philosophy of Science,* 79-94.

240. Haack disputes this: "On the contrary, it is the knowledge possessed by individuals that is primary" (ibid., 85).

241. Ibid., 83-84.

242. Andrew Ross, "Reflections on the Sokal Affair" (October 1996), in *The Sokal Hoax,* 245.

243. Brenkman, "*Lingua Franca* Roundtable, May 1997," 258.

244. Ibid.

245. Ellen Willis, "My Sokaled Life: Or, Revenge of the Nerds" (1996), reprinted in *The Sokal Hoax,* 133-38, citing 135.

246. Special issue on "Science Wars," *Social Text* 46/47 (1996); Brenkman, "*Lingua Franca* Roundtable, May 1997," 268.

247. Brenkman, "*Lingua Franca* Roundtable, May 1997," 253.

248. Ibid.

249. Alan Sokal and Jean Bricmont, "Comments on the Parody," appendix to *Fashionable Nonsense,* 259-67.

250. Alan Sokal, "Transgressing the Boundaries: Toward a Transformative Hermeneutics of Quantum Gravity" (1996), reprinted in *Fashionable Nonsense,* 212-58, citing 213.

251. Kathy Pollitt, "Pomolotov Cocktail" (1996), reprinted in *The Sokal Hoax,* 96-98, esp. 97 (originally published in *The Nation*).

252. Bruce Robbins and Andrew Ross, "Response: Mystery Science Theater," in *The Sokal Hoax,* 54–58.

253. For a particularly indigestible but brash exemplar, see Richard Biernacki, "Method and Metaphor after the New Cultural History," in *Beyond the Cultural Turn,* ed. Victoria Bonnell and Lynn Hunt (Berkeley: University of California Press, 1999), 62–92.

254. Alan Sokal, "Revelation: A Physicist Experiments with Cultural Studies" (1996), reprinted in *The Sokal Hoax,* 50–51.

255. Aronowitz, "Alan Sokal's 'Transgression,' " in *The Sokal Hoax,* 200–204; Robbins and Ross, "Response: Mystery Science Theater," 54–58; Stanley Fish, "Professor Sokal's Bad Joke," in *The Sokal Hoax,* 81–84.

256. Albert, "*Lingua Franca* Roundtable, May 1997," 264.

257. Brenkman, "*Lingua Franca* Roundtable, May 1997," 265.

258. Paul Boghossian, "What the Sokal Hoax Ought to Teach Us," in *A House Built on Sand,* ed. Koertge, 23–31, citing 25.

259. Ibid., 26.

260. Evelyn Fox Keller, letter, in *The Sokal Hoax,* 59.

261. John Brenkman, letter, in *The Sokal Hoax,* 64–65.

262. Aronowitz, "Alan Sokal's 'Transgression,' " and Fish, "Professor Sokal's Bad Joke."

263. Kathy Pollitt, "Pomolotov Cocktail," in *The Sokal Hoax,* 97.

264. Keller, letter, in *The Sokal Hoax,* 59.

265. Stanley Fish, "Professor Sokal's Bad Joke," 81.

266. Barbara Epstein, "Postmodernism and the Left" in *The Sokal Hoax,* 214–29.

267. Hacking, *Social Construction of What?* 30.

268. Ibid.

269. Andrew Pickering, letter of June 6, 1997; cited in Hacking, *Social Construction of What?*

270. Latour, "Is There Science after the Cold War?" in *The Sokal Hoax,* 126.

271. Alan Sokal, "Why I Wrote My Parody," in *The Sokal Hoax,* 128–29.

272. Marco D'Eramo, rejoinder to Sokal, in *The Sokal Hoax,* 118.

273. Marco D'Eramo, "Academic Insult in Greenwich Village," in *The Sokal Hoax,* 116.

274. Paul Boghossian, "What the Sokal Hoax Ought to Teach Us," 27.

275. Steven Weinberg, "Sokal's Hoax," reprinted in *The Sokal Hoax,* 148–59.

276. Steven Weinberg, *Dreams of a Final Theory.*

277. Weinberg, "Sokal's Hoax," 153.

278. Ibid., 154.

279. Ibid., 155.

280. Ibid.

281. Michael Holquist and Robert Schulman, "Sokal's Hoax: An Exchange," in *The Sokal Hoax,* 159–62, citing 161.

282. Michel Callon and Bruno Latour, "Don't Throw the Baby Out with the Bath School!" in *Science As Practice and Culture,* ed. Pickering, 343–68 353.

283. Norton Wise, letter to the *New York Times,* in *The Sokal Hoax,* 163–66.

284. Weinberg, rejoinder to letters, in *The Sokal Hoax,* 168.

285. Ibid., 169.

286. Harding makes a botch of things by thinking that they are. See Sandra Harding, *The Science Question in Feminism* (Ithaca, N.Y.: Cornell University Press, 1986), 43; and the rebuttal: Gross and Levitt, *Higher Superstition,* 127–31.

287. Thomas Gieryn put it precisely: "Sociologists of science will never be in a position

to adjudicate the ontological accuracy of scientific (or any other) beliefs" ("Relativist/Constructivist Programmes in the Sociology of Science," 293).

288. Michael Bérubé, email to Sokal, in *The Sokal Hoax*, 141.

289. Ibid., 142.

290. John Searle, *The Construction of Social Reality* (New York: Free Press, 1995).

291. Bérubé, email to Sokal, in *The Sokal Hoax*, 142.

292. Ibid., 147.

293. Ibid.

294. Ibid.

295. Alan Sokal, "What the *Social Text* Affair Does and Does Not Prove," 9–22.

296. Alan Sokal and Jean Bricmont, "Intermezzo: Epistemic Relativism in the Philosophy of Science," in *Fashionable Nonsense: Postmodern Intellectuals' Abuse of Science* (New York: Picador, 1998), 50–105, citing 61.

297. Ibid., 51.

298. Sokal, "What the *Social Text* Affair Does and Does Not Prove," 17.

299. Kitcher, "A Plea for Science Studies," 32.

300. Ibid.

301. Ibid., 34–35. Kitcher, in his footnotes, reckons with the radical outliers in philosophy—Rorty, Putnam, and Goodman prominent among them—who would not concede these claims were so "uncontroversial." Yet his position is eminently sensible for the purposes of empirical inquiry in the human sciences, and that is his point.

302. Ibid., 36. This moderate position would appear—taking Gross and Levitt and Sokal at their word—to be "uncontroversial" for the critics of science studies, but I wonder how true that is.

303. Ibid., 37, 45.

304. Ibid., 37.

305. Ibid., 38.

306. Ibid., 39.

307. Ibid., 40.

308. Ibid., 41.

309. Ibid., 42.

310. Boghossian, "What the Sokal Hoax Ought to Teach Us," 28.

311. Satya Mohanty, "Us and Them: On the Philosophical Bases of Political Criticism," *Yale Journal of Criticism* 2 (1989): 1–31; Clifford Geertz: "The Uses of Diversity," *Michigan Quarterly Review* 25 (1986), 105–23; and Rorty, "On Ethnocentrism: A Reply to Clifford Geertz," *Michigan Quarterly Review* 25 (1986), 525–34.

312. Kitcher, "A Plea for Science Studies," 43.

313. Ibid.

314. Ibid., 44, 55 n. 40.

315. Rachel Laudan, "The 'New' History of Science: Implications for Philosophy of Science," *PSA 1992*, vol. 2, 478–79.

Conclusion

1. Ian Hacking, *Representing and Intervening* (Cambridge: Cambridge University Press, 1983), 45.

2. Ibid., 130.

3. Dudley Shapere, "Reason, Reference, and the Quest for Knowledge," in *Reason and the Search for Knowledge*, Boston Studies in the Philosophy of Science 78 (Dordrecht: Reidel, 1984), 402–3.

4. Ibid., 405.

5. "[I]t was assumed that the rules for justification are discoverable a priori by rationalistic logical analysis" (Bonnie Tamarkin Paller, "Naturalized Philosophy of Science, History of Science, and the Internal/External Debate," *PSA 1986*, vol. 1, 258).

6. Ronald Giere, "Philosophy of Science Naturalized," *Philosophy of Science* 52 (1985): 331–56, citing 336.

7. Rorty drew the proper personal conclusion, leaving the discipline of philosophy to pursue "conversations" in the wider humanities.

8. "If the philosophy of science is naturalized, philosophers of science are on the same footing with historians, psychologists, sociologists, and others for whom the study of science is itself a scientific enterprise" (Giere, "Philosophy of Science Naturalized," 343).

9. Philip Kitcher, "The Naturalists Return," *Philosophical Review* 101 (1992): 63.

10. Paller, "Naturalized Philosophy of Science," 258.

11. Giere, "Philosophy of Science Naturalized," 331.

12. Paller, "Naturalized Philosophy of Science," 259.

13. I confess to having been bitten with the virus of the *Zeitgeist* along with my contemporaries, for I have thronged my readings of Immanuel Kant with talk of turns.

14. This campaign achieved its contour in *Philosophy and the Mirror of Nature* (Princeton, N.J.: Princeton University Press, 1979) [hereafter cited parenthetically as "PMN:(p)"], and in a series of essays that appeared in that context, notably his exchange with Charles Taylor and Herbert Dreyfus in *Review of Metaphysics* (R. Rorty, "A Reply to Dreyfus and Taylor" and "Discussion," *Review of Metaphysics* 34 [1980]: 39–46 and 47–55, respectively); in his essay "Method, Social Science, and Social Hope" (1981; reprinted in *Consequences of Pragmatism* [Minneapolis: University of Minnesota Press, 1982], 191–210); in *Contingency, Irony and Solidarity* (Cambridge: Cambridge University Press, 1989); and in the various essays that compose *Objectivity, Relativism, and Truth: Philosophical Papers*, vol. 1 (Cambridge: Cambridge University Press, 1991) [hereafter cited as "ORT:(p)"].

15. Herbert Dreyfus, "Holism and Hermeneutics," *Review of Metaphysics* 34 (1980): 3–23.

16. Rorty, "Pragmatism without Method," in ORT:76. This is a variant of the "web of belief" approach developed by Willard Quine. See Willard van Orman Quine and J. S. Ullian, *The Web of Belief*, 2d ed. (New York: Random House, 1978).

17. Rorty, "Inquiry As Recontextualization," ORT:102.

18. Rorty, "Method, Social Science, Social Hope," 197.

19. Ibid., 201.

20. Ibid., 203.

21. Rorty, "Texts and Lumps," ORT:83.

22. Rorty does acknowledge that E. D. Hirsch has discerned a unique aspect of texts as against lumps in authorial intention. The upshot of his reflection, though, suggests that it is a trivial uniqueness ("Texts and Lumps," ORT:87–89). For a later insistence on the meaninglessness of the sorts of distinctions involved with Quentin Skinner and Hirsch—this time disputing distinctions proffered by Umberto Eco—see Rorty, "The Pragmatist's Progress," in *Interpretation and Overinterpretation*, by Umberto Eco, with Richard Rorty, Jonathan Culler, and Christine Brooke-Rose, ed. Stefan Collini (Cambridge: Cambridge University Press, 1992), 89–108, esp. 93ff.

23. Rorty, "Texts and Lumps," ORT:84.

24. Charles Taylor, "Understanding in Human Science," *Review of Metaphysics* 34 (1980): 24–38, citing 26.

25. Willard van Orman Quine, *The Ways of Paradox and Other Essays*, rev. ed. (Cambridge, Mass.: Harvard University Press, 1976), 229–30.

Index